Photos on corners of odd numbered pages may be flipped to give the appearance of circling the face from above. This is the closest approximation of a three-dimensional image.

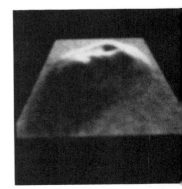

# THE
# MONUMENTS
# OF MARS

## A City
## on the Edge
## of Forever

# RICHARD C.
# HOAGLAND

North Atlantic Books
Berkeley, California

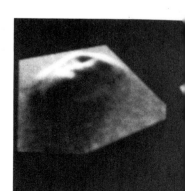

**The Monuments of Mars: A City on the Edge of Forever**

Published by
North Atlantic Books
2800 Woolsey Street
Berkeley, California 94705

Credits: Mark J. Carlotto of The Analytic Sciences Corporation, Reading, Massachusetts, for the use of his photographic reproductions of NASA data tapes; M.I.T. Press for the Paolo Soleri drawings (reproduced by permission of Gail King and Paolo Soleri of the Cosanti Foundation, Scottsdale, Arizona); R. L. Walker of the U.S. Naval Observatory, Flagstaff Station, Flagstaff, Arizona, for the drawing of the Great Pyramid; James Channon and Kynthia Lynn for original artwork (as noted in the text); Daniel Drasin and The Mars Mission for overlays, composites and measurement updates in photograph section; Erol Torun for graphics, data and mathematical analyses; National Space Science Data Center for photographs. Also, please note individual credits assigned to specific illustrations in text and photograph section.

Cover art by Nancy McIntosh-McNey
Cover and book design by Paula Morrison
Printed in the United States of America

**The Monuments of Mars: A City on the Edge of Forever** is sponsored by the Society for the Study of Native Arts and Sciences, a nonprofit educational corporation, whose goals are to develop an educational and crosscultural perspective linking various scientific, social, and artistic fields; to nurture a holistic view of arts, sciences, humanities, and healing; and to publish and distribute literature on the relationship of mind, body, and nature.

**Library of Congress Cataloging-in-Publication Data**

Hoagland, Richard C., 1945-
    The monuments of Mars.

    Bibliography: p.
    Includes index.
        1. Mars (Planet)—surface  2. Life on other planets.
3. Cydonia (mars)  I. Title
QB61H63   1987      999'.23     87-10813
ISBN 1-55643-118-X

To Mom and Dad, who encouraged us to Dream . . .

To those heroes of my childhood, who inspired the grandest Dream of all—

"To seek out new life and new civilizations . . ."

In memory of a Special Dreamer who, one day, became a colleague and a friend—
Gene . . .

And to the Next Generation . . . especially—Lindsey, Angela and Matthew . . .
Who will Know.

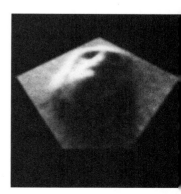

## ABOUT THE COVER

Those who are familiar with the ongoing, multi-disciplinary investigation into the "Monuments of Mars," will recognize that we have changed the cover on this updated version of our continuing work. Kynthia's strikingly predictive analog clay model of the "Face on Mars," which formed the basis of the original cover, has moved inside (see illustrations section)—where we can now directly compare it to the completely separate, digital 3-D computer reconstructions subsequently carried out by Dr. Mark Carlotto.

The cover for this new edition was designed by Nancy McIntosh-McNey, who has been a major source of inspiration and support during the critical research and political breakthroughs which have occurred since *Monuments* was originally published in 1987. Her synthesis—which began on a Fall day in 1988 as some idle doodling on a table placemat sketch, during a conversation about the "meaning of Cydonia" in a small cafe just down the road from the President's famed western Maryland retreat, "Camp David"—symbolizes now the extraordinary connection between what we've found on Mars . . . and . . . what we've also found on Earth.

For, although we didn't know it when Nancy made the sketch, we have now not only *confirmed* the reality of "the Monuments of Mars"—we have successfully "unlocked the code" to their fundamental meaning . . . elegantly symbolized by the ghostly *tetrahedron* linking Mars, the "Face," the Sun . . . and Earth itself.

And . . . wonder of wonders . . . we have now found that *same* "code"—laced repeatedly around the Earth—

Even in the exquisitely artistic, but totally mysterious, "tetrahedral circle" which graces the back cover of this book—a "glyph," part of an equally baffling fifteen-year phenomenon—appearing suddenly one July morning in 1991, in a field just across from a famous English landmark, Barbury Castle.

Whoever executed this stunning, two-dimensional, now blatantly *tetrahedral figure "in the crops,"* laid out in eerie parallel to Nancy McIntosh-McNey's own cover for this work (and only a month after it was decided that Nancy's painting would *become* the cover)—also, somehow, "knows" the identical, highly sophisticated geometry and "physics" contained in our (till now unpublished) decoding of "the Monuments of Mars!"

To find out how, read on . . .

# CONTENTS

# AUTHOR'S PREFACE

This is the space where it is customary in these matters to acknowledge all those who made special contributions; writing a book is not the easiest of undertakings, and behind every author there usually stands an unseen "legion" of family, friends and supporters—without whom, at some point, he (or she) would have the good sense to give it up and become perhaps a plumber!

It is also customary for readers, at best, to skim this section quickly; after all, who really gives a hoot about a group of faceless people, to whom the author claims he'll "forever be indebted?"

In this case, I hope that doesn't happen.

There are some remarkable individuals included here, who have in fact made this Whole Investigation (as well as this book) possible. Many provided their time and talent in very special ways in an effort to help us understand the inexplicable. What originally began as one man's curiosity (following on the curiosity and insight of two others), has now evolved into a diverse and loosely-organized multi-disciplinary Investigation—exploring evidence for something which, if established, will transform human history.

Among the many attempting to understand this problem—which includes at this point physicists, anthropologists, artists, architects, image-processing computer specialists, and even *novelists*—some individual contributions stand above the rest . . .

Initially . . . when there was just me . . . a set of images . . . and over-whelming questions—Rich Blumenthal appeared (literally!). Rich was "destined" to become my "Lord Carnarvon" (but without nearly his resources!).

Without Rich—nominally, a "high-tech sales engineer," but in reality a pioneer and highly talented photographer (who would have taken knock-out shots of Cydonia, if I could have figured out the small detail of getting him a ticket!)—there would have been no "Mars Project."

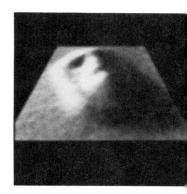

Rich is a man of most uncommon vision—to risk an entire life savings on a 5-inch photograph . . . and one somewhat enthusiastic science writer. But more than that, he is a man of special loyalty and patience—to stick with me through years of what turned into seemingly endless technical details (and at times some rather formidable problems), while I attempted to understand what I had found . . . and then organize an appropriate inquiry to doublecheck the work.

Rich Blumenthal, however, did not just become the financial "angel" of my efforts to figure out Cydonia; through our long, eventful (and still unfinished) odyssey into the Unknown, I discovered that Rich is something even rarer than evidence for an extraterrestrial civilization:

He's what "friendship" is truly all about.

There is another "Richard" (Richard Grossinger—my editor and publisher) who also deserves a word or two on behalf of his unique contributions to this effort (if not for having the simple courage to tell me when I was going wrong). Our editorial relationship—on a book dealing with particularly difficult material—saw perhaps its "finest hour" in our joint struggle with the most intractable problem of any we confronted: how to present, credibly, potential evidence for a so-called "terrestrial connection."

Richard's anthropological expertise, and his experience with the slippery slopes of historic myth and legend, were invaluable in helping frame my arguments; needless to say, I take full responsiblity for the data presented and the arguments themselves. If, however, they are (as I hope they will be) thoughtfully considered, I feel it will be due in no small measure to Richard Grossinger's highly insightful and anthropologically-based editorial suggestions.

Remarkably, Grossinger's sensitivity to the central question of this work—"Why is there an apparent *human* face lying on another planet?"—seems traceable to an eerily prescient prediction, made years before the staggering epistemological problems presented by "the Face" had ever come to light.

Observed Grossinger . . . in 1981:

"Whether we fully appreciate it or not, biology (life science) means to us the living forms on this planet only . . . It does not include, for instance, life forms on Mars; even the simplest of these creatures, if they exist, are not protozoas in a terrestrial biology. They are something else . . . *We would be most shocked to be visited by spacemen who resemble us* (italics added), for if this did not indicate that matter has in it a template leading to the human shape, it would mean that the combination of environmental, pregenetic, and genetic conditions leading to man is duplicatable (and perhaps the only chain capable of intelligence); that we did not originate on this world; or that we are the lineal ancestors or descendants of people on other worlds . . ."

As fitting to me as our eventual editorial relationship, is the fact that, at the time I selected Richard Grossinger (or, he selected me . . . ), I had no idea of this "prediction"; it was only later, in perusing his own epic work *The Night Sky,* that I found his fascinating observation (which even Grossinger apparently had long forgotten!).

Another central figure in our continuing efforts to understand this resonant enigma, has been anthropologist Randolfo Pozos.

To Randy (and to his wife, Kathy—also an anthropologist) must go credit for being the first to recognize the *scientific* nature of this data— and then helping me to organize "The Independent Mars Investigation." I also wish to express my sincere appreciation to both Kathy and Randy for the many personal kindnesses extended during the course of our association; again, what began as a professional relationship, has grown . . .

I would be extremely remiss if I did not single out for inclusion here Dan Drasin—a "generalist's generalist."

Dan's extensive talents, in areas ranging from the philosophy of science to techniques of photography, made real differences in every facet of this Investigation. I have not always agreed with what Dan has said re these Martian "anomalies,"—particularly, their overall significance to the "human condition" if they should prove "real." But I have valued his willingness to have definite opinions and to share them freely; only this type of vigorous public debate, regarding both the evidence itself and its general implications, will assist in the eventual solution of this major scientific problem.

Dan extrapolated several key aspects of the data—among these, the "1:2:4:8" ratios of distances along the solstice sightline (see back cover). His active presence thoroughly "Drasinized" many aspects of the Mars Project—to the benefit of this Investigation.

Then, there's Tom Rautenberg.

Tom's deep involvement is amply related in the pages that will follow. I would like, however, in the brief space I have here, to single out perhaps the most important quality of Rautenberg's association with this Investigation: his courage.

Tom took major risks in championing an, at times, very controversial inquiry—and in the

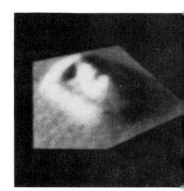

face of substantial ridicule from many of his former (!) academic colleagues. His willingness to persist, in spite of continuing advice to the contrary, deserves both my gratitude and lasting admiration. It is my sincere hope, for Tom's sake as well as all the rest of us, that he doesn't have to see how this comes out "from his rocking chair . . . "

Architect Dan Liebermann first recognized the habitational and structural engimas presented by the objects at Cydonia; subsequently he placed at my disposal his entire architectural studio. Later, he introduced me to the many complexities of historical architectural analysis (particularly as applied to ancient Sumer and Egypt), and shared his own unique designs and research into the relationship of human beings and their constructed environments (see "The Horizontal Cathedral: The Future Metamorphic Human Ecosystemic Community" and "The Theory of the Relativity of Spacial Perception," Liebermann, D., 2728 Benvenue Ave., Berkeley, CA., 94705, 1983). In my opinion, Dan's appreciation of the architectural significance of the "anomalies" clustered at Cydonia stemmed directly from his lifetime of contribution to the true frontiers of Earthly architecture . . .

And while I'm on the subject, I would like to extend my appreciation to another architect—about a thousand miles further east: Harry Jordan. Harry and the Architectural Design Class of the Creighton Preparatory School for Boys, in Omaha, Nebraska, were the first to produce complete blueprints and a scale model of Cydonia for use in our on-going analyses. Their efforts not only exemplified the ideal "democratization" of the Mars Project, they catalyzed significant new recognition of the relationships exhibited by the remarkable objects at Cydonia, as related elsewhere (see Appendix II).

Another individual who made a singular contribution to our eventual understanding of the significance of Cydonia, is Elizabeth Schrank, also a "generalist" (Elizabeth's many professions have included motherhood, textile engineering, and journalism). Elizabeth's enthusiasm and excitement in those first few weeks, as we jointly pored over the photographs and tried to come to grips with what they might contain, will be a vivid and treasured memory forever. But beyond a shared experience. Elizabeth's lasting contribution will lie in her recognition of a pivitol aspect of "the Face"; she was the first to recognize what may turn out to be the most important clue of all: that at certain sun angles, "he" truly does look *simian*!

Likewise, sculptor Kynthia Lynne "pushed beyond her limits" to create the first meticulous clay model of "the Face"; she finally fulfilled my long-term dream (to see "him" from "the City"), and thus also discovered additional aspects of the deepening mysteries latent in Cydonia (see Appendix II). Kynthia's model became the cover of this book. For her pioneering work and her loving friendship, I am deeply grateful.

I would also like to take this opportunity to thank the following persons for a variety of contributions which have resulted in partial verification of our initial work:

Ren Breck—who provided, at some personal cost, the electronic "nervous system" of the Independent Mars Investigation; Paul Shay—who had the vision to recognize the scientifically worthwhile . . . and the political courage to support it; Lambert Dolphin—for his many archaeological comparisons and insights, his technical support of key aspects of the Independent Mars Investigation—

And for his confirmation that, indeed, "Cairo" does mean Mars!

Mark Carlotto—for an unusual combination of scientific integrity and artistic appreciation, whose work has now forever put to rest the lie that "it was all just a trick of light and shadow" (see Appendix II); and finally, in this special catagory, Jim Channon—for recognizing that "he" was truly an awesome work of art.

I would like to express belated thanks (by about ten years!) to Eric and "Billy" Burgess—for gracious hospitality and endless patience, during our unforgettable "Viking Summer." Without you guys, I wouldn't have known which questions were the important ones to ask; the same goes, Eric, for our shared "pre-Cydonia experience" with messages . . .

My gratitude to Robert Jastrow—who understood the true complexities of Viking all along (and who encouraged me to continue asking questions); to C. West Churchman—for recognizing the quintessential epistemological problem; and to Arthur Young—for providing, in a "sea" of New Age nonsense, the Institute for the Study of Consciousness.

Thanks to Roger Keeling, for sharing the determination that this work must be "democratized" as soon as possible, and for his friendship; to Stan Schmidt—for the first editorial in support of the only logical conclusion: for God's sake, let's go back and take a better look!; and to David Woolcombe and Anya Kucharev—for volunteering to become a "spacebridge," this time literally between two worlds!

To Alise Agar, my love for simple loyalty and friendship (and for introducing me, a couple of lifetimes ago, to Arthur Young . . . ); and to David Pasion and Ken Irwin, mutual respect

*Author's Preface*

and a promise to return . . . for, if nothing else, enough cake to last a lifetime! A special thanks to Charlie Boyle: "Virginia, there are no such things as straight lines . . . "; to Brook—for telling me she would never speak to me again, unless I finished this (my new number is . . . ); to John Hewitt—who, in addition to those months of careful measurements, reminded me that Karl Gauss once proposed constructing a huge geometric symbol *out of wheat,* to let "the Martians" know there was intelligence on Earth; and to George Gaboury—for not only being another generalist, but for saving me the center seats . . . in the third row.

My thanks to "Kitty" Carr, Roger Smith, Tom Nursall, Carmine Buzzelli, Robert Judd, "Smitty," J. Thompson, and E. Daley—for proving once again that people are basically extremely fair, if given half a chance.

And to all those who, but for my execrable memory, also deserve to be here—my sincere apologies.

Richard C. Hoagland
Tryon, NC
January, 1987

*The Monuments of Mars*

# AUTHOR'S PREFACE—ADDENDUM

In the four years since *Monuments* was originally published, a lot of things have happened.

The hallmark of true science—as opposed to "all that other stuff"—is when independent lines of research (if not researchers) converge on highly similar, verifiable interpretations of a phenomenon in question; that has now occurred in the continuing Cydonia Investigation—with spectacular results.

So much, in fact, has taken place in these four years—on both the research and political fronts—that to do justice to all we now know (or strongly suspect) really demands a follow-up to *Monuments*—an entirely new book. However, the press of several upcoming events, initiated by these same rapidly escalating developments, make that all but impossible at the moment.

These activities, besides managing the expanding pool of serious researchers, engineers, and technicians—including those who have joined us in exploring several imminent *technologies* now directly stemming from this ongoing Cydonia research (!)—also include perhaps the even more important short-term goal of *politically* insuring that the *now-officially-scheduled verification of Cydonia's reality*—which we have apparently secured via NASA's now tacit commitment (see below) to new *Mars Observer* Cydonia images in 1993—*actually takes place.*

One key means of doing this, we have decided (after considerable deliberation), will be to implement the nationwide launching of what we have chosen to call "The ENTERPRISE Mission": an innovative "21st Century educational program" we have planned, developed, and now tested as a pilot project in an inner-city school in Washington D.C. Not only has the Project been nominated (by NASA!) for one of the prestigious White House "Point-of-Light Awards" for "innovative public service"—both NASA Administrator, Richard H. Truly,

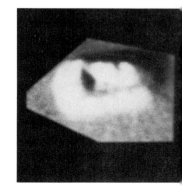

and First Lady, Barbara Bush, have visited the Project and endorsed its revolutionary means and goals for "upgrading overall math and science education in America."

"ENTERPRISE" (named after another mission "boldly going where no one has gone before . . .") is designed ultimately to link schools directly with NASA, and with each other, coast-to-coast and worldwide, via *satellite* . . . allowing kids of all ages to see—*as they come in*—the stunning new Cydonia images in 1993. This above all, we now feel, will go a long way toward keeping the politics of the "Cydonia verification" honest. Contact The ENTERPRISE Mission (through *The Mars Mission,* see address, below) for information on how *your* school can also "join the Fleet."

But all of this takes up a lot of time. So, against this backdrop of important tasks we feel *must* be completed before 1993—which center on the crucial *Mars Observer* mission—we have compromised:

This "updated" version of *Monuments* presents an overview of the major research breakthroughs (and consequently, crucial political developments) which have taken place since the original editions of this book were published—including an astonishing new complexity in the rapidly expanding *terrestrial implications* of these breakthroughs: the discovery (and verification now) of not only *identical* "Cydonia Geometry" here on Earth, amid several major terrestrial archaeological sites and artifacts, but the discovery of equally identical geometry amid the major new "enigma" in our midst:

The *"Crop Circles"*—

Those baffling, mysterious, and increasingly blatant *geometric* "messages" which, since 1975, have been appearing in the fields of England . . . and all around the world.

(And no, they were NOT all carried out with the aid of "string and board" by two "tipsy gentlemen" out of an English pub—as we shall vividly demonstrate later in this work.)

Also included in this new edition are some of the surprising *astrophysical* applications we have now discovered in the solar system—planets (if not the Sun!) actually "operating" directly according to the "Cydonia Geometry" we have painstakingly (and apparently successfully) decoded. As a direct corollary to these major "energy" discoveries, we conclude this recitation of new research results with mention of several recent terrestrial *engineering* demonstrations of these energy-related aspects of the Cydonia breakthroughs—and at least touch on their "explosive" economic and political implications for the near future, if they are properly pursued.

this "attempt"—including direct communication to *The Mars Mission* from the Manager of the *Mars Observer* Project—are included in our new Epilogue.

I say "apparent willingness," because the politics of simply testing our "intelligence hypothesis" over the past eight years have been anything but "clear and simple." The major breakthrough for this crucial aspect of our work—that fundamental concept of a *fair test* of all the data—only came in 1989, in a series of successful meetings on Capitol Hill with key congressional and committee representatives, regarding the vital need now for *government verification* of the preceding, compelling Cydonia research results.

NASA—as the congressionally-mandated U.S. government agency with the virtual monopoly on civilian launch vehicles (like the Shuttle), unmanned interplanetary spacecraft (such as *Mars Observer*), deep space communications systems (such as NASA's worldwide Deep Space Network of tracking and data reduction computers, etc.)—obviously has a stranglehold on all U.S. means to confirm the "intelligence hypothesis" on Mars itself. Yet, even after NASA suddenly (and rather mysteriously) decided in 1986 to add a very sophisticated camera to the only U.S. mission funded to go back to Mars within this century—*Mars Observer*—(as we reported in the initial Epilogue to *Monuments*), for years it still refused to give any *public assurances* that *Cydonia itself* would be reimaged with that camera—a camera capable of relaying images fifty (50) times sharper than the original *Viking* photographs.

Official spokespersons for NASA merely temporized repeatedly:

> "At no time in the six-year investigation and analysis period associated with the [original NASA] *Viking 1* and *2* explorations did NASA ever believe the 'faces' [sic] were anything but a curious formation caused by wind processes. *Nor, by the way, does the agency have any special plans to photograph these features on the **Mars Observer** Mission . . .* (emphasis added)."
>
> —NASA public affairs officer, Charles Redmond, October 1988.

So, in 1989 we requested—and received within three days—a face-to-face meeting with then Chairman of the House Committee on Science, Space and Technology (which directly authorizes NASA's entire fiscal budget), the Honorable Robert Roe. Also at the meeting were several other colleagues involved with the Cydonia Investigation.

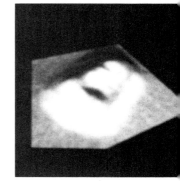

*Author's Preface—Addendum*

The details of that 1989 briefing—and the subsequent political exchanges *vis-à-vis official* NASA verification of Cydonia in 1993, which have taken place since this pivotal face-to-face meeting with Chairman Roe that April morning—will be reported fully in the Epilogue; these details, however, must now include recent, disturbing new developments: a remarkably ambiguous three-page "technical letter," sent to *The Mars Mission* late in 1991—direct from the Project Manager of *Mars Observer,* at JPL—telling us in essence why "NASA [despite its previous assurances to Roe] *cannot guarantee* new Cydonia images . . . !"

Such continuing attempts to evade simply rephotographing the "Monuments of Mars," in our opinion, now underscore the urgent need for a new *grassroots political movement*—before the 1992 elections—*specifically designed* to maintain the rising Congressional interest *we have spurred in new* **Mars Observer** *Cydonia images*, regardless of NASA's changing whims, or inevitable personnel changes on the relevant House and Senate NASA Oversight Committees.

You can now directly participate in this essential "Cydonia insurance policy"; among other things, by officially joining our research and "lobbying efforts," through:

**The Mars Mission** c/o The Society for the Study of Native Arts and Sciences, 2800 Woolsey Street, Berkeley, CA 94705

It is increasingly clear, despite NASA's hurried "assurances" to Congress after our initial 1989 meetings, that only the *active* involvement of significant numbers of Americans can even hope to guarantee full long-term NASA compliance with the wishes of its citizens, and of this key congressional committee—most especially on *this* issue!

The stakes for *direct* political action—an entire, verified "ET civilization"—were never higher . . . as you'll discover in the subsequent pages of this work.

This contention is strongly indicated by *another* set of events that began to mysteriously unfold—even as our "Capitol Hill conversations" were developing: a rash of sudden, almost inexplicable (given the "official" NASA reception to our work) "official NASA Center invitations" . . . to present our eight years of painfully developed Cydonia research, now, to literally *thousands* of NASA engineers and scientists!

From a status where the "kindest" things that some NASA spokespersons could say about us over the past eight years were, "Well, they mean well . . . ," beginning in December, 1988 (significantly, even *before* we spoke with Roe!)—we began suddenly to be invited by separate NASA Centers to present "the Intelligence Hypothesis and data" to thousands of rank-and-file NASA personnel and scientists around the country!; the first breakthrough in this direction came with a *formal invitation* to address the NASA-Goddard Space Flight Center Engineering Colloquium.

We'll describe in the Epilogue, and in some detail, the most provocative of this "wave" of recent official invitations—to make a Center-wide Address to NASA's Lewis Research Center, in the Spring of 1990—and then, the almost bizarre story of what this invitation apparently triggered back in Washington . . .

For, if some within the far-flung Space Agency are now finally *listening* to what we have discovered, others (back in Washington . . .) seem more determined than ever to *suppress* all "serious Cydonia discussion."

We'll detail *specific* evidence now of NASA's overall, *deliberately suppressive attitude* toward our continuing Cydonia investigation: the Space Agency's decision in 1990 (through the previously mentioned NASA-Lewis Center) to produce—then *cancel*—then (after receiving "several thousand angry calls and letters from concerned Americans") to once again *reschedule* (but in highly edited form) . . . *an official NASA television broadcast on our work,* for PBS—to be called by NASA (not by us!)—

"Hoagland's Mars."

If nothing else, NASA Headquarters' now highly visible, suppressive political actions *vis-à-vis* "Cydonia"—through its sudden cancellation of this program—is a cautionary tale for 1993 . . . and the fact that *nothing* can be taken as "assured"—until we actually *see* those crucial Cydonia images from *Mars Observer.*

*       *       *

I cannot conclude this Addendum without citing at least some of the major people who have recently joined forces in this major effort—important folks who have volunteered for *The Mars Mission* team since the first editions of *Monuments* were published. Several have made stunning scientific contributions to our growing insights *vis-à-vis* Cydonia, not only in confirming the mathematical reality of what we have discovered, but in assisting us towards the beginnings of an understanding of the profound *meaning* behind those key mathematics . . . even before we have returned to Mars in 1993.

These new "important folks" include: Erol Torun, my friend and colleague, at the Defense Mapping Agency. Erol's addition to this research in 1988 resulted in a quantum leap, not only in terms of specific breakthroughs *vis-à-vis* my original "Cydonia relationship model," but in terms of his exquisite geometric insights on the meaning of that model. You'll read the highlights of his story—the beginnings of his subsequent major contributions to this work—in the updated Epilogue.

Another "player" who didn't really know what he was getting into in 1989—when he picked up one of the first editions of this book for his son's science project on "the planets," and then called me—is now another valued friend and colleague, David Myers. David's background includes a classical Greek education at Stanford, a stint as an "exec" in the U.S. Navy (including time on a destroyer off Vietnam), and now full time Director of Operations for *The Mars Mission,* in addition to serving as Editor of its official Journal of ongoing research results and membership reaction: *Martian Horizons.*

One of David's other hats is managing the growing file of mathematically verifiable *geodetic* relationships we are discovering on Earth, which now compellingly support a *testable* "terrestrial connection" for Cydonia. This includes integrating his own remarkable geodetic discoveries with those of other pioneers, such as Carl Munck—who, unknowing of the Cydonia work, has also been measuring quietly for several years, "rediscovering" the Cydonia Grid . . . right here on Earth!

How do I "properly" introduce Stan Tenen . . . ?

Stan and I actually met face-to-face only after the initial edition of this book was published, at a gathering at Ren Breck's Marin home in 1987. Years earlier we had been made aware of each others' research efforts through a mutual friend and colleague—Lambert Dolphin (see previous Author's Preface, and chapter VIII). But, at the time, I can remember vividly asking myself (after Lambert invited Tenen to SRI for discussions of his work): "Why is Dolphin involving himself in *that* digression?!"

I could not have been more wrong.

Stan Tenen's twenty-one years (as of 1991) of painstaking and (until recently) completely separate efforts at resolving his own "resonant mystery"—which involves his discovery and successful decoding of explicit and redundant *tetrahedral metaphors* in a score of "ancient texts" here on Earth—turns out to be parallel, if not *central,* to our own successful "decoding" of Cydonia! But Tenen's mathematical decodings lie in some of the most sacred documents on Earth—including Genesis; "ours" lie in the striking geometric configuration of a set of "alien ruins" . . . and on another planet.

The implications of these now strikingly parallel discoveries are profound.

Unfortunately, the developing *interconnections* between these two highly controversial research programs are simply too vast to be ade-

*The Monuments of Mars*

quately dealt with here—for the geometry that each is based upon is not just "parallel" . . . it is *identical;* and, if that were not enough, this explicit geometry (and its underlying *meaning*) is now echoed in what Myers and others have discovered—also independently—in the geodetic "tetrahedral codings" of equally *sacred sites* around this planet. Until *The Cydonia Papers*, we can therefore only hope to provide enough detail later in this work to inspire others in "picking up the trail" . . .

(Information and tapes on *Meru* "textual" discoveries which relate specifically to our Cydonia efforts may be obtained through *The Mars Mission*—see address above.)

Then there's Marty Arant.

With a background in high technology and, specifically, education and computers, Marty first called because he'd encountered a copy of *Monuments* in 1989. His initial area of interest was in "democratizing" (for a wider audience) the work to date. As new research and political developments took place, Marty's involvement also rapidly expanded: ranging from initiating the first national, public computer conference on the ongoing Cydonia research—as Systems Operator for Mars Mission Issues, a section of the Issues Forum on the international, 500,000-subscriber computer information system, CompuServe—to developing, with me and Nancy McIntosh-McNey, the prototype "ENTERPRISE Mission" school demonstration in Washington D.C.

Marty is currently serving as Administrative Director of "ENTERPRISE"—a pivotal role in "democratizing" the truth about Cydonia, as we expand the program to school systems across the nation, if not beyond.

Keith Morgan picked up *Monuments* in the summer of 1988, showed it to his producers at ABC News, and the result was the first discussion of the "Monuments of Mars" coast-to-coast on NIGHTLINE. Keith then engineered a two-and-one-half hour briefing by us, for a producer designated by Ted Koppel to follow the progress of the continuing research. Eventually, NIGHTLINE may do something "even more significant." When they do, Keith Morgan—by simply refusing to give up—will be the one who made it happen.

In addition to his determination to expand public awareness of this work at the network, Keith's vision and persistence have been invaluable in our establishment of "ENTERPRISE" in Washington, at Dunbar Senior High. His broad-based background in computers and electronics, which has "saved" us more than once in this unprecedented project, has been matched only by his dedication to finding out

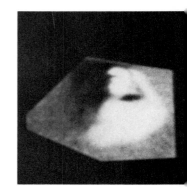

*Author's Preface—Addendum*

what's lying at Cydonia—then using that system to expand the awareness of an entire generation of imperiled inner city youth. Keith even managed to recruit into setting up the "U.S.S. DUNBAR" (which is what the student-crew call their "experiment in education . . .") his dad—a master electrician of uncommon vision and ability, who saw exactly what we wanted, then made it happen for "mere pennies"; and his brother John, an equally dedicated and visionary professional with architectural experience, who helped transform our own vision of a "training vessel of Starfleet Academy" into a very solid, inner city "starship"—"homeported" a stone's throw from Capitol Hill . . .

For his initial efforts on our behalf at ABC, for helping to convince Howard University that we were serious about "ENTERPRISE" in the beginning, for his constant good humor and ingenuity in creating the U.S.S. DUNBAR itself, and for enlisting the extraordinary efforts of his talented family in that effort . . . but, most of all, for his simple steadfast friendship and respect . . .

Thanks, Keith; we're still working on 'the curve.'

Then, there is Mark Dwane . . .

Mark is a young composer in the Midwest. After reading an initial version of *Monuments* in 1987, he was so inspired by the possibilities that he began a series of compositions to Cydonia . . . which resulted in the first record album created specifically in honor (as far as we can tell) of a scientific work—*The Monuments of Mars: the Music* (Trondant/Orbian Music)!

Needless to say, I was "blown away" by this development; the music is so eerily appropriate to what we have discovered, I only wish, as I've now told Mark repeatedly, "I'd had "*Monuments*" to listen to . . . when I was writing *Monuments*! I do now—for this update. Mark is now working with Colette Dowell (who first discovered it—see below) and Erol Torun, on a new "twist" to the Investigation: "tetrahedral music." More details in *The Cydonia Papers* . . .

I cannot leave this subject of appropriate acknowledgments without mention of another valued friend and professional colleague in these areas, Dr. John Wilson.

John is a world-renowned psychologist—globally recognized now for his pioneering work on "Post-Traumatic Stress Disorder." PTSD, as it is professionally termed, is the lingering psychological difficulty (which can overwhelm anyone) following a major physical or psychological trauma: from airline crews, passengers and rescue workers involved in catastrophic aircraft accidents; to those suffering private personal tragedies; to members of the Armed Forces involved in deadly conflicts. In recent years, the disorder has often been termed simply "the Vietnam Syndrome."

Because of John's courage and persistence in initially recognizing, and then fighting for appropriate treatment for this then highly controversial (and often ignored!) psychological condition over twenty years ago, Vietnam (and now Desert Storm) veterans, and victims of a wide range of stressful, catastrophic, or other highly traumatic personal situations, are now receiving essential counseling and treatment.

John's unique role in the Cydonia Investigation has been to assist us to recognize, then implement across the culture via many mechanisms, "pre-traumatic stress preventatives"—*before Mars Observer* shocks us to the roots of our collective psyche, with vivid images of "vast, ruined, alien architectures at Cydonia . . ." in 1993.

In that vein, I should say a special word regarding Alan Shawn Feinstein. Alan—ever since someone dropped him a copy of my book in 1987—has become a very courageous, innovative and consistent supporter of this effort—not only through his own publications, "The Insider's Report" and the "Eye to the Future" newsletters (which are preparing up to 500,000 readers to confront the inexplicable in 1993), but through his pioneering and persistent underwriting of our creation and development of "ENTERPRISE" in Washington . . . which, in Alan's view, will do the same for future generations—our kids—who must be properly prepared to live with Cydonia's reality . . . if not its implications.

It was Alan who convinced two entire *governments*—first, the government of Sierre Leone, then that of Grenada—to *officially* memorialize the Cydonia Investigation, through the first official stamp series dedicated to the "Coming Exploration of Mars," including the resonant mystery posed by the presence of "Cydonia."

The Souvenir Sheets of these unique philatelic editions specifically memorialize the haunting central feature of this work (see illustrations section)—The "Face" itself!

There are many others we should take the time to cite, for their contribution to the progress we have made in these four years:

Bob Roe, of course, former Chairman of the House Committee on Science, Space and Technology—for understanding what he was seeing "in the photographs and studies," and then *acting* on it; Valerie Thomas, at NASA's Goddard Space Flight Center, who literally "opened the door" with that initial, surprising invitation to address a Goddard Engineering Colloquium, regarding some very controversial work . . . and John Klineberg, former Director of NASA's Lewis

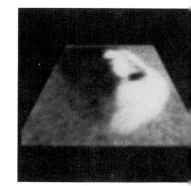

Research Center (now head of Goddard), for following up and extending his independent invitation for a critical, center-wide address to all the scientists and engineers of NASA-Lewis on the same subject . . . and Joyce Bergstrom, head of John's NASA-Lewis ALERT Program, for taking the time and care to see that it so successfully came off—and then going beyond the call of duty in personally responding to the flood of requests (from all over the country) for video copies of the NASA-Lewis program.

And to my long-time friend and NASA colleague, Charlie Boyle— so recently of NASA Headquarters, and now (thanks to John Klineberg) "back where he belongs," at Goddard—a very special, heartfelt "thank you."

Charlie, as I knew he would, *instantly* recognized in "ENTERPRISE" the vast potential for spurring youthful interest in science and math education in our ailing schools. So he turned his considerable talents—both inside and outside NASA—to seeing that take place . . . through appropriate *national* implementation of the program.

It is Charlie Boyle who was ultimately responsible for getting the NASA Administrator himself, Rear Admiral Richard Truly, "piped aboard" the only "starship" currently in Washington—the U.S.S. DUNBAR. And it was Charlie who introduced us to someone who was to become a key "outside player" in the Project: Carolyn Harris.

Because of Charlie Boyle's unique blend of long-range vision and Washington "insider" experience institutionally, it's now "ahead Warp Factor one" for "ENTERPRISE."

And while we're at it, a similar "well-done" to the remarkable, pioneering Bridge Crew of the U.S.S. DUNBAR herself—the students of Dunbar Senior High:

"First Officer" Herman Smith, and all the other initial "Bridge personnel"—John, Robert, Alex, James, Eula, Nancy, Sabrina, and the rest.

I'd "ship out" anywhere with you guys now!

And another "well-done" to Dunbar Pre-Engineering Coordinator, Judy Richardson—an educator of uncommon vision—and to all the other staff and faculty at Dunbar Senior High: my deep appreciation to all of you, for launching this new "ENTERPRISE" upon her own very special "five-year mission . . ."

And to you, Carolyn—who, with this sterling "crew"—took the time (at considerable personal sacrifice, and without institutional financial backing) literally to "write the book" . . . on how to *really* go "where no one has gone before . . ."

All of you have an extraordinary course "laid in" for the next few years, if what we have discovered at Cydonia is true . . . With your example, your assistance, and participation, soon . . . we'll *know.*

(Again, those wishing further information on "ENTERPRISE," and how it can be implemented in schools in your area *in time for the Mars Observer Mission*, should contact *The Mars Mission.*)

A special mention should also be included here of producer Cliff Curley, winner of five Emmys, who, after NASA Headquarters "killed" Lewis' version of its planned PBS broadcast, *Hoagland's Mars*, made it possible to independently release my Lewis Address to a worldwide audience—as the home video version of *Hoagland's Mars: the NASA-Cydonia Briefings, Vol. I.*

We should also include Val and Doug Thompson, for their many kindnesses and conversations through the years; Lois Lindstrom and her husband Tal, who arranged our critical meeting on "the Hill" with Chairman Roe; Walter Gelles, "Billy" Cox, and Bob Kiviat, journalists of uncommon integrity who knew a *real* story when they saw it—and then followed up; Donna Gaal, for friendship and for arranging our first "out of town tryouts" in Charleston, West Virginia; Colette Dowell and Lisa Fisher, for providing another "window" on the magnitude and *context* of what we have discovered; and especially Colette, not only for her friendship through personally difficult times, but for successfully enduring that experience and going on to "do the numbers" on the *other* vital aspects of The Message: "the *circles*" and "the *music*"; to "Cherry" Warren, for her warmth, her loyalty and friendship; to Jayne Smith, for all her many kindnesses "before we 'knew,'" for her personal "reconnaissance" in Egypt which gave us priceless "ground-truth" on the Pyramid, and for tracking down Lockyer's "tome" all the way from Sedona to the Mississippi!; to Jean Hunt, for her patient years of simply "tracking the Flood survivors"—It now looks, Jean, as though you were really on to something, all along . . .; thanks also for founding a first-class institution which can finally put some "teeth" into the entire subject of a possible "Terrestrial Connection"—the Louisiana Mounds Society.

Deep appreciation also to Susan Karaban, for instantly recognizing what this is truly all about—and then for efforts "beyond the call of duty" to bring the basis of a *real* "new world order" to the world: at none other than where it is now daily being born . . . the *United Nations*; to David Percy, for his own unique understanding, from the beginning, of what we've found, and then for his deep personal commitment to "getting out the word"; to Scott Roberts, for "tackling the big one"; and to Robert Watts, for making possible, in a very spe-

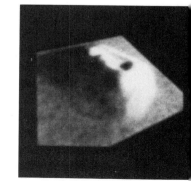

cial way, this project's special vision . . . the People's Right to Know.

Thanks to Gary Gunderson and Butler Crittendon, who have now provided us the tools to make it happen . . .; to Kent Watson, for introducing us, and for your own very special "puzzle piece" and unique vision: "We'll make yours happen too!"; to Ed and Georgia Pearson, for introducing us to Don Kelly; and Don Kelly, for introducing us to Bruce DePalma . . . and to Dr. Bruce DePalma, for doing the critical experiments *twenty years ago* that now let us know *we're* truly "on to something"; to Troy and Evelyn Reed, for your pragmatic engineering . . . and persistence; Arthur Thiel, for your independent fifteen years of also proving—via a totally revolutionary route—the reality of the "Cydonia Equations"; Gerry Franklin, for beginning to tackle the tough mathematical modeling that lies ahead; to Chuck Harder, Kent and Laura Phillips, and all the other "For the People" staff—and the growing and highly involved "For the People" audience coast-to-coast—for their national support for simply "finding out."

To all the other media folks over the past four years who have heard us out, and then, rather than snickering at something "so absurd," overwhelmingly took the time to listen . . . and to wonder.

And a "thank you" to Dick Criswell—who, understanding the crucial role of television in this story very well—freely offered his major communications efforts to "simply finding out."

Finally . . . to my family—and to the families of all those involved in one way or another with the Cydonia Investigation—we appreciate your love, your patience, your caring and support for all these many years. As we do all the many others who have helped along the way, whose names (if I have not) I should remember . . .

I would like to close with a very special acknowledgment . . . on this unfolding odyssey . . . to Nancy: who not only co-created "ENTERPRISE," whose cover not only perfectly encapsulates the essence of what we now believe we know about this wondrous subject—and who was "there" for three very crucial years during these extraordinary new developments "on Mars," through the exciting *and* the difficult—but whose love, pioneering spirit, discernment, and artistic abilities are exceeded only by her almost infinite caring for the "loving furry aliens" in our midst (there, guys, I got you in!).

To Nancy—and to everyone who has sustained us on this journey— "the real voyage has only just begun . . ."

Richard C. Hoagland
November, 1991

*The Monuments of Mars*

# PUBLISHER'S FOREWORD

by Richard Grossinger

For a reader first discovering the matters in this book, the topic may seem bizarre and obscure. Monuments on Mars? A sphinxlike human face among the craters and sand?

Is this a science-fiction story?

Is this a new attribution to the already-ubiquitous "ancient astronauts?"

In truth, at first glance the Face on Mars seems more like the invocation of a theosophical cult, or a bit of astral sightseeing, than what it is: a concrete object apparently carved from a Martian mesa in precise alignment to the sun and to surrounding structures. The Face and its adjacent "Monuments" are not the imaginings of flying saucer abductees or paintings from the cover of a science-fiction magazine; they are physiographic protuberances photographed by NASA, our American space agency, in 1976, and yet not discerned . . . because they were not supposed to be there, because they would have been too outrageous to believe.

Outrageous even to the New York publisher who first purchased the rights to this book and then tried to reclassify it as science fiction. Finally it was deemed too far out to publish . . . as fiction!

Now it would be foolish for me to tell you that for sure and certain there is an artificially constructed statue of a human face on the surface of Mars, surrounded by the outcroppings of a buried city. I don't know that as a fact, nor does Richard Hoagland. This is, however, a bona fide puzzle for us as a culture and as a species to resolve over the next two or three decades.

However, I can say safely that the objects photographed by Viking on the surface of Mars, particularly in the Cydonia region, have many intrinsic characteristics of artifacts that go beyond (and beneath) the mere subjective appearance of a face and a city, and that these characteristics are

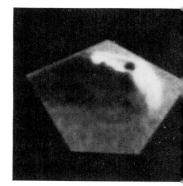

the outcome not of idle Man-in-the-Moon stargazing but of the scrutiny of experts in a variety of fields after eight years of critical study. As Hoagland himself has said many times—in this book and elsewhere—if these remarkable objects and alignments turn out to be natural features, then they are a trivial waste of time . . . but if they were made by someone, they absolutely cry out for an explanation—and that explanation, whatever it is, will change the human race's sense of its own identity and destiny.

From my own first knowledge of this phenomenon I have realized, with growing respect, that it is the Face itself which weights the "message" so heavily. To discover life on another world would alone be a fantastic event in our human history. But to discover a Face on another world calls into question the very nature of who *we* are. Either our history is different from what we have traditionally assumed, or our biology is, or both are. Another civilization elsewhere can have its own origin and *raison d'être* without impinging on our separate terrestrial destiny, but the Face is our face, and it *cannot* have an exogenous or trivial explanation. If someone made it, then "they" picked the one object that would absolutely compel us to come there, for "they" have held a mirror up to our entire planet.

From its sudden unexpected appearance on a NASA data-tape, the Face has been the victim of one debunking after another. NASA didn't even give it the respect of looking twice; in the official collective mind, it had to be a mirage. The space agency squandered its major opportunity to go back and re-photograph the site while the on-board cameras were still functioning. At the time, NASA was considered virtually infallible, and its integrity was unchallenged, so the images fell by the wayside.

The first serious papers on the phenomenon, years later, were ridiculed, and their authors were subjected to cruel teasing and unconscionable innuendo about their motives. They still suffer the professional and personal scars of those encounters. The astronomer with the largest lay audience and the widest personal acceptance, Carl Sagan, has spoofingly compared the Face to a variety of mirages in natural objects ranging from hemispheres of the Moon to a tortilla chip.

The press coverage has been uniformly abysmal; occasional tabloid headlines about the "Monkey Man on Mars" and "Lost Martian Civilizations" have served more as confirmations of Sagan et al. than as rebuttals to them in defense of a significant finding at Cydonia. A humanoid visitation from outer space is a territory that Erich Von Daniken has so abused that it can no longer be invoked seriously by other researchers.

*The Monuments of Mars*

Von Daniken cried "Wolf!" so loud that we may now be immune to recongnizing the artifacts of actual aliens even in the context of compelling evidence. In recent years the Face has even been the butt of uneducated debunking in the countercultural press (as if writers accused of being gullible about everything intend to prove their critical faculty by setting up and knocking down this one straw man).

We are, as a culture, peculiarly immune to the Face on Mars. Perhaps we have seen too many episodes of *Star Trek* and been through too many "Star Wars," "Wars of the Worlds," and "Close Encounters" to appreciate the meaning of actual contact with the "Other." A shocking percentage of the supposedly educated public in the West has little appreciation of, for example, the real distance from here to Mars, or from this Solar System to another star system. The availability of a clever omnipresent technology and super-animation has dulled us to the actual universe. The Industrial Light and Magic Company may enact spectacular special effects for movies, but it has misguided us as to the *actual* effects of gravity and the scale of the stars and planets. (One editor who rejected this manuscript did so because, in her words, we had already sent men to Mars and, if there were a Face there, our astronauts would have reported it. The misconceptions of both fact and scale here boggle the imagination.) If we think about the aliens of Hollywood, or even the scientific reconstructions of PBS, the Face on Mars may seem tame. But in the real night under the actual sky, its possible presence is chilling.

This is not a movie we have already seen.

Richard Hoagland believes that fear as much as indifference keeps us from acknowledging the Face, i.e., from even *confronting* the dilemma of whether it is natural or artificial. Something in our nature doesn't want to "face" ourselves in this way; the reflection, after all, is just as potentially damaging to the rigid priesthood of Western science as it is to the officials of Western religion. If a human face is there on Mars, it blows just about every orthodoxy wide open.

But then maybe we're ready to have everything blown open. As a species we are teetering on the precipice—of an Armageddon war, of the limited resources and fragile biosphere of our world, and, perhaps most significantly (especially if we expect to solve any of these problems), of meaning itself: our values, beliefs, and sense of self have no ground anymore; we drift among stars and atoms we have conceived as drifting

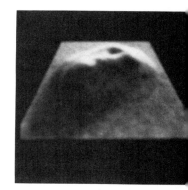

aimlessly among one another. The Face is not a cure to all our ills, but it is a signed, sealed, and delivered paradigm shift, out of which a real human transformation may come.

Already the Face suggests some important things: It tells us we may not be alone and our destiny may be tied to the destiny of Others. It tells us that we are one genetic message, culminating in intelligence. (More than a decade ago, photographs from Moon-travelling ships told us that we were one world, one breathing organism and circulatory system without boundaries.) The Face warns us that however we come to its riddle (or oracle) we do so as a species, not as separate warring nations. [Already, the possible confirmation of such an artifact virtually begs for a joint Soviet-American expedition, both to avoid a deadly competition between us and to put the "face-to-face" meeting at the disposition of the whole planet. (In such an undertaking it little matters that the superpowers do not really represent the whole planet and that they will not go as real trusted friends: symbolically, their collaboration would stand for a potential planetary unity and reconciliation, and the men and women on the mission would be true co-workers; sometimes it is most important to act decently without complete faith . . . then faith may follow.)]

I am reminded of the letter Hoagland received from a woman after appearing on one radio talk show; she wrote, "In all the years since I realized what nuclear weapons were and to fear them, nothing seemed as though it would be big enough to rival them, not in my lifetime. Now, I know this sounds strange, but this is the first thing that has given me hope . . . the idea that someone would have gone through all that trouble to build this image of us on another planet. I don't know why, but it just might be the antidote."

At another time Hoagland described the Face as perhaps the one droplet of matter into solution needed to precipitate its crystal. A Face on Mars is such an unlikely thing at this stage of our history and in our present crisis that its confirmation could literally crystallize us in a new way, perhaps not all at once but over many years, even generations. And then of course . . . there is its message . . . its unknown message . . . assuming it has one other than the very fact of its existence.

The author of this book, Richard C. Hoagland, is, by career, a science writer as well as a consultant in the fields of astronomy, planetarium curating, and space-program education. Although his apprenticeship was in the hard sciences, biology and physics, he has studied widely in so many diverse fields that his approach transcends conventional academic definitions. Hoagland is by no means a subjective dabbler. As a longstanding member of the space community he upholds rigorous scientific requirements in all his endeavors. He is at once a quantifier and a synthesizer—the sort of throwback independent scholar spawned by the Space Age, a curious combination of [*Star Trek* creator] Gene Roddenberry and Mr. Spock himself. His professional activities (and hence his unique qualifications for this investigation) become understandable only when his career is reviewed in detail:

In 1965, at the age of nineteen, Richard Hoagland became Curator (possibly the youngest in the country) of the Springfield, Massachusetts Museum of Science where he designed and produced an elaborate commemorative event to coincide with Mankind's first visit to the planet Mars, a fly-by of *Mariner 4* on July 14. A live 2000-person audience in Springfield was linked across the country to 5000 press and NASA scientists gathered in Pasadena, California, at the Jet Propulsion Laboratory.

In 1966 Hoagland served as NBC consultant for the historic soft-landing of a U.S. spacecraft on the Moon—*Surveyor 1*. Later, he appeared on "The Tonight Show," explaining the significance of the landing to Johnny Carson. The following year he directed an exhibit and radio special on the first Russian soft-landing on Venus and the second U.S. fly-by (with *Mariner 5*); the program included a visit by Esther Goddard, widow and biographer of Dr. Robert Goddard, and a live trans-Atlantic interview with Sir Bernard Lovell of the Jodrell Bank Radio Astronomical Observatory.

In 1968 Hoagland became Assistant Director of the newly-constructed multimillion-dollar Gengras Science Center and Planetarium in West Hartford, Connecticut, where he directed a Premiere Show centered around the projected NASA "Grand Tour of the Planets." At Christmas of that year he was asked to become a consultant to CBS News, where he designed space simulations and served as an advisor to Walter Cronkite on the science of NASA's epic journeys to the moon in the Apollo Program.

Hoagland's relationship to astronomy has always been a marriage of the grand conception to the human scale, as he has sought to translate the vast dimensions of the universe into events familiar to us. In 1969 he conceived a project to demonstrate strikingly the ultimate meaning of the *Apollo 11* landing on the Moon; he arranged a rendezvous between the *Queen Elizabeth 2* and the 5000-year-old Egyptian boat being sailed

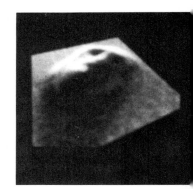

across the mid-Atlantic by Thor Hyerdahl. Using both NASA and CBS equipment, he planned a major media event, with satellite navigation and communication. Hyerdahl, however, cancelled at the last minute so as not to compromise the "scientific integrity" of his voyage, and Hoagland reduced the link to a short-wave radio interview with Cronkite from New York.

In 1970 Hoagland's major project was the network television for the "Eclipse of the Century" in March. He designed each remote along the 5000-mile path of totality, including a special airborne unit using a borrowed Air Force KC-135, from which he coordinated joint ABC/CBS coverage. At the same time he served as Science Advisor to the Committee for the Future, a non-profit pro-space organization, and he designed the first private-enterprise mission to the Moon. Called Project Harvest Moon, it proposed to make use of "surplus" Apollo hardware. The Committee's submission of this concept to NASA and the White House led to a House of Representatives Resolution calling for future NASA cooperation with private-enterprise endeavors.

Among Hoagland's most valued contributions to history and science is the conception, along with Eric Burgess, of Mankind's First Interstellar Message in 1971: an engraved plaque carried beyond the solar System by the first manmade object to escape from the sun's influence, *Pioneer 10*. Hoagland and Burgess took the idea to Carl Sagan, who successfully executed it aboard the spacecraft, and acknowledged their creation in the prestigious journal *Science*.

In 1971 Hoagland was involved in a number of unusual projects. Unabashed by his position outside the formal scientific cadre of the space community, he proposed an ad hoc *Apollo 15* experiment to observe the first total eclipse of the sun *from the Moon,* inducing NASA to form a team to carry out observations on this and remaining Apollo flights under the direction of Dr. Thornton Page. He also proposed that two objects, one light, the other heavy, be dropped simultaneously on the airless lunar surface, recreating Galileo's classic "Leaning Tower" legend. During *Apollo 15,* Astronaut David Scott dropped a hammer and falcon feather together before a worldwide TV audience, proving Galileo right!

At that time Hoagland also served as project scientist (along with Astronaut Phillip Chapman) for British aviatrix Sheila Scott's attempt to become the first woman to solo non-stop across the North Pole in a light aircraft. Hoagland "borrowed" NASA satellites to track the flight and communicate with experiments aboard the aircraft. His prediction that cold air above the North Pole traps pollutants from around the planet was borne out when Scott found a high concentration of $SO_2$ emitted by power plants burning coal and oil.

Since 1971 Hoagland has held a number of editorial, managerial, and consulting positions in the space-science world. During 1974 and 1975 he was the Coordinator of Special Projects and Public Affairs at the American Museum of Natural History's Hayden Planetarium in New York City. For five years after that he served as a consultant for the Goddard Spaceflight Center in Greenbelt, Maryland, producing a series of Direct Broadcast Satellite Demonstrations, including the pre-Bicentennial program featuring Carl Sagan, Jacques Cousteau from the *Calypso,* and Raphael Salas, Director of the United Nations Population Fund. During this same time, Hoagland was the founder and first president of the High Frontiers Foundation (based in New York) and a contributing editor and then Editor-in-Chief of *Star & Sky* Magazine. In 1981 he became science advisor to Cable News Network in New York where he produced field coverage of the *Voyager 2* fly-by of Saturn, serving as on-camera commentator and co-anchor.

During these years Hoagland was most noted for his innovative public events, science-cruises, space campaigns, and daring articles. On one cruise, aboard the Holland-America Lines luxury ship *Staatendam,* he brought together Isaac Asimov, Hugh Downs, Robert Heinlein, Norman Mailer, Katherine Anne Porter, Carl Sagan, and many others. During its ten-day crossing of the Caribbean, the conferees witnessed the departure of the last Apollo Mission while "parked" offshore from Cape Canaveral, and visited the world's largest radio telescope at Arecibo Observatory in Puerto Rico. Hoagland's "Voyage to Kohoutek" cruise aboard the *Queen Elizabeth 2* featured participants form Carl Sagan to Burl Ives. His demonstrations at Goddard included Arthur C. Clarke and William (Captain Kirk) Shatner, who "beamed down" via NASA satellite to hand-held "communicators."

On the seventy-fifth anniversary celebration of the consolidation of New York's five boroughs, Hoagland wrote, produced, and presented a complex multimedia event in the Hayden Planetarium entitled "The City and the Stars: 75 Years of New York Looking Up," for which he received a citation of excellence from the City.

In the years following his stay at the Hayden, Hoagland was active in both the effort to enlist public support in requesting then-President Gerald Ford to christen the first Space Shuttle after the popular *Enterprise* of the *Star Trek* tele-

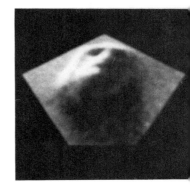

vision series (in an effort to "democratize" the space program), and The Halley Fund to support NASA plans for a fly-by mission of Halley's Comet in 1986.

Hoagland's most notable contribution to the American Space program since the *Pioneer 10* Plaque and the "Galileo Experiment" has been his Europa Proposal: that pre-organic material, or even simple life forms, exist in a satellite-wide ocean under a crust of ice on the second moon of Jupiter. When published in January, 1980, in *Star & Sky,* the idea created a whirlwind of reaction, including newspaper headlines. Scientific reaction was equally dramatic, ranging from outright dismissal to plaudits for identifying "the first new plausible location for life in the Solar System in ten years . . . ," as Terry Dickenson, then the editor, put it. Dr. Robert Jastrow, former Director of Goddard Institute for Space Studies, and one of the founders of NASA, called it the most compelling and startling model to arise from the Jovian system of moons. Perhaps the most fulfilling response came from Arthur C. Clarke who wrote in *2010: Space Odyssey II,* "This quite brilliant concept . . . may provide one of the best motives for the projected Galileo mission [back to Jupiter]."

In a July 1982 story in *Science Digest* Hoagland focused national attention on the mystery of "The Thing in the Ring," a fascinating enigma found by *Voyager* to orbit Saturn in the rings. Just before becoming involved in the later Martian enigma (of a far different nature), Hoagland was organizing efforts for radio telescope observations of Saturn's rings.

For the last ten years, as chronicled in this book, Hoagland has devoted his life to the mystery at Cydonia—and its ramifications on Earth. He has coordinated diverse groups including physical and social scientists, architects and electrical engineers to refine and analyze the NASA images. Whether or not one agrees with his lines of inquiry and interpretations, it is clear that he has provided, thus far, the most significant breakthroughs of conceptualization and contextualization in the quest to evaluate and "decode" Cydonia. He has travelled the country educating the public on the possible implications—cultural and technological—of Martian artifacts. These "travels" have included over a hundred radio talk shows, lectures, televised news appearances, and press conferences during which he has countered misconceptions, debated often uninformed opponents, and even addressed the personnel at NASA facilities under official auspices. At the same time, he has written the original edition of this book and several scientific papers (to appear in the forthcoming book *The Cydonia Papers*) and has also helped write and produce an audio-tape version of this book and a separate video *(Hoagland's Mars)* detailing his NASA appearances and elucidating the new "evidence" derived from analysis of the original photographs.

As both editor and publisher of this book, I have had a major role in shaping the material presented here. In fairness to the author I should point out that I have served as "devil's advocate" and led him into a book somewhat different from the one he initially envisioned and that he might have written on his own. The tug of war between us originally resulted in a number of compromises of both content and interpretation. In this new edition the author has reclaimed his own book.

Although editors often provide a critical function, because of the controversial and ultimately historical nature of the claims in this volume, I will point out two major areas in which Mr. Hoagland and I have disagreed:

1. I myself would tend to deemphasize, first, the *certainty* that the Monuments, City, and Face are artificial and, secondly (and more substantially), the hypothesized terrestrial connection. I think the mystery of the raw giant Face in rock stands by itself, beyond statistics and without requiring ancient Earth cultures, Egyptian counterparts, etc. The entire proposition seems incredible enough to require rephotographing for verification—and even were the data tapes made by a subsequent satellite to suggest artificial construction yet show anything short of actual hieroglyphs and infrastructure, I would still consider "extraterrestrial intelligence" a long shot as an explanation.

However, since the first edition, the computer work of Mark Carlotto has sharpened the images and confirmed their anomalous relationship to the surrounding geomorphology; meanwhile, continued investigations by Hoagland, Torun, Myers, et al., have uncovered at least the hypothetical parameters of an entire alien science. These extraordinary occurrences have gone a long way toward making me "doubt" my own doubts. I finally must conclude that I do not understand statistics, algebra, and trigonometry well enough to have a gut conviction one way or the other about whether the Cydonian code lies on the terrestrial or Martian side of the lens—or both.

Numbers lie—we've been shown their tricks many times. But our whole world-view is based on the fact that numbers also uniquely tell the truth. Which "numbers" are we being offered from Cydonia? If they turn out to be carefully arranged universal constants not repeated on

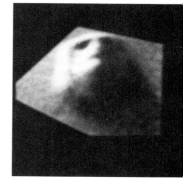

random selections of equivalent planetary surfaces, I would finally have to agree that the bias of evidence has shifted strongly toward those supporting Intelligence on Mars.

At one time I objected to the inclusion of any material on a "proposed terrestrial connection." My bias is that reconstructions based on myth and legendary history are subject to at least three distortions: of the psychological levels of the material itself, of transmission over long aeons, and of the subjectivity of the modern investigator (especially in our era of Hollywood-reinforced projections of the everyday millennial). However, I was ultimately convinced by Hoagland's argument that, if the Face should prove to be an artifact, exploration for a terrestrial connection becomes all but inevitable. To avoid preliminary wrestling with the potential textual evidence from ancient cultures was sheer timidity.

Since then, as Mr. Hoagland has reminded me many times, he took my objections to heart and sought evidence which would specifically override all three objections. The trigonometry of Cydonia that he and Torun have derived from the intrinsic angles and extrinsic positions of the artifacts and the subsequent derivation of a possible energy "paraphysics" from those numbers certainly trumps conventional sacred geometry or legendary history while continuing to require a terrestrial connection, albeit at subtler, less obvious levels. Even more provocatively—in fact, provocative at the ultimate level of megaconspiracy— the proposed connection between Cydonia and the current crop circles (and other snow and field markings) takes the terrestrial connection out of a shadowy archaeological past 500,000 years ago and explodes it potentially into tomorrow's headlines. It is one thing for there to be a proposed consanguinity between eroding old "Sphinxes" and Pyramids on neighboring planets; it is another thing for the architects—or at least the architecture—of that relationship to remain active (or to have become active again). With the "crop circle" connection, we have moved from a kind of moderate *2001* scenario in which ancient aliens left a message for our evolution to an imminent "Close Encounter" with someone (or some thing) which—now that we have met ourselves at Cydonia—refuses to let us turn away from the mirror.

Whatever else we may learn from the mystery at Cydonia, even its apocryphal message is a powerful one.

It is telling us we are very, very old. It screams mutely of an event that occurred in this Solar System before the beginning of time. It is instructing us that our human facsimile is the link, the clue. It is drawing us to look closely this time. It hints that the Egyptian Sphinx might be something else entirely. It says, Third Planet, again, and again, and again—

by its massive piece of statuary, its lines of sight to the heavens, its angles and universal constants, its redundancies. If it speaks at all (if it is artificial), then it utters slow and didactic syllables, making sure we understand them in their precise, prelinguistic syntax.

It is an oracle on another world.

The harder we look the more we see: not ourselves but our shadow; not civilization but primal intelligence; not only what must have been (once upon a time) but also what must inevitably be again. Here merge King Arthur, the Easter Island heads, Osiris, Quetzalcoatl, and the Dreamtime stones of the Australian Aborigines. Here Mount Rushmore, Mona Lisa, and the skull of Pithecanthropus are bound in a single crypt. Sarcophagus deathmask, mermaid figurehead, wild monkey, coin of Caesar, Jivaro shrunken skull, DNA threads. . . scale is irrelevant—there is only one biological trail leading back through the mist.

Hoagland feels that the present *active* force of the cosmic imprint visible at Cydonia has been powerful enough to shatter the Soviet Union, that ideological megalith of the late Twentieth Century. If so, we are at the true edge of history. If not, our present imagination of the Other is invoking something almost as crucial in the human spirit.

In any case, the "argument" between author and editor at the core of the original text has been blown out of the water in this new edition.

I yield to Mr. Hoagland. If tomorrow's verdict proves him right, may it be not another conquistador behind a Trojan Horse but a guest come with compassion and grace for all humankind.

2. A more substantial disagreement of epistemology between Hoagland and me has been that, whereas I feel the fundamental mechanisms of human consciousness (for instance, as described by Buddhism, Freudian and Jungian psychology, shamanic practices, linguistic philosophy, etc.) are immune to transformation by the Face and, in fact, precede it in the absolute chronology of the universe, Hoagland feels that the Face may well turn out to be a major precipitating factor in the history of consciousness itself.

Elsewhere I wrote:

"The original ground on which we stand is this world, the familiarity of our body/minds, language, and civilization, however much in crisis that might seem and no matter that we judge ourselves to have made a mess of it. It is still what we are, by karmic law what we have chosen to be, a materialization of our profoundest de-

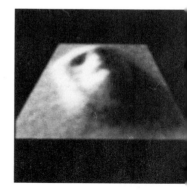

sires. We shouldn't be looking anymore for a literal biblical god, or his fundamentalist equivalent in a super alien biologist. There are probably billions of possible manifestations in the universe and we could transmute among them authentically for innumerable lifetimes, all manner of planets and dimensions and subatomic particles. That's not yet our destiny; even unenlightened, we live, experience wondrous things, breed, and die, and we should not stand ready to cede that to aliens, to submit it to their judgment, or to deem them automatically superior to us—not out of false pride but because we exist only as the legitimate outcome of the forces we express. We cannot be anything else, and I would argue that this is true even if we *are* someone else's experiment—and I will get right to the point on this one: even if we are the genetic experiment of a scientifically advanced race, the experiment can only be a manipulation of existing life forms; it does not include the invention of our whole existence. Our genes can be synthesized, or rearranged; the biological frame of our life can be altered, but biology itself cannot be invented out of nothing by something that is itself a product of the evolutionary universe. . . .

"Put another way—genetic manipulation (which is just an intense form of natural selection) may determine what range of phenomena we deal with as organisms, but it cannot invent the original phenomenology, the impulse for something to become in a universe where nothing might as easily have been. . . . Our creators are not going to be any more enlightened than we are; they will just have advanced technologies. They know no more about how the universe came to be, why it is, and what anything in it is. Their missions and experiments should not trap us and our civilization in the meanings they give to them and the ostensible goals they set (any more than we can change the existential fact of a genetically-altered rabbit)." ("Giving Them a Name" from *The Night Sky,* revised edition, J.P. Tarcher, Inc., 1988).

That the Face itself could also have an esoteric or theosophical meaning is certainly possible, but it will not redefine our entire gnostic and dharmic heritage; it will simply translate terrestrial terms into interplanetary or transgalactic terms, perhaps replacing astral symbology with actual cosmic geography. But the Face (as artifact) at very least must redefine the *exoteric* history and "anthropology" of our planet.

I am reminded of other words I wrote after my first meeting with this creature:

*The Monuments of Mars*

"When I was driving home I fiddled with the radio and happened on an FM station on which, the announcer said, they were going to play a very old version of "Silent Night"—how it might have sounded as a Germanic dialect of proto-Indo-European with tribal instruments. I turned up the volume. Then I heard the bells, the panpipes, the lute . . . and an ancient Christian pre-Christian melody from the snowy north. I imagined were-lights, polar stars. . . . Then the words. When I reached my driveway, I sat there in the car listening. I opened the envelope and took out the pictures. Then the real oldness of the Face struck me—even before the Ice Age, our image in stone, on the Martian tundra. The Shroud of Turin was not so wrong. For being both human and prehuman, for being stone and suggesting compassion and sentience, the Face was archetypally Christos, bringing our fragmented and warring planet together in a single mask, in a unity beyond our history, outside of ordinary time. "Silent Night/Holy Night" for sure, but on another world, perhaps even another dimension of creation, the deep night inside us as well." ("Interview with Richard Hoagland," *Planetary Mysteries,* North Atlantic Books, 1986).

Now on the radio Paul Simon is singing "Graceland": out of South Africa and in an accounting of our troubled times he still affirms—
　　"These are the days of miracle and wonder. . . ."

The Face is one of the wild cards in our present deck. Albeit on a different scale and for different reasons, it has the potential, like AIDS, like revolution in Africa, like the breakup of the Soviet Union, like currency crises and ghost particles, the capacity to transform us.

Among the present voices vying for our attention, it is truly (Paul Simon, again) "the long . . . distance call."

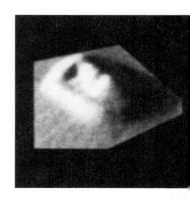

*Publisher's Foreword*

# SECTION I
# DISCOVERY

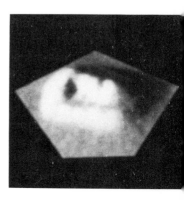

# A MODERN MARTIAN CHRONICLE

> "If Mars is empty . . . we will fill it. But still the
> voice of Mr. Burroughs calls out on nights when
> we pace our lawns and eye the Red Planet: "All
> the evidence is not in! Maybe . . . . "
>
> Ray Bradbury, on the initial survey
> of Mars by Mariner 9, 1971.

One of the most memorable books of my growing up was Ray Bradbury's
*Martian Chronicles*, a haunting collection of vignettes depicting the colonization of the planet Mars. Long before the current wave of respecta-
bility for science fiction and science-fiction writers, this work was hailed
as a masterpiece of literature even by mainstream critics. It therefore takes
a lot of nerve (I'd prefer a term more like "awed respect") to begin a
book with a chapter named after such a classic. But, knowing Mr. Bradbury, I felt the occasion was appropriate: for new evidence *has* been
painstakingly uncovered; evidence which once again may tip the scales
in favor of the kind of Mars depicted in those tales.

Because . . . there once may have been "old Martians," living in cita-
dels amid the Martian deserts. Long before men looked at Mars and
dreamed of going there one day, someone may have looked at Earth and
watched it rise, green and sparkling, before a Martian dawn. We have seen
the evidence—a collection of enigmatic artifacts
lying in the reddened Martian sands—and it is
staggering: the possible ruins of a City, crumbling
for all too many years back into the windswept
wastes of the fourth planet of the sun.

And lying to the east of this collection of

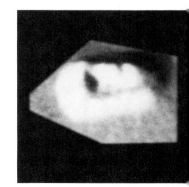

eroding "structures," something even more extraordinary—a likeness of a human face—etched perfectly into a mile-wide mesa where no human face has any business being.

On Mars.

\*　　　　\*　　　　\*

It is the summer of 1976, the American Bicentennial. NASA has just placed two unmanned spacecraft—Viking Orbiters 1 and 2—in Martian orbit, prelude to a far more ambitious step: the actual landing of a robot craft on Mars. This momentous feat is accomplished July 20, 1976, when the first Lander successfully descends from its high flying "mother craft," Viking Orbiter 1. The Orbiter, freed of its precious cargo brought safely to the sands of Mars from Earth, continues its orbital survey of the planet, snapping thousands of unprecedented photographs from its sweeping vigil, which on every orbit dips as close as 1,000 miles to Mars' almost airless deserts. It is on orbit 35, as the spacecraft is cruising a thousand miles above a barren region called "Cydonia," about 41 degrees above Mars' equator in the northern hemisphere, that it takes a picture of an object which will be later referred to as, "The Face." The image, along with additional historic views recorded by Viking's cameras on the surface and hundreds of other pictures flashing back from orbit, will make its way through the maze of antennas, recorders, and computers comprising NASA's world-wide Deep Space Network, and into the imaging area to be examined by scientists with all the others.

Toby Owen, a member of the imaging team on this Sunday afternoon is on his hands and knees with a magnifying glass, looking at a set of mosaic polaroids for a safe site for Viking Lander 2—still a month away from Mars aboard its orbiter. Suddenly he finds on one frame, 35A72, a very peculiar-looking mesa. It looks remarkably like a human face. In the words of Gerry Soffen, the Viking project scientist, Owen exclaims softly, "Oh my God, look at this!"—a perfectly reasonable reaction. One does not expect to find a human face on Mars, needless to say, and certainly not one a mile long. However, after a few moments of, "Gee whiz, isn't that weird," etc., etc., the strange mesa is very quietly forgotten. It obviously cannot be "real." It just doesn't fit any paradigm that one could whomp up, even in the wildest science fiction, for mankind's first brush with an extraterrestrial intelligence.

I was a member of the JPL press corps covering the Viking mission for *American Way* magazine that afternoon. There were about a thousand of us there—from all over the world. Everybody who was anybody: Italian television, BBC, even the Russians. The Japanese occupied one whole quarter of the press room. Everyone was clamoring for data—Life on Mars, the Search, the daily drama, the press conferences, the reports on the experiments: it was a wonderful three-ring circus, as these things usually are. And this one had a special cutting edge because NASA acknowledged that this was it: THE SEARCH FOR LIFE ON MARS.

Gerry Soffen got up in front of us and showed us this quirky face and said, "Isn't it peculiar what tricks of lighting and shadow can do." And then he said, "When we took a picture a few hours later it all went away; it was just a trick, just the way the light fell on it." Those of us who were there accepted this. Gerry Soffen was a very open, very careful, engaging project scientist who typified the spirit around Viking, which was a multidisciplinary, open, American approach to probing the unknown. He went out of his way to try to make people feel comfortable with unfamiliar material. His job was basically to keep eleven teams and several hundred scientists all happy and working together, but he established a credibility among the press by *including* us in the process. In several cases he even allowed interdisciplinary squabbles to be publicly paraded and written about—as part of the process of inquiring into something as unpredictable and unprecedented as looking for life on Mars. So when he said that nothing was there—that it was just a trick of light and shadow—his credibility was overwhelming and certainly dissuaded anyone from doing any hindsight checking. We believed him. The "Face," the mesa, the frame (35A72), were soon forgotten.

Not to emerge from this obscurity until over three years after the events I've just described.

*     *     *

Vincent DiPietro is an electrical engineer with fourteen years' background in digital electronics and image processing. He first saw "the Face" in a magazine—and promptly dismissed it as a hoax. The "magazine" purported to be a journal of *extraterrestrial archaeology!*

Two and a half years later, however, while leafing through the archived NASA photos in the National Space Science Data Center, at the Goddard Spaceflight Center, Greenbelt, Maryland, he was turning over the glassine envelopes—and once again came face to face with the curious image of "a man on Mars." This time he realized it must be something real—or NASA wouldn't have put it, as he later said, "boldly in the file." Now very curious to learn more about this enigma captured by a television camera several million miles from Earth, he sought additional scientific data on the object—and found that there was nothing. Other than noting in its description of the photograph that it was "an oddity of light and shading," NASA had done no further research on the subject.

This was not surprising, when you consider what was occurring at the time the original photograph was taken. Viking scientists associated with the "imaging team" were like kids in the proverbial candy store—up to their hips in extraordinarily detailed shots of Martian geology, ancient climatic features, and those first-ever scenes taken by the Landers. To them, busy with solving the "real" problems presented by the planet, to have wasted time on what was obviously just a trick of lighting would have been absurd.

The image of the curious "face" in a NASA file also intrigued a friend and colleague of DiPietro's, Gregory Molenaar. Molenaar, a computer scientist with a background similar to DiPietro's, was soon intrigued enough to suggest that the two of them together undertake a bit of private space research: a project to improve the quality of the features in the NASA image by a technique known as "computer enhancement."

As images are taken by a spacecraft camera, they undergo several processes en route to your newspaper or your home television screen. First, they are transmitted to the Earth as a set of coded numbers, each number corresponding to a shade of grey in the original scene. After reaching Earth and being intercepted by extremely large and sensitive antennae, the numbers are recorded as magnetic impulses on long reels of tape. Later, these reels are duplicated on smaller reels of tape. It is from these duplicate magnetic tape recordings that images can be "reconstructed" at any future time, by running the tape recording through an appropriate computer equipped either with a TV screen or another means of creating an actual photographic negative or print.

Computer experts like DiPietro and Molenaar, once they had the tape, could play the same kinds of "games" with the numbers as NASA has with countless images from other planets over the past twenty or so years. Such "computer enhancement" of images has become standard technology, an electronic way of altering contrast, removing errors in the original transmission back to Earth, and even improving detail to a limited degree. This manipulation of the original data can dramatically improve the quality of the "raw images" transmitted from the spacecraft (as has been seen in countless NASA publications throughout the years). The only difference here was that two private computer scientists, not NASA, were bringing these skills to bear upon the image of the Face on Mars.

After much painstaking trial and error, after reconstructing the image of the Face in several steps—each one designed to improve some aspect of the image—DiPietro and Molenaar became convinced that standard

*The Monuments of Mars*

techniques of computer enhancement would not suffice. They devised a new technique, different from the advanced mathematical manipulations of digital data NASA has applied to spacecraft images. They called this process "Starburst Pixel Interleaving Technique" [SPIT]. (A "pixel," by the way, is the smallest dot, or *picture element*, in a computer-reconstructed image.)

Each step in this exacting process brought out correspondingly more detail in the eerie human likeness on the Martian desert. When the final SPIT process was applied to the mile-wide mesa with the humanoid countenance, it was obvious that the object was essentially symmetrical: where the other "eye" should be, there was, indeed, a suggestion of a shadow, partially backlit by light reflected from the desert; the "brow" seemed also to conform to the geometry of a human-like depiction (as did the "nose" and "mouth").

At the extremely low sun-angle of the photograph (10 degrees) the desert lighting in the shadow was too dark, however, to allow enough enhancement of these details to be absolutely certain of the symmetry. And it was on such symmetry, unlikely to be found in nature, that initial conclusions as to the ultimate origin of this intriguing object were based.

One of the reasons, apparently, for NASA's initial dismissal of "The Face on Mars" was the perceived lack of any corroborating evidence. There seemed to be no subsequent pictures taken of the structure by either Orbiter—or so the Viking Project Scientist, Dr. Gerald Soffen, indicated when he was quoted as saying that images made of the area "a few hours later" revealed only an ordinary mesa. When DiPietro and Molenaar searched through the Viking Data File for such images, that was true, none existed. If the Viking Orbiter had taken pictures of the region of the Face on a subsequent orbit "a few hours later," no trace of them could now be found.*

However, science does not rest upon asserted authority, but on the

---

*Frame 35A72 was taken at approximately 6:00 p.m., Local Time, resulting in the lengthy shadows seen in the resulting image. An image "a few hours later" would, by necessity, have been taken in total darkness—thus belying Soffen's initial assertions regarding the existence of such an image. In other words, the physics of the situation definitely reveal that the Viking Project Scientist did not, in fact *could not have*, checked the Viking picture file, but relied instead on the overwhelmingly favored presumption: that the whole thing was just a quirk of sunlight. This tendency to presume, rather than to check, would manifest itself again and again in the course of events surrounding investigation of this enigmatic landform.

discovery of evidence. One of the things Vince and Greg should be credited with is not believing Gerry Soffen. They looked through the entire Viking data set, the file, number by number and frame by frame by frame; and they finally found, misfiled, a second set of pictures taken over this area thirty-five *days* later, at a different sun angle. When they blew up one of those frames, which was 70A13 (seventieth orbit, "A" spacecraft, thirteenth frame), lo and behold "our friend" was still there at a lighting change of twenty degrees. This second picture was also taken from a different spacecraft angle—allowing comparative stereo of the mesa and the surrounding areas.

SPIT processing of the second photograph brought out many additional details, making it evident that the Face was no mirage of light and shadow. A distinct chin materialized. The visible eye cavity was sustained, while the "hidden" one began to emerge. Where the apparent hairline was visible on one side it was mirrored continuously on the other.

With the availability of a second photograph and the stereo viewing that afforded, the surface relief on the mesa making up the facial features could also be examined. The researchers found appropriate cavities where the eyes and mouth appeared, and a raised area conforming to the brow and nose (as opposed to a mere change in the surface reflectivity of the mesa in these areas). That, and a striking symmetry between the left and right halves, made it hard to ignore the possibility that something other than a completely natural phenomenon was on the sands of Mars.

DiPietro and Molenaar considered the full range of geological and meteorological events that could have given rise to this bizarre object and concluded that neither wind nor tectonics could reasonably be considered the cause.*

In the course of investigating the surrounding area, searching for additional examples of "wind-faceted" mesas with a curious resemblance to humans, the two computer experts scanned many additional images made by Viking of the area. Between the two orbital sweeps previously identified—the one which took the first picture of "the face," and the subsequent pass 35 days afterward—DiPietro and Molenaar turned up at least 10 photographs of the "immediate countryside"—for a total of approximately 4000 square miles to be examined! In all that real estate, they found nothing even remotely similar. But their search did turn up something else, equally intriguing. In 70A13 there was an immense pyramid, roughly a mile by 1.6 miles. In their report they commented that

---

*DiPietro V., Molenaar, G., *Unusual Martian Surface Features*, Mars Research, Third Edition, Glendale, Maryland, 1982.

*The Monuments of Mars*

"there appear to be four sides that go down to the surface at sharp angles. The corners exhibit symmetrical material, almost as if they were being buttressed. One would expect erosion along a natural pyramid-shaped mountain at the center of the wall rather than at the corners."[1]

Pyramids on Mars were nothing new. Mariner 9, the spectacular mission by the United States which had preceded Viking, had first discovered "pyramids" during its historic orbital surveys of the planet in 1971. And Carl Sagan, in his call for further unmanned exploration of the Red Planet—with a vehicle capable of roving across the Martian surface—had painted vivid images of a robot rolling up to the Mariner 9 pyramids one afternoon, relaying back live television images as scientists learned simultaneously with the watching public what the famed "pyramids of Elysium" actually were . . .

No one, not even Sagan, believed the pyramid-shaped objects detected on those Mariner frames were *real* Pyramids.

Now, far from Elysium (which is actually on the other side of the planet from the region of "the face"), Viking had apparently found another pyramid-shaped object, one that DiPietro and Molenaar noted was aligned with the spin axis of the planet. It was startling to them to find this "artifact" only ten miles from the Face.[2]

This observation was part of a growing list of "coincidences" which this region was beginning to exhibit. Ironically, a search for the Viking frames of the Elysium pyramids revealed another case of misfiling of the Viking data; the pictures listed in the catalogue turned out to be of a region in the *opposite* hemisphere from Elysium. Thus, if Viking also photographed Carl Sagan's "pyramids," which became famous in his series *Cosmos*, they (the photographs) have been mislaid.

After four months of image processing, computer enhancement, and data searches through the Viking File for additional confirming photographs, DiPietro and Molenaar decided to announce their findings. They called a press conference for 10:00 AM, at a Ramada Inn in Lanham, Maryland. Two reporters and a couple of NASA planetary types showed up. The continuing hostage crisis in Iran was apparently of greater interest than the possibility that two independent researchers had discovered the first credible evidence of extraterrestrial artifacts on another planet—even if they didn't make any outright claims to that effect.

A month later, the climate surrounding this "discovery" had changed —seemingly for the better.

Invited to address the prestigious American Astronomical Society's June Meeting, DiPietro and Molenaar soon were inundated with requests from countless numbers of the press for copies of their work. Over a thousand astonomers from all regions of the country attended the subsequent meeting, and more than a hundred actually stood in line to see a demonstration of the three-dimensionality of the famed "Face on Mars" as well as its marked bisymmetrical aspect. Of the many questions thrown at DiPietro and Molenaar from the scientists in attendance at their talk, many concerned the geology of the entire region—which is noted for its unusual erosive qualities—and the probabilities of wind-carving of both the Face and pyramid. But DiPietro and Molenaar retorted that "the non-uniformity of alignment between adjacent pyramids and the bisymmetry of the face leave doubt that nature was totally responsible."[3]

Following the AAS meeting, DiPietro and Molenaar continued to apply new techniques to the existing images of the curious likeness on the mesa. Because of the very low contrast of the original images, they decided to play some more "games" with the computer: they replaced the original tonal values (grey levels) with a range of colors, from red for the brightest intensities seen in the image, grading to blue for the darkest details. They were shocked by what they saw.

There, in the deepest shadows of the photograph, was the unmistakable presence of . . . an eyeball, with a pupil!

But by then nobody was listening.


The central problem lay in the nature of the "evidence."

DiPietro and Molenaar carefully compiled their, by now, impressive mound of data on "the face," spending considerable money to have three increasingly comprehensive versions printed—the last one in four colors! They then sent copies of this Report to as many geologists as they could interest in this unique phenomenon lying in the desert of another planet.

Their objective was simple: a hope that others in the scientific community, moved by the weight of evidence that "something" unusual existed on Mars would undertake a call for a return mission, to verify with better pictures the true nature of "the Face."

DiPietro and Molenaar did not foresee that this strategy—trying to interest members of the geological community in returning to the Red Planet based on even a suggestion of some form of *life*—would be exactly inappropriate. Even before Viking, there had been a basic disagreement between members of the *geological* community and those scientists representing the *life sciences*—about the nature and priorities attached to Viking's purpose. And because Viking had been presented publicly as "a search for life on Mars," and because that search (with the Landers on the surface) had returned conflicting and inconclusive data, the geological

*The Monuments of Mars*

community now felt that too much priority had been given to the biological sciences with Viking. The last thing they wanted to hear, in the midst of budget battles over a proposed equally expensive radar mission to the planet Venus to map its geological details in 1980, was that two outside researchers—and not even geologists!—thought we should send another expensive mission back to Mars—just to check out a controversial contention (no matter how conservatively phrased) that something on the Red Planet left "doubt that nature was totally responsible."

This was the situation, then, which persisted for another three years—from May of 1980 to July, 1983—until this writer, through a series of serendipitous events, became entangled in the story. And suddenly realized that he was staring at the vital missing puzzle piece in DiPietro's and Molenaar's fascinating tale . . .

The City.

## Notes

1. DiPietro, V., Molenaar, G., *Unusual Martian Surface Features*, Third Edition. Mars Research, Glen Dale, Maryland, 1982.
2. *Op. cit.*
3. *Op. cit.*

# II

# DISCOVERING THE CITY

> "Why, sometimes I've believed as many as six
> impossible things before breakfast . . ."
>
> The White Queen
> *Alice's Adventures
> Through the Looking-Glass*
> (Lewis Carroll)

In 1981 a group of us travelled from all around the country, to the small college town of Boulder, Colorado. We came to this idyllic setting in the westering shadow of the "Flatirons," to discuss the planet Mars—and the ways in which we might one day return, to explore the myriad enigmas presented by the Viking missions.

The get-together had been given the provocative (and somewhat hopeful) title, "The Case for Mars." Its organizers were a group of undergraduates from the University of Colorado, who managed to gather, despite their inexperience, some of the most thoughtful conferees ever to assemble for discussions of the Red Planet.

Experts from other universities, several NASA centers, and (of course) keenly-interested members of the science-journalism community had been invited. I was fortunate to be among the latter. The main idea was to share analysis of the outstanding "problems" remaining after Viking, as well as the technological (and political!) constraints preventing us from returning for the answers.

There were the usual array of technical discussions regarding current Martian geology, meteorology, climate—and their impact on future missions to the planet, such as: "Could a manned mission 'live off the land?'" We knew (from Viking) that everything from water to nitrogen was available (although in small amounts) in both the atmosphere and soil. This would make a mission back to Mars most unlike a lunar mission, perhaps more resembling an expedition to our own South Pole. It was in the midst of these dissertations that I chanced one night to wander into an event not even listed on the program.

It was an almost clandestine meeting in a back room of the University Inn where the conferees were staying; and, looking back, I realize I almost missed it—as I came in when the "formal" presentation was concluded, the handful of hangers-on standing around asking questions of the speakers, as they packed up their slides and booklets. I glanced at one of the booklets on the table . . . And came "face to face" with the enigma in the Martian desert.

That was how and where I first met Vincent DiPietro and Gregory Molenaar.

I was intrigued by their improvements in the quality of the "human-looking face," and admired their persistence. I even managed to snag a copy of their booklet. But for some reason, the problem didn't "click." Perhaps it was the limited quality of the photo reproductions in the booklet; perhaps it was the inherent ambiguity of the situation itself; was the Face simply a fascinating example of Martian erosion or was it something else . . ?

Even with their exhaustive dissection of the original NASA image, and their pixel-by-pixel reconstruction, the enigma remained just that— an enigma. For it was always possible that remarkable geological and meteorological factors had conspired on the deserts of the Red Planet to create the most improbable natural wonder in the solar system—a mile-wide "human" head!

With the data presently in hand, there was just no way to know.

I remember taking the booklet home, reading . . . and rereading it . . . and feeling frustrated.

In addition to "the Face," DiPietro and Molenaar had also included reproductions of "the Pyramid," noting its orientation (north) and location (10 miles to the southwest) of the Face itself. In the brief text there was made passing mention of "several other pyramids in the area." But examination of the half-tones did little to clarify their locations (which were not specified), or the criteria used to differentiate them from the half-dozen suspicious pyramid-shaped hills that dotted the surrounding landscape.

One curious "structure" did catch my eye, however: a strange-looking straight-edged object located several miles west of the Face itself. Its strangeness was highlighted by two features apparently not shared by any

*The Monuments of Mars*

other objects in the photograph: two "walls" that met at almost a right angle at the southeast corner; and an inexplicable "keep"—a decidedly square-looking dark space—contained by these outer walls. For a moment I fantasized that I had found some ancient "castle" or a "fortress," its linear "battlements" built so as to afford the occupants with an uninterrupted view out across the desert of the Face. But a moment's calculation was enough to shatter this stray thought; the "walls" were almost a mile in length, the overall dimensions of the "castle" totally dwarfing any counterpart on Earth. No, I decided, the "keep" was merely the triangular shadow of another pyramid-shaped hill located slightly west of the line-like "walls." And they, merely the result of faulting in the landscape. With the small scale of the reproduction in DiPietro's and Molenaar's booklet, it was barely possible to trace the outlines of the hill I suspected of casting the suspicious shadow.

*(Still, for a moment my mind had conjured up a vision . . .*

*(Alien Martian armies, pennants flying, sallying forth from a massive fortress squatting on the rust-hued Martian desert, off on some ancient and inexplicable campaign, one of—how many?—carried out beneath the upturned gaze of the enigma lying out across the sands . . .)*

My next conversations with DiPietro and Molenaar would not take place for two years. In the interim I had become preoccupied with many things, including another mystery turned up by NASA's "Golden Age of Planetary Exploration"—the existence of "a thing in Saturn's rings."

It was in the course of following up new leads to that continuing enigma (a point-like, extraordinarily powerful radio-source orbiting within the center ring of the most beautiful celestial object in the solar system) that DiPietro and Molenaar would once again come to my attention. For, on the edges of the rings is a potential clue to their formation: a collection of small satellites, barely discernible on NASA's published versions of the Voyager photographs transmitted back by the robots sent to Saturn. Contemplating those fuzzy images of the flotsam whirling around the outside of the rings, I recalled DiPietro's and Molenaar's impressive image-sharpening technique—and promptly called them.

My plan had been to locate a private source of funds for enhancement of the Saturn images, for which (I told DiPietro and Molenaar) I would need samples of their work. They responded by sending several copies of the latest version of "the booklet," as well as a collection of actual photographs of the region of the Face, including SPIT-processed blow-ups of the Face itself. In addition, DiPietro threw in several original Viking photos of the region, processed by the National Space Science

Data Center, at the Goddard Spaceflight Center, Greenbelt, Maryland. And it was when the actual pictures came and I sat there looking at the glossies, particularly one approximately five inch square full frame enhanced and processed version of 35A72, that the whole resonant mystery sort of came to a point. . . . And I realized that I was looking at something that was either a complete waste of time, or the most important discovery of the twentieth century if not of our entire existence on Earth.

There is no middle ground. It either is or is not artificial. If it's not, it is not worth worrying about. If it is, it is imperative that we figure it out, because (it comes back to the "Face" itself) *it does not belong there.* Its presence, if it was made by someone, is trying very hard to tell us something extraordinary.

But more was needed—like finding where the "people" lived who made the Face, if that's indeed what happened.

In the back of my mind was the vaguest stirrings of a thought: if the Face had been "made," then it must have a purpose. Find the purpose, the thought whispered, and you will find additional evidence supporting the preposterous notion that it was, indeed, an artifact of intelligent design.

That afternoon so many years ago at JPL, as a group of press passed around copies of the "head" photograph and laughed, someone had jokingly remarked that "the head is to tell us where to land." Now, looking at the image, I realized it wasn't a joke at all—it was actually my opening wedge in getting past its mere existence.

If someone made it, with the purpose of attracting our attention, there was a certain logic to a face. What better way to call attention to a specific place on Mars as a site for further exploration? Place something strikingly familiar on the landscape, a resemblance to the dominant intelligence on the planet most likely to one day send explorers—Earth!

Even with primitive robotic spacecraft, and cameras, the existence of such a mile-long marker could hardly be missed—as Viking's historic photograph—35A72, the 72nd frame taken by the "A" Orbiter on its 35th swing around the planet—amply demonstrated.

All right, so perhaps it was a marker. But marking what? What were we supposed to find in this otherwise desolate region of the planet, at 41 degrees north latitude, 9 degrees west longitude (as terrestrial astronomical convention has mapped the surface of Mars)? Thanks to Vince DiPietro, I had a collection of actual photographs of the entire region. Somewhere, in those frames encompassing thousands of square miles, there might be something Vince and Greg had missed . . .

My eyes were drawn once again to the "fortress" located a few miles west of the immutable Face. On the actual photograph (as opposed to the half-tone I'd looked at years before) there was a host of marvelous detail—all of it, unfortunately, very very small, as the print I had been loaned measured about four and a half inches on a side. This translated to about 6.6 miles (35,000 feet) per inch.

The actual resolution of the Viking image was supposed to be about 150 feet (meaning that, from its thousand mile altitude, the Viking camera was theoretically able to differentiate between two spots only 150 feet apart). If I was going to detect the smallest details in the photograph, I was going to have to use a magnifying glass capable of enlarging a tiny portion of the image to where I could see two pixels about a hundredth of an inch apart!

And I hadn't the foggiest idea what I was looking for.

Once you allowed yourself to cross that "magic" line—between dismissing the Face as merely a remarkable phenomenon of nature, and seriously considering the possibility that someone made it—all bets were off. When you'd crossed over into "the suspension of disbelief," at least three extraordinary possible explanations for the existence of the figure immediately came to mind:

1) The figure had been created by indigenous "Martians" (whatever that means).

2) The Face was the product of a designer from beyond the solar system who, at some time, paid a visit both to Earth and Mars—and left a "calling card" behind to mark the visit.

3) In the dim prehistory of our race, a previous technical civilization had developed, gone to Mars, and left the monument as a message "to the future"—whomever would come after.

There were immense problems with all these possibilities.

Oddly enough, the third—that somewhere down the cavernous halls of geologic time a previous technical civilization had developed on the Earth—was at first the least unsettling to me.

*     *     *

Contrary to popular opinion, we know almost *nothing* about the past history of our own planet, save for details on the scale of several million years. A paleontologist once compared the problem of trying to reconstruct an accurate history of the Earth to trying to reassemble the famed Library of Alexandria after it was burned in 415 A.D. "We have a few scrolls—in most cases merely scraps of parchment—out of the million or so volumes in the library. With these pitiful remnants, then, we try to assemble an accurate picture of the original contents of the library—plainly an impossibility."

If all the plant and animal fossils ever found were stacked within one room, they would probably not fill an auditorium (the human remains would not cover a billard table)[1]—that, out of the countless numbers of creatures and hundreds of millions species which have lived on planet Earth over the last half billion years. As we attempt to probe both the origins of our own planet and the origins of the myriad creatures which have evolved upon it, our accuracy becomes increasingly uncertain as we "go back in time." Of events which happened within the last few thousand years, within "recorded history," we appear quite certain (which has always bothered me, considering the flagrant inaccuracies in even a simple story—with pictures—on the evening news). But when we venture into "pre-history," before the written word, we embark upon a voyage filled with circumstantial evidence, educated guesses, and outright speculation. The further back in time we try to grope, the more the record is distorted by inevitable gaps caused by Earth's inexorable destruction of all "artifacts"—be they bone or metal, and by our own preconceived ideas.

A case in point is the current lively scientific discussion over the origin of the immediate precursors to ourselves, an ancestor once termed "Neanderthal Man" (after the first fossil recovered in the Neander Valley, near Dusseldorf, Germany, in 1856). An even more intriguing mystery is how these humans vanished, and were replaced by so-called Cro-Magnon Man (named after the region in France where its first fossil was discovered). Cro-Magnon, in fact, appears to be nothing less than ourselves—but removed in time by about 40,000 years.

The popular conception of the Neanderthals—as slow, brutish creatures with essentially sub-human qualities—has undergone radical updating as a result of recent discoveries in paleontology and anthropology. In fact, measurements of the volume of their skulls (which is an indication of the size of the brain) imply somewhat greater capabilities for Neanderthals than for ourselves! (Though the experts quickly caution against extrapolating mere brain-case volume into something as abstract as intelligence.)

The fact is that, from the (meager) fossil record, *Homo sapiens neanderthalensis* appears about 100,000 years ago, "reveals a human population complex with a special pattern of anatomical features that extends without interruption from Gibraltar across Europe into the Near East and central Asia . . . down to 40,000 to 35,000 years ago (depending on the locality),"[2] only to vanish "abruptly."

The two parameters limiting our new knowledge of the Neanderthals and their relationship to the humans who came after, are the existence of relatively few specimens which anthropologists can examine, and the uncertainty in dating them. The "40,000 to 35,000 years ago" referred to earlier is on the brink of reliable carbon-14 dating; some specimens of Neanderthal (such as those found in the Shanidar Cave, in Iraq) could be much earlier. Therefore, the key problem for modern anthropologists—finding the transition forms between Neanderthal and Modern Man—seems to come down to accurate dating of a handful of fossils—in a period of Earth's history extending back at least 100,000 years.

[More archaic hominids, such as pre-Neanderthal fossils termed *Homo Erectus*, are totally excluded from these speculations. Not only is there an absence of circumstantial cultural evidence indicative of an intelligence necessary to produce a "high tech" civilization (deliberate burial of the dead with ceremony, "advanced" stone tools, cave paintings, etc.), the skull remains of *Homo Erectus* are extremely primitive, with braincase volumes rarely exceeding 900 cubic centimeters—compared to our average 1300 cubic centimeters, and Neanderthal's 1500.][3]

If one were to suggest seriously that an advanced technical civilization had evolved somewhere in those 100,000 years—or even earlier—the first reaction of the anthropological community would be amusement, followed by a lengthy discourse on what we "know" about the progressive evolution of the period. The clear lack of any evidence of such an extraordinary civilization would be cited immediately as proof of the absurdity of such a claim—particularly the absence of even one artifact of high technology. The only "tools" identified from this period are traditional flakes of flint. (The tools from the even later period, after 35,000 years ago, are viewed as "advanced" only because they use a different form of flint—long, narrow blades struck from the core, which could then be worked into a greater variety of tools. The innovation was also more economical in its use of flint, sometimes a scarce material in the Upper Paleolithic.)

But would we recognize a truly advanced (one of "high technology") artifact existing in this period even if we found it? Since the prevailing paradigm insists such developments have never happened, wouldn't any archaeologist or palaeontologist dismiss a "printed circuit" or "beer can" found in some Paleolithic level as a prank by some college student on the "dig," or with some equally trivial explanation?

But there is far more substantive evidence against the existence of a heretofore "undiscovered sophisticated human technical civilization" existing somewhere within the last 100,000 years: the absence of large amounts of "detritus" from such a civilization.

All human populations leave behind one item almost impossible to obliterate totally: their garbage. Because some of it is in the form of pottery (or those stone tools referred to earlier), the forces of destruction are often less effective in wiping out this evidence of human habitation than they are in crushing and dissolving the human skeleton. Then too, the *numbers* of such artifacts usually exceed by far the number of inhabitants who used them, thus the statistical chances of some of these artifacts surviving even immense periods of time are higher.

This is why we can identify cultures by their tools of stone, even when we can't find any remnants of those who made them. There were apparently once a lot of them, discarded when they wore down, to be replaced by new ones—vaguely prophetic of our own cultural practice of buying a new car each year! Even in a million years (it is argued by analogy) future archaeologists won't have much difficulty in discerning ample traces of our "high" civilization. All they'll have to do is find a junk yard.

Unless, of course, the steel, plastic, and assorted other metals and materials rapidly corrode, in which case all they'd find would be a site unusually rich in certain chemicals and alloys!

The truth is, we don't really *know* how durable the "detritus" of our high-tech culture is, or how rapidly its forms would vanish if our civilization were suddenly struck down by something as devastating as nuclear annihilation. How many "beer cans" would actually survive being buried for a hundred thousand years?

I now raise this argument, concerning the possibility of a previously undiscovered, advanced technical civilization here on Earth—who might have gone to Mars and left the Face—not because I believe that it once happened but because the "evidence" through which it would be dismissed is almost as tenuous as the idea itself.* In going through the arguments

---

*Such "evidence" includes, for instance, "Portalano" charts ("to guide navigators from *port* to *port* . . . ") of the early Renaissance, and maps from the Middle Ages, which suggest that "someone" might have mapped the entire earth thousands of years before Egypt. Some of these Fourteenth-Century replicas, created from much more ancient originals (now lost), appear to show Antarctica with its coasts free of ice, yet from a perspective which requires *spherical*

for and against this admittedly, far-out possibility, I managed to convince myself that something as complex as the evolution of an entire technical society—capable of reaching Mars—could not have happened so "recently" in terrestrial history without leaving abundant "stuff" on Earth for us to find, so widespread is the refuse from our culture. (This may be an example of "throw-away society chauvinism," but so be it.)

The exercise led me to rethink the nature of the artifacts we find from cultures. Rare is the *deliberate* monument or tool constructed for the future. Rather, the archaeologist usually uncovers some prosaic implement used in the day-to-day existence of the people who created it. Exceptions to this "rule" are tombs and monuments dedicated to the rich and powerful, or those monumental architectural examples built for some religious purpose—and all of these are characterized by a scale, uniqueness, or grandeur which requires considerable resources to construct.

Exactly like the Face on Mars!

If the reader is bothered by the application of terrestrial examples to an "artifact" located on another planet, I might offer the reminder that "terrestrial" logic has been a hallmark of discussions surrounding the search for extraterrestrial intelligence (SETI) for over twenty years—from the "economic" argument which states that it is far more likely that a distant culture would send radio messages across the galaxy than spaceships, to those who contend that if extraterrestrials really were ex-

---

*projection*—as seen from high above a *global* Earth! Other maps using this projection depict the Americas, the Arctic, and the Black Sea. This entire collection of remarkably sophisticated cartography—apparently far ahead of its "time"—has been authenticated by extensive mainstream scholarship. The source of the highly accurate information on which the maps were based is currently unknown. Charles Hapgood, of the University of New Hampshire, who for the last thirty years or so has performed the most extensive analysis of this material, concludes the following:

"The evidence presented by the ancient maps appears to suggest the existence in remote times, before the rise of any known cultures, of a true civilization, of a comparatively advanced sort, which either was localized in one area but had worldwide commerce, or was, in a real sense, a worldwide culture. This culture, at least in some respects, may have been more advanced than the civilizations of Egypt, Babylonia, Greece, and Rome. In astronomy, nautical science, mapmaking and possible ship-building, it was perhaps more advanced than any state of culture before the 18th Century of the Christian Era . . ."[4]

*Discovering the City*

ploring Earth today (as "UFOs") they would more likely "land on the White House lawn" than skulk around some backwoods swamp. Both are examples of "terrestrial chauvinism" at its best.

More relevant, carefully chosen terrestrial examples which have a basis in universal principles of physics—such as the amount of work required to create a mile-wide monument on *any* planet—can be used judiciously in any context—provided one is always wary of too much extrapolation from too few facts.

The likeness of the Face to the colossal monuments of Egypt—including an eerie physical resemblance to the deathmasks of the Pharaohs—imbued the photograph I held. Both were the product of tremendous effort (if the Face were, in fact, constructed), requiring the excavation and shaping of several million tons of limestone (in the case of Egypt's pyramids) and, perhaps, ten times that much material scooped out of the Martian mesa to form appropriate eye cavities, a brow and nose having suitable three-dimensional relief, and the mouth. That much effort, even for "advanced technology," could not reasonably have occurred "of one afternoon."

Which brought me back to purpose: why expend such prodigious energies (as would be required to "sculpt" a mile-wide mesa into such a striking likeness) unless the reason was of almost overwhelming importance to the builder?

Suddenly, the idea that someone had created this extraordinary piece of engineering as "a message to the future" seemed less likely than the alternative: that it had been done for reasons vital to the culture which had built it . . . and it had lasted long into the future just by the good fortune of being on a planet where erosion is now almost nonexistent. [Close examination of the Face itself reveals evidence of some erosion; the "nose" appears to be broken off. Where the tip should be, there is a suggestion (on the SPIT images) of a flattened area. The "tear" that DiPietro and Molenaar identified in their color-slices of the grey levels I believe to be the fragments of the "nose," which apparently rolled down the "cheek"—one piece landing below the "eye" and the other continuing until it came to rest on the desert floor itself beside the "hair" which frames three sides of the figure. In addition, there is evidence of a small (several hundred foot) crater on the face, slightly below the eye line on the bridge of the "nose" itself. Such erosive features suggest some age to this impressive structure—under current Martian atmospheric conditions—at least several million years.]

Didn't the level of energy expenditure merely to create this monument tell us something vital about the society which built it (assuming still that it was "built")? That the culture which created it was either rather primitive, or very far advanced, but not somewhere in between? (Careful! Am I extrapolating from too little information? Follow . . . )

A mid-level culture (such as ours) would probably require an economic or other short-term incentive in order to undertake such effort. In a truly democratic structure, where individuals must reach consensus about discretionary uses of the society's wealth, getting the resources for a project which costs a lot, yet will take a long while to get results, is very difficult. That's why, in our society, the space program itself is in such trouble; we are trying to convince an entire culture, with very short-term interests, to commit a percentage of its wealth (and a *small* percentage at that) to exploration of a new domain which has the potential for extraordinary payoff—in the time frame of a generation.

Only a military project could command the level of effort required to complete a project like the Face on Earth today, and there just didn't seem to be any obvious "military" application to a mile-wide figure in the desert.

Which left "primitive" and "very far advanced" societies as explanations.

Extrapolating from current trends on Earth we can see (nonetheless) that it won't be very long before individual non-governmental institutions, and even wealthy individuals themselves, command the kind of technology and power required to create something like the Face. Simple advances in robotics, computer technology, spacecraft design, and the associated wealth inherent in harvesting the riches of the solar system will eventually enable even one person to create "works of art" on such a prodigious scale. Even now, artists such as "Christo" are able to run miles of fabric across landscapes, wrap entire islands in pink plastic—such is the economic and technological level of the society in which they live. And the social freedom which is essential to their opportunity to carry out such projects in the first place.

Is the Face an example of an extraterrestrial "work of art" created by some interplanetary "Christo" with immense technical resources?

While raising all sorts of fascinating questions (for instance, "Why did the artist choose a *human* face to illustrate . . . and on the planet *Mars*?"), such a hypothesis has little that is testable—and is thus of negligible worth in evaluating the basic question, "Is 'the Face' really an artificial construct or 'merely' an extraordinary work of nature?"

(There is also the correlative thought: Mars could be littered with the faces of the artist—or of "the Martians"—and we would never know it! Only the striking resemblance to *us* made us look twice at this otherwise totally unprepossessing piece of desert . . . )

Having attempted to use logic as a tool for finding out just *what* to look for in the photographs scattered before me on the desk, I now remembered the other item DiPietro and Molenaar had pointed out:

The "pyramid."

If it was a pyramid, it was immense: a mile along the short side, by 1.6 miles long. There it rested, a few miles southwest of the face. And there, just as they'd described them, were the remarkably rectangular four corners—complete with the "buttressing" they'd talked about.

In the original booklet I'd seen in 1981, the authors had talked about erosion as the cause of the distortion evident in the Viking photographs. With the availability of the actual photographic prints, including a SPIT-processed enlargement in another 4.5 inch format, the distortion seemed less the result of Martian sandstorms than the product of two "direct hits"—meteor impacts.

One impact had evidently taken place near the top of the pyramid on the eastern side; the other had occurred much lower down on the flank, but on the same side of the "structure." Both were evident from the characteristic crater-like appearance at the impact site. Material thrown out of the craters as "ejecta" (so termed because it is "ejected") could be traced out across the floor of the valley to the east of the pyramid, as a multi-lobed discoloration on the surface. Additional ejecta could be seen on its western front—evidently flying over the top and cascading down the western side as streams of lighter "stuff" (thinly covering the darker material of the mountain on which the pyramid was "built") flowing outward in radial streams from the pyramid's base.

DiPietro and Molenaar were right; this was the most interesting version of a pyramid yet found on Mars, far more compelling (as an actual construction) than the three-sided affairs seen in the Mariner 9 photography. Yet again, it was unique. One could always argue that the strange permafrost conditions of the planet, coupled with the winds, were able to create exotic "structures" unlike any seen on Earth. After all, both these enigmatic objects—the "pyramid" and "face"—were within the region of the planet interpreted by the geologists as "stripped plains." This designation simply translates: the northern hemisphere of Mars is most weathered and eroded, by a combination of running water and the atmosphere itself—sometime in the yawning geologic past of this strange planet.

Almost any unexpected landform, no matter how improbable, might therefore be created under these unpredictable conditions. Or, given the

4.5 billion years that Mars had been around, even a four-sided mountain—with ruffles at the corners!—might somehow form.

But the existence of the "pyramid" had a perceptible impact on my train of thought; what were the odds against *two* "terrestrial-like monuments" on such an alien planet—and in essentially the same location? Moreover, if construction of the Face was not merely an exercise for an afternoon—even for advanced technology—what about the creation of a pyramid measuring a mile, by two, by half a mile in height (as determined by a simple calculation, based on the sun-angle and the length of shadow)?

For, no matter what its purpose, the Martian pyramid contained over a cubic *mile* of "earth"—piled up to a 30 degree angle to the surface! Even under "Martian gravity," which is about 38% the surface gravity of Earth, such a feat was staggering. But even more, if the pyramid was the result of a gargantuan project in civil engineering, it provided the crucial clue to what I should be looking for by way of confirmation . . .

Where the "people" lived while they were building it.

It was becoming obvious; if you accepted the premise that the face and "pyramid" were too much of a coincidence—strikingly familiar and in the same location—to be natural formations, then, by elimination of hypotheses, you were left with only one reasonable alternative: whoever made them had a long-term purpose and (from the engineering represented by the "works" themselves) would have required a considerable period of time for their completion.

Ergo: where did they live during the construction?

Disregarding for the time the obvious environmental problems posed by the currently inhospitable Martian atmosphere and climate (which will be dealt with in greater detail in Chapter III), I found myself being forced by the evidence into accepting what I originally had not wanted to believe:

That there were, indeed, "Martians"—if not now, then sometime in the long and varied history of this enigmatic planet.

With this series of almost inexorable clues, it took me about half an hour, scanning the small photographs before me—to find them. Or rather . . . where they might have lived.

The "City."

Ironically, it wasn't on the SPIT-processed versions that I found it, but on the higher-contrast montage Vince and Greg had assembled from the prints furnished by the National Space Science Data Center. For the critical details were not very small; they were very *large*—and showed up best in the large-scale mosaic.

The "City," once you knew what you were looking for, almost leaped off the page—a remarkably rectilinear arrangement of massive structures, interspersed with several smaller "pyramids" (some at exact right angles to the larger structures) and even smaller conical-shaped "buildings." The entire gathering measured something like 4 by 8 kilometers (2.5 by 5 miles); a strikingly rectangular pattern created by numerous features at right angles to each other, including aligned corners and even "streets" running roughly north and south.

In the center of the "City" was a highly unusual arrangement: a rectangular layout of four structures—with a fifth one in the middle.

While the objects on the perimeter were oblong, the fifth object in the center was exactly circular. And, no, the feature did not appear to be an "eroded crater"; the breaks in the "walls" were too extensive and too exact. A line "north" and "south" laid across two of the structures and across the center of the configuration exactly bisected another line running "east" and "west." And where these two lines met, in the center of the arrangement of structures, was also the exact center of the fifth. The entire pattern looked very artificial; at each right angle appeared one of the four "buildings," set precisely on the circumference of the pattern.

The arrangement reminded me most strongly of a city square!

Then I noticed something even more intriguing; the entire "city" was located such that the "inhabitants" would have had a perfect view of the inexplicably human-looking object lying a few miles to the east—the Face. The object I'd fantasized as a "fort" those years before was, in fact, the easternmost structure in this entire complex; its strikingly linear "walls" adding to the growing impression that it too was part of a deliberate plan . . .

An entire City built so as to have an unimpeded view, out across the desert, of the "monument."

I took out a ruler and laid it on the photograph, at right angles to the central axis of the Face.

It passed directly across the "mouth" of the upturned enigma lying in the Martian sands, west—precisely to the central building of the "city square!"

Underneath that line, a short distance east of the "square" itself, was a faint track etched in the surface—running directly toward the distant face. It looked like a mound of earth, with a subtle shading of shadow, as if deliberately constructed to mark some specific direction leading out of the city square—as if some ancient sightline had been laid out to highlight the importance of that particular alignment.

It was part of a growing gestalt forming in my consciousness about the City: that the entire set of massive structures, associated smaller ob-

*The Monuments of Mars*

jects, and repeating right angle patterns were arrayed parallel to that same track . . .

As if it were the "main avenue" aimed from the center of the City toward the one inescapable point of interest (for a human being . . . ) on the entire landscape.

The Face.

That all of this could merely be "coincidence" was rapidly becoming more improbable than the alternative: that I had, indeed, found some kind of artificially constructed Martian "complex"—and all that such a phenomenal discovery implied.

*      *      *

Extraordinary excitement and outright disbelief began to war within me that night . . . which has continued through this moment, as I attempt to relate exactly how I came to discover the first evidence that I would begin to call "conclusive," that there were, indeed, "Martians."

An entire city laid out—on Mars!—with the precision of a Master Architect, and on a scale as amazing as its apparent purpose: To afford a unique view of the equally astounding monument lying to the east.

That the two formed a unit—the City and the Face—was undeniable. Each reinforced the other; without the existence of the Face, I wouldn't have looked twice at that particular piece of Martian desert. Without DiPietro's and Molenaar's refusal to let the enigma fade back into the long solar system night, none of us would have looked twice . . .

The Face itself would always have remained just that: an enigma.

But, the City and the Face, and the inescapable geometrical relationship which linked them (as I was to discover), formed an argument that compelled further investigation.

For already my mind was racing, snatching at fleeting thoughts and hypotheses to explain what I was seeing.

If there once were Martians . . . where were they now? How could they possibly have ever lived—much less evolved—on such a "hostile" planet? And finally (but certainly not least!): why did they expend such prodigious energies in laying out a city, much less carving the Face, itself perhaps a representation of a race of creatures who would arise on another planet, Earth—with whom they could never come in contact!?

From all the available evidence gathered by Mariner and Viking about the physical make-up of the planet, including its geological and climatological history stretching back countless aeons to the past, the "Martians" who had created this amazing complex had to have disappeared ages before the human race—or even its ancestors—evolved.

How could a race of extinct beings create a monument to someone who would not appear in the same solar system—let alone on the same planet—for several million years . . . ?

## Notes

1. Johnson, D. and Maitland, E., *Lucy: The Beginnings of Mankind*, Simon & Schuster, 1981.
2. Trinkaus, E. and W.W. Howells, "The Neanderthals," *Scientific American* Vol. 241, Number 6, 1979.
3. "The most important item of Dubois' original discovery [of *Homo Erectus*] was a calvaria of *Homo Erectus*, which, in the very flattened frontal region, the powerfully developed supra-orbital ridges, the extreme platycephaly, and the low cranial capacity (estimated at about 900 cc.), presents a remarkably simian appearance." Le Gros Clark, W., *The Fossil Evidence for Human Evolution*, Chicago, 1964.
4. Hapgood, C., *Maps of the Ancient Sea Kings*, Chilton Books, 1966.

# III

# THE ONCE AND FUTURE MARS

"Mars ain't the kind of place to raise your kids
"In fact, it's cold as Hell.
"And there's no one there to raise 'em, if you
dig . . ."

Elton John
*Rocket Man*

Elton John is right—now.

But was Mars always "cold as Hell"? That is the kind of question which has haunted planetary scientists and "astrogeophysicists" since the "Martian revolution" of over twenty years ago—July 14, 1965.

That was the night the first American Marsprobe—Mariner 4—flew a looping course of over two hundred fifty million miles, climaxing its spectacular interplanetary odyssey with a fly-by a mere 6000 miles above the fabled Martian deserts. What Mariner 4 relayed back to Earth that night was to change the way scientists (and thus the rest of us) forever looked at Mars . . . or so we thought. For the Mars that Mariner 4 discovered, through its 22 primitive television "photographs," and by passing its own radio signal through the planet's atmosphere, was a far more hostile planet than anyone had contemplated even months before.

I have maintained ever since that summer July evening (and the first published close-ups of Martian craters to appear on the front page of the prestigious *New York Times* the following dawns) that the American space program began to die that night, even as Mariner 4 shattered our last "Lowellian" dreams of finding any "Martians."

Lowell? Who's Lowell?

Percival Lowell, member of a well-known, upper-class family from

Boston ("The Lowells talk only to Cabots, the Cabots talk only to God . . . "), was probably the single individual most influential in shaping our pre-Mariner conceptions and expectations. He believed in the existence of intelligent beings on the planet Mars, and in 1894, he founded an observatory at Flagstaff, Arizona, to prove his belief.

1894 was the year of a particularly close "opposition" of Mars, from the vantage of Earth. Because of the "race-track" nature of the two planets' orbits, Earth "lapped" Mars about every two years—catching up and passing the slower-moving outer planet. Because of the elliptical (egg-shaped) Martian orbit, all close-approaches were not equal. At times Earth would pass Mars at a distant sixty million miles; at other times, the difference in the radii of the two orbits brought Mars within thirty-five million miles. (These close passages inevitably took place when Mars was directly opposite the sun—hence the term "opposition.")

The spacing of these particularly favorable oppositions, in contrast to the intermediate distances of Earth's "fly-by" of Mars about every two Earth years, was more like once every 17 years. And in 1894 was the closest opposition of Mars since the last such "close-encounter"—in 1877.

1877, as planetary scientist Bill Hartmann once described it, "was a banner year for Martian studies."

It doesn't take much scientific background to guess why: the closer Mars could come to Earth, the more detail astronomers could see upon its surface. In 1877, because of certain subtleties associated with that egg-shaped orbit, Mars came closer to Earth (or, we to it) than for almost a hundred years. This fortuitous circumstance was also well timed; technological progress—in the form of better telescopes—had spawned some truly fine observatories around the world, coupled with a generation of astronomers who knew how to use them.

The result, at the opposition of 1877, was a series of "landmark discoveries" about the famed Red Planet, including the discovery of its two diminutive moons, Phobos and Deimos, by the American astronomer, Asaph Hall; the identification of high-altitude "white spots" hovering above the Martian sunsets and dawns as "condensation clouds;" a total renaming of all identifiable features on the planet after classical references in the history and mythology of the Mediterranean region (to put an end to the previous confusion caused by astronomers naming various features after other astronomers—some still living!), and one thing more . . .

The "discovery" of "canals."

Actually, these most infamous of Martian features—supposed line-like markings covering vast areas of both the bright regions and the dark ones—were detected decades before 1877, by an Italian Jesuit, Father Secchi. And even he acknowledged that previous observers had seen some of them.

But in 1877, in the general high interest surrounding both the close approach of Mars and the new telescopes arrayed to greet it, the social atmosphere was "ripe" for popularization of the existence of "the canals." And so, when Giovanni Schiaparelli, a former classical scholar who, in fact, had taken the initiative and renamed the previous Martian features after famous places in and around the Mediterranean, wrote in addition about "the canali" or "channels"—his descriptions were picked up by the popular press, and "canali" became mis-translated into English as "canals."

The rest, as the cliché goes, is history.

The "canals", and the presumed existence of someone on Mars to dig them, would become the focal point of a scientific and popular controversy which would rage over sixty years—until the Mariner 9 spacecraft proved they didn't exist—by photographing almost the entire planet from orbit.

But in between, the controversy over the existence of canals (who could see them, whether they suddenly became "double" between the two-year oppositions of the Earth, which regions of Mars had them— "bright" or "dark") carried on. Including, of course, the general implication of a network of geometric "works of engineering" on the planet:

Who constructed them?

And in this area, it was Percival Lowell who led the fight for civilized Martians.


Popular opinion (not to mention scientific assessments) of the habitability of Mars has swung wildly back and forth. At one time, in the early 1800s, the scientific community's picture of the Red Planet was that of a world of oceans (the dark regions), lakes, dry land (the reddish bright areas), clouds, polar ice-caps, and a day slightly over 24 hours long—in short, a planet not too different from the Earth.

By the end of the nineteenth century, scientific opinion (and thus public perception) was in the process of transition; Mars was increasingly revealed to be far more desiccated than had been assumed even a few years before.

The failure to detect "sunglints" from the presumed "oceans" pretty much eliminated large expanses of water as the explanation for the extensive dark regions of the planet. Later, when "canals" were reported to have been seen *crossing* some of the dark regions (which obviously couldn't happen if they were, in fact, large reservoirs of open water) the picture of a dry, increasingly desertlike planet came into favor. This growing impression was buttressed by simple physical calculations: For instance, the amount of atmosphere Mars possessed could be estimated in one of two ways, by assuming Mars had as *massive* an atmosphere as the Earth, only thinner at ground level due to the lower Martian gravity (which wouldn't compress it as much as Earth's higher gravitational field); or, by assuming that Mars had been given an atmosphere in proportion to the mass of the planet. Since the latter was only about 11% that of Earth, for an equivalent percentage atmosphere, thinned out by the reduced gravity, the ground-level pressure on those "Martian deserts" worked out to barely 14% that of Earth—or the equivalent for a person on Earth, of trying to breathe at about 50,000 feet.

Water, under these conditions, could not long remain liquid on the planet; if present in large, open reservoirs—such as oceans, it would rapidly evaporate. In fact, the only stable source of water on the entire planet—according to this picture—would be the polar caps . . .

Which is precisely where Lowell thought the "builders of the Martian canals" had gone, in their desperate attempt to divert the last remaining reservoir of water to the equatorial regions—where they battled "a dying planet."

Because of the theoretical uncertainties in pinning down the exact pressure of the atmosphere, and the almost total lack of any observational data on this important point, the "new Mars" of the Victorian Era hovered on the brink of habitability. While physical calculations continued to make Mars more and more unlikely as a place where life—let alone intelligent life—existed, Percival Lowell began a vigorous public campaign in defense of his own theory:

That not only did the Martians exist, but that their works of "super engineering," visible even on Earth as a planet-spanning canal network, implied advances far beyond ours—not only in terms of scientific accomplishment but in terms of social progress.

(There is a valuable lesson here, about too much extrapolation from limited geometrical evidence . . . )

Lowell's conception—of a dying, increasingly desiccated planet, on which the inhabitants were eternally engaged in a never-ending battle for survival against the inevitable—caught the popular imagination like nothing else in the science of the period. But his extension of these ideas—that "the Martians" had unified their planet into a single political entity before the implacable "real enemy" (Mars itself) was, somehow, lost. Another well-known popularizer of the period, H. G. Wells, in 1898 turned

Lowell's portrayal of a race of valiant Martians into a race of "imperialistic and ruthless Martians" who would stop at nothing, including an invasion of the Earth, in their pursuit of water and a "new beginning" for their dying civilization.

Thus did Wells' *War of The Worlds* become the template for innumerable Grade B movies of the 1950s, as well as the very term "BEM"—"bug-eyed monster." This xenophobic distortion of Lowell's original conception, grounded in part in the blatant imperialistic exploitation of what we now term the Third World by Britain and almost every other "civilized" country of the period, was to shape popular perceptions of extraterrestrial life for decades that would follow—a classic example of accumulating extrapolation from almost nonexistent facts!

Lowell's imaginative and changing canal network, amplified in several books[1] he published describing his observations of the Martian deserts from (ironically) the clear air of a terrestrial analog in Arizona, were ultimately shown simply not to exist. Or, as Carl Sagan once succinctly put it, "There was never any question about the 'canal network' having been created by 'intelligence.' The only question was, 'On which side of the telescope did it exist—on Earth or on the planet Mars?'"[2]

Successive Mariner missions to the planet, beginning with Mariner 4's historic fly-by that July night in 1965, would demonstrate that the canals had been an optical illusion—the combination of random small detail actually on the planet, and the eye-brain tendency to attempt to organize that detail into a comprehensive pattern.

Thus died the Martian legend . . . and "the Martians."

The planet that these epic space voyages revealed[3,4] was orders of magnitude more "hostile" than even the most severe Lowell critic of that era could have imagined.[5] The atmospheric pressure was finally measured as less than 1% that of Earth—similar to an altitude 100,000 feet above our planet. Similarly, temperatures—except for a limited region straddling the equator—never exceeded the freezing point of water. And water itself, although thought to be a scarce resource, was determined by a combination of Mariner and Viking orbiters, and then the Viking landers, to be a *vanishingly* scarce resource; if all the water available in the atmosphere and in the polar caps were condensed from all over the entire planet, it wouldn't be enough to fill Lake Erie—on a world with a surface area equal to the combined continental area of Earth!

And without abundant liquid water, so said the biologists, life on Mars was very dubious, at best confined to microorganisms.

Another telling argument against the existence of some form of "ad-

vanced life'' was the total absence of free oxygen. Instead, over 90%
of the almost nonexistent atmosphere was $CO_2$—carbon dioxide, the stuff
which makes the "fizz" in soft drinks. If life did currently exist on Mars,
so observed some astronomers, why wasn't it turning some of the carbon dioxide into oxygen—as the process of photosynthesis does all the
time on Earth? (The idea that extensive biological activity could take place
on a planet, and *not* be dependent on the sun in some significant way,
was inconceivable—mainly for energy considerations).

Then there was the "ozone argument."

Without oxygen in the planetary atmosphere, in the form of the tri-
atomic molecule which has achieved a certain "celebrity status" back
on Earth—ozone—any organisms present would quickly succumb to lethal
ultraviolet radiation—sunburned into non-existence.

In this space-age portrayal of a world—not merely "dying" but quite
dead—there was one additional factor which, in almost everybody's mind,
"sealed the coffin":

Craters.

A few far-seeing astronomers, like Opik, Baldwin, and Tombaugh,
in the decades before we sent our spacecraft, had predicted the possible
presence of such features—the result of random asteroid impacts. But,
in the wake of finding "crater piled on crater" on the Moon—which was
confirmed as definitely "dead" by a variety of robot spacecraft—the
numbers (and the state of preservation) of the Martian craters was both
startling . . . and quite depressing.

For, by a visual analogy with the unending scene of barren craters
on the Moon, this Mariner-relayed landscape more than anything con-
veyed a visual impression of another utterly "dead" planet. And in a
sense, this was an accurate impression; for the preservation of so many
craters implied almost no erosion—no soft rains before the Martian mornings, no wild hurricanes spawning tornados in a Martian "monsoon season." But even more fundamentally, the presence of so many ancient
landforms (for almost all the Moon's craters were formed in the "wee
hours" of the history of the solar system—nearly 4 billion years ago)
implied that Mars was "lifeless" geologically as well as biologically; we
would find no great chains of mountains rippling across its surface, no
raised "continents" above abyssal "ocean basins,"—as the dynamic internal process of our own Earth have created . . . and destroyed . . .
countless times in the history of the solar system.

Those craters—first seen in a few television images transmitted back
by Mariner 4—were a giveaway; the preservation we saw was the preservation of the Tomb—for Mars had apparently died long ago. Stillborn.

This was not the picture envisioned just a few short years before by Lowell—a slowly "dying" world, "ahead" of Earth (in terms of evolution) because its once abundant atmosphere had slowly leaked off into space. Rather, it was a Mars which never had a substantial atmosphere—because it lacked the internal heat sources necessary to "boil off" the more volatile gases (as had happened here on Earth).

The presence of those craters, and the absence of any other familiar geological features formed by internal heat—such as volcanos—was the clue.

In 1969, the same year American astronauts in triumph reached the "dead, grey Moon," additional American robots to Mars confirmed this dismal picture, adding the one planet in the solar system where we had really thought we'd find something (if not "somebody") waiting, to the growing list of places in the solar system that were lifeless.

Thus did advancing American technology obliterate the last vestige of "the Martian legend."

For if the Mars that we were seeing—a small, glaciated world, an "eternally-frozen desert"—had always been like this . . . then hope was forever gone that once there could have been a spark of life upon the planet. From the best evidence, not only was any living form impossible right now—it had always been impossible.

The effect this profound realization had upon the scientific community was considerable. But it was less than its effect upon the public, which would exert far-reaching "political" repercussions.

For, if one looks at the amount of funding spent by government on science—particularly the space program itself—one can't help but notice a dramatic falloff after 1969—the year all hope of Martians officially was banished.

Particularly hard-hit was NASA's unmanned planetary program—evident in the fact that, following Viking (which was already planned, as of 1969), no further missions back to Mars have been authorized by Congress, or even requested by the space agency itself for almost ten years. Further, with the exception of the Galileo program—to orbit Jupiter and probe its atmosphere—no new explorations of the solar system *of any kind* have been funded by the government of the United States, since the equally dismaying (to the public) "failure" of the Viking landers to find a trace of life, from their unique position on the actual surface of the planet.

Without too much exaggeration, one could cite the currently dismal state of the once brilliant U.S. planetary program as an excellent example of "killing the messenger"—the very tools through which we learned that we inhabit the third planet of the solar system (apparently) in solitary splendor.

To an entire generation raised on the timeless possibilities contained within that phrase, "the Martian Legend"—from the fantasies of Edgar Rice Burroughs, to the haunting images evoked by Bradbury's own immortal tale, of Men from Earth striding, silent, through the now-deserted Cities of the Martians beside the moonlit canals—the messenger indeed brought bitter news . . .

That the Martians were not there.

Long live the Martians! Who are no more . . . and never were.

But what if the evidence, whereby "the Martians" had been "tried"— and found nonexistent—was slightly incorrect? What if a door which had been firmly closed, was once again reopened . . . if only just a crack?

In 1971, precisely such a remarkable thing happened.

The occasion was the achievement of placing humankind's first artifact in orbit about another planet—Mars. Mariner 9, the successor to the spacecraft series which preceded it, was that artifact. It would become the first manmade device placed as an artificial "moon," circling "forever" around another world (although, its orbital lifetime was actually put at "more than 50 years"; an attempt to prevent the unsterilized spacecraft from contaminating Mars until we totally eliminated all possibilities for some form of life. Even in science, there is such a thing as hope . . . ).

Imagine, then, after the dismal portrayal of Mars reported back by three successive spacecraft, the shock it was to find out that Mariner 9 was in effect orbiting a *different planet* from the one we thought we knew. For, from its vantage point in orbit, the details in the television images that streamed back from this spacecraft were of a place almost the complete opposite of the Mars in those brief "snapshots" taken as we'd flown past with Mariners 4, 6, and 7.[6]

Oh, the atmosphere was just as bleak (from the perspective of something that might attempt to breathe it), and the temperatures were just as drastically extreme—from a "balmy" 50° F. in the noontime sun on the equator, to a frigid −200° F. at night. No, it was the ability to examine *photographically* almost the entire planet (as contrasted to the mere 10% that was imaged by the previous fly-by spacecraft) that caused planetary scientists almost totally to reassess their "first impressions."

By an extraordinary coincidence, each of the previous missions had flown past Mars and photographed the same hemisphere—the "dull" one! In the Mariner 9 mosaics of the surface, as its orbit took it completely around the planet twice each day, image after image revealed literally "a whole new side" to this continuing enigma of the Red Planet.

For Mars was apparently divided into two extremely different hemispheres, the one we'd had the bad luck to examine first (the one with all the craters) and the other one, which contained a host of extraordinary features, unlike (in terms of scale) those found on any other planet. The first of these was quite literally the "largest shield volcano found in the entire solar system"—convincing proof (at last!) that Mars wasn't "dead"— just "sleeping."

For the existence of such a huge amount of lava [the volcanic mountain measures over 27 kilometers (16 miles) in height, and is over 500 kilometers (300 miles) across ], implied a substantial source of heat internal to the planet. In fact, the feature (to be known as "Olympus Mons"— the Mountain of Olympus) quickly became only one of four such "shield volcanos" Mariner 9 would discover in this one region. The others, lying on a ridge-like elevation running roughly northeast to southwest, would all be comparable in height. Gravity measurements (made by tracking the orbit of the spacecraft as it repeatedly passed over this "planetary bulge") would reveal that this "Tharsis Ridge" seems to be the result of an enormous ejection of lava onto the external crust of the planet. The fact that such a "bulge" existed at all (in seeming contradiction to a tendency for such large masses to depress the crust) introduced an even more extraordinary concept into the on-going discussions over the past history of Mars:

That Olympus Mons was, in fact, among the youngest features on the planet, its vast lava flows so recent that they had not yet depressed the planetary crust beneath their mass.

In this view, Mars was not only far from "dead"—it was just now "heating up," as the age implied for the last eruptive phase of Olympus Mons was "only" a hundred million years!

While the geologists on the Mariner 9 team were attempting to sort out the implications of this remarkable discovery, Mariner 9's cameras were sending back additional surprises—such as strange pictures of "poker chip" terrain around the Martian poles. Visible in the most detailed images (which had the highest resolution), this peculiar "layered terrain" was revealed to consist of alternating "light stuff and dark stuff," apparently laid down only in the planet's polar regions. Initial speculation as to what caused the enigmatic layering centered on sequential deposits of frozen carbon dioxide "ice," the darker layers consisting of this "dry ice" mixed with dust.

It wasn't long after this discovery that geologists proposed a fascinating explanation for this multi-layered geological formation: that it

*The Once and Future Mars*

represented episodic deposition and/or erosion of ice-layers responding to drastic changes in the Martian climate. The arrival of the spacecraft during the height of a planet-wide duststorm in November, 1971, had sensitized everyone attached to the Mariner 9 team to the effects of such enormous storms—which had literally blotted out all features to both Earthbound telescopes and Mariner's own cameras for several months. Now, the possibility that such duststorms varied in both frequency and severity as a function of planetary climate, leading to the deposition of layers of ice on the polar caps containing different amounts of dust, opened up another fascinating prospect:

That, somehow, Mars might have had a *different* environment in the past—including one with a substantially greater atmospheric pressure, and (wonder of wonders) one which might have made possible liquid water on the surface.

As these discoveries and implications were being pondered by the scientists in charge of analyzing the pictures coming back, Mariner made perhaps its most critical discovery—which was to color the interpretation of both the "poker chip" terrain and the ages and duration of the volcanism mentioned earlier.

It found the channels.

No, not the "canals"—channels. Natural, erosive patterns in the now-arid ground which looked for all the world (in many cases) like the meanderings of "once mighty Martian Mississippis!"

The discovery touched off a firestorm of controversy among geologists, which has not abated much even at this writing—over fifteen years after they were found. For in the wake of the previous Mariner "evidence," of an arid, hostile Mars lacking in so much as a cup of liquid water on the contemporary planet, the growing suggestion that *liquid water* could have carved the features seen wending their way across more and more pictures of the surface, was a devastating reversal. Many tried to "explain" the features as the result of some other process—from fractures in the surface, to erosion caused by another kind of liquid (an "exotic hydrocarbon" was suggested). One scientist jokingly remarked, "About the only thing *not* suggested was champagne . . . and only because there's so little of it in the Universe!"

And all because the very presence of the channels—if they were indeed carved by liquid, running water—implied a totally different kind of Martian environment from the one we see at present—including a much denser atmosphere, actual rainfall (!), and temperatures above the freezing point of water over a good deal of the planet.

Shades of a Lowellian Mars!

The door had been reopened . . .

At the close of the "extended mission" of this remarkable member of the "Mariner family," and for many years thereafter, opinion in the scientific community regarding Mars was divided: between those who readily—eagerly—accepted this startling evidence of a "different kind of Mars" sometime in the geologic past; and those who remained stubbornly attached to the "old Mars"—the dry, extraordinarily bleak world revealed by Mariner 4.[7]

To these latter geologists, the persistence of "ancient cratered highlands" on Mars refuted the more exuberant claims of the group now favoring a previous, wetter epoch in the Martian climate. For (the "dry" geologists maintained), how could you reconcile the persistence of so many features from the earliest ages of the planet, if the Martian atmosphere had ever literally been "dripping wet?" Water—and freely flowing water even more so—is well-known for its erosive properties on Earth. Yet, the geologic evidence from the myriad ancient craters still abounding on the Martian landscape argued forcefully for a planet with a miniscule if any atmosphere—and certainly a very dry one—for the majority of Mars' 4.6 billion years existence as a planet.

Or, as Dr. Bruce Murray, member of the Mariner 9 team (and one day to be appointed head of NASA's prestigious Jet Propulsion Laboratory) once phrased it, "If the channels were created by rainfall, it would seem that one must postulate two miracles in series: one to create the earth-like atmosphere for a relatively brief epoch and another to destroy it."

(As we'll see later, Dr. Murray may not only have been correct in his assessment of a need for a "miracle" on Mars, but eerily prescient.)

Fueling this controversy was the mystery of "layered terrain" at both the Martian poles.

While preliminary dating of the channels (via the well-known crater-counting process—tally up the number of craters of a given size in the area represented by the channel, then look up the calibrated "age" for that many impacts in a given time-period as estimated for a similar area on the Moon) gave evidence that some of them were very old, similar crater-counting methods when applied to the peculiar "poker chip" terrain revealed one of two things: either the layered terrain destroyed craters almost as fast as they were "born" from random impacts, or the terrain itself was relatively new to Mars—something like 100 million years.

If the multiple layers in this unique geological formation were, in fact, the result of alterations in the Martian climate, then the "newness" implied relatively "recent" changes in the atmosphere. Could a climatological version of "Lowell's Mars" have existed as recently as a 100 million years ago? Several scientists associated with the Mariner project began calculations in an effort to find out.

The basic factor in changing any climate is variation in the amount of absorbed energy delivered to the atmosphere. Since for Mars that energy derives from sunlight (as it does on Earth), the investigators were looking for one of two possible events: an overall change in the amount of sunlight emitted by the sun itself; or a change in the distribution of sunlight on the planet Mars. Why distribution?

It was reasoned that to create a drastic change in planetary climate, the increased heat from sunlight would have to "thaw" something—either carbon dioxide or water ice. Mariner 4 had indicated, Mariners 6 and 7 had confirmed, and Mariner 9's instruments had confirmed again that the current Martian polar caps were frozen $CO_2$—dry ice. The suspicion among some members of the Mariner team had been that, beneath the "yearly" polar caps—the ones that wax and wane as seen for centuries from Earth—there lurked "residual" polar ice caps—composed (in their model) of even greater depths of frozen $CO_2$.

If you could somehow "warm up" the polar caps—if only just a bit—some of this "atmosphere in waiting" would come out, they reasoned, adding to the current atmosphere.

The mathematical investigations focused on changes in the Martian orbit (as evidence for a solar increase, which would have produced certain changes in the climate of the Earth, was lacking). Specifically, the scientists were looking for a way periodically to release additional reserves of frozen $CO_2$ through some subtle increase in the amount of sunlight reaching both the polar caps. Initially, a combination of two factors—the slow, but predictable "precession" of the polar axis (caused by the sun's slight gravitational "tugs" on Mars' equator) and an equally predictable precession of the entire Martian orbit around the sun itself—seemed to conspire to bring the necessary additional sunlight to the polar regions to thaw out a percentage of the "hidden" $CO_2$.

First predicted in 1966 (even before the discovery of the layered terrain by Mariner 9 in 1971), this mechanism would increase or decrease the amount of sunlight available at the Martian poles by about 1% over about 50,000 years.[8] According to the authors (Drs. Robert Leighton and Bruce Murray), this should affect the yearly duststorm process, accounting for the "stripes" of light and dark material within the "poker chip" terrain. The overall climatological effect, however, was recognized as very slight, and could not account (according to the authors of the theory) for the "stacking" of tens of individual layers in even larger "plates." It was these plates, eroded backward from an even frontal edge, which gave the layered terrain its distinctive "toppled poker chip" appearance. The existence of these "poker chips" required an additional, longer-period climatic change on Mars.

In the pursuit of the expected, science often runs headlong into the unexpected. Thus it was with explanatory models to account for changes in the Martian climate.

During the course of the investigations, one of the authors of the previous calculation (Murray) and two additional colleagues (Ward and Yeung) came across what they themselves termed "previously unrecognized long-term periodic variations in the solar insolation (sunlight) reaching Mars, which can be expected to regulate the growth and disappearance of perennial $CO_2$ deposits and probably also the production of planetwide dust storms."

According to Murray, Ward, and Yeung, "These insolation variations arise from *variations in the eccentricity of the orbit of Mars* (italics added)."[9]

Their description, derived from a decades-old paper on celestial mechanics, prepared by two astronomers named Brouwer and Van Woerkom treating subtle changes in the now-eggshaped Martian orbit (induced by the gravitational fields of primarily the Earth and Jupiter), was discovered by Murray, Ward, and Yeung to have substantial impact—if viewed in the context of climate changes on the planet Mars.

The authors discovered that the "eccentricity" of Mars' orbit (its deviation from a perfect circle) varied with two superimposed periods: a "short-term" variation every 95,000 years; and a "long-term" periodicity of about 2 million years. The orbit, in its excursions from the "norm," could vary from almost ciruclar to even more egg-shaped than it is at present (the actual numbers ran from .004 to .141).

It doesn't take much scientific insight to project the consequences of varying a planet's distance from the sun; the closer it is (on average) the warmer it will get. The farther away, the lower the temperature will fall. If these changes occur over the course of a single year (as they will if the orbit isn't circular), the effects upon the seasons can also be readily imagined; the hemisphere (and thus the pole) which has "summer" when the orbit is closest to the sun, will develop unusually warm summers, while the winter in the opposite hemisphere will form somewhat more mildy than its norm. Half an orbit later the situation is reversed: the winter in the first hemisphere will now take place when the planet is farthest from the sun (and will be *very* chilly!), while the summer hemisphere is experiencing a reduced amount of sunlight (and thus heating) because of the increased distance.

What Murray, Ward, and Yeung discovered was that, because the actual degree of this variation in distance from the sun *varied*, its effects on the overall temperature of Mars could also vary—producing potential variations in both the long-term climate and the deposition of frozen $CO_2$ at the polar caps. Or, in their own words, "This pattern of insolation variations bears a striking resemblance to the long-period and short-period fluctuations that seem to be recorded by the layered deposits."

But even in this remarkable discovery there lurked the possibility for additional surprises.

Or as one of the authors (Ward) warned in a footnote to the paper, "It was found (in Brouwer's and Van Woerkam's calculations) that the inclination of the orbit of Mars could vary from 0 degrees to 6.5 degrees . . . The effect of this variation on the obliquity (tilt) of Mars is not immediately apparent."[10]

The importance of the additional subtlety was simply this: a change in the tilt of the overall planetary axis to its orbit, created by some kind of "dynamic coupling" between its changing eccentricity and the ever-present precession (wobble) of that spin axis, could substantially alter the amount of sunlight reaching those vital (climatologically speaking) polar regions. Imagine if the current tilt of the Earth's axis—about 23.5 degrees off the vertical—were altered; polar ice caps would have an increasing tendency to melt (if the tilt were increased), more water vapor in the atmosphere would increase the number and severity of storms, and last (but certainly not least!), the oceans would have a nasty tendency to rise—inundating cities like Paris, London, and New York!

Over time (according to these same celestial mechanical calculations) the tilt of the Earth's axis to its orbit has only varied by "a few degrees." But in the Martian situation, continued Ward, "it is possible that *obliquity variational effects* dominate those described in this report (italics added)."[11]

The reason had to do with water.

One of the "raging controversies" before the Mariner 9 mission was the amount—and location—of whatever water still remained on Mars. Mariners 4, 6, and 7 had confirmed a theory put forth in the last years before these missions: that the present polar "icecaps" were, in fact, not like those on Earth, but composed of frozen $CO_2$. Temperature measurements made of the southern cap by Mariners 6 and 7 confirmed the ultracold temperatures required of the "dry-ice theory." Since these latter spacecraft had also detected a trace of water vapor in the atmosphere (confirmed by ground-based terrestrial observations), a logical question arose: where was the small amount of water thought to be on Mars currently "hiding?"

During the Mariner 9 mission, one potential answer developed from a peculiar observation.

Mariner arrived during "summer" in the southern Martian hemisphere, when the polar cap was waning. Scientists had looked forward eagerly to being able to chart the regression of the dry-ice fields day by day, to fit the amount of ice that was melting ("subliming"—going directly from a solid phase into a vapor—would be a more accurate description) into a comprehensive model of the relationship of this "dry-ice cap" to the current atmospheric pressure.

Mars is currently unique; nowhere else in the entire solar system have we encountered an entire world where the predominant component of the atmosphere can literally freeze out—and fall as snowflakes on the ground. (Titan, the largest moon of Saturn, comes closest; methane, a substantial component of its nitrogen atmosphere, does freeze out—because of the exceedingly low Titan temperatures: near $-300°$ F.)

Earth, with its abundant water vapor, doesn't even come close. The snow and ice of our world represent a "trace constituant" of the atmosphere itself. Now, for the terrestrial situation to match the Mars we know, the dominant gas—nitrogen—would annually have to freeze out and thaw—something that would have a major impact, needless to say, on the entire terrestrial environment, as the freezing point of nitrogen is almost $-300°$ F.!

But Mars is very different. There, the carbon dioxide does freeze—and falls in icy drifts upon the polar ground. Because of the remarkably low atmospheric pressure currently observed on Mars, however, some scientists predicted that even in the dead of winter the *amount* of $CO_2$ thus frozen out was relatively small—only enough to cover up the ground to a depth of a "few meters at most" (in stark contrast to the miles-deep icefields here on Earth, composed of a frozen "trace constituent").

Thus, when Mariner 9 arrived in time to monitor the way the southern icecap disappeared, it was predicted that *all* of it would go—as even on Mars the summer temperatures at the poles climb far above the freezing point of $CO_2$, a frigid $-185°$ F.

Imagine the surprise among the members of the Mariner 9 team when, after a few weeks of a predictable retreat, the southern ice fields slowed . . . then stopped their melting, remaining unchanged for the remainder of the summer season. This startling behavior could only mean one thing: that something else—besides dry-ice—was underneath the "yearly" polar cap—revealed when the overlying layers of frozen $CO_2$ had melted!

There was only one abundant compound which could account for such behavior: water. The "residual icecap" on the southern pole of Mars was apparently a reservoir for the element that had once existed as a liquid on the surface, and had once created channels.

All that would be needed now . . . or sometime in the past . . . would be a way to "warm up" Mars—until that water melted. For, if its vapor could be added to the reserves of $CO_2$ that were suspected to exist elsewhere on the planet (perhaps absorbed in the very soil around the polar regions), the resulting "greenhouse effect" could dramatically increase the average temperatures over the entire planet, by trapping solar radiation.

And a "Martian spring" would come . . .

Even as Viking was landing two spacecraft in 1976 and placing two Orbiters on patrol above the planet, Dr. William Ward was proving how such a "spring" could happen.

In a series of papers published from 1973 to 1981, Ward and several coauthors were investigating the effect upon the Martian climate of the previously unsuspected changing tilt of Mars to its own orbit—

And discovering that Mars—in all the solar system—is unique: it alone can change its axial direction in space by as much as 24 degrees (plus or minus 12).[12]

This radical departure from the normal obliquity of the planet (which is about 25 degrees) would provide the crucial mechanism for totally transforming Mars by tilting the ice-bound polar caps far over, towards the warming sun—from the bleak, glaciated planet Viking was capturing in its spectacular new photographs even as Ward was working, to a softer, warmer world sometime before . . .

A world in which the Origin of Life could, indeed, happen, a world in which there could now be Time . . . Time perhaps for whomever built the City and the Face to have been born.

## Notes

1. Lowell, P., *Mars*, Boston: Houghton, Mifflin & Company, 1896; Lowell, P., *Mars and its Canals*, New York: Macmillan, 1906; Lowell, P., *Mars as an Abode of Life*, New York: Macmillan, 1908.

2. Sagan, C., *The Cosmic Connection. An Exterrestrial Perspective*, New York: Doubleday, 1973.

3. Leighton, R.B. et. al., "Mariner IV Photography of Mars: Initial Results," *Science* 149:627, 1965.

4. Leighton, R.B. et. al., "Mariner 6 and 7 Television Pictures: Preliminary Analysis," *Science* 166:49, 54, 1969.

5. Wallace, A.R., *Is Mars Habitable?*, London: Macmillan, 1907.

6. *Mars as Viewed by Mariner 9*, NASA SP-329. Washington D.C.: U.S. Government Printing Office, 1974; Hartmann, W.H. and Raper, O., *The New Mars: The Discoveries of Mariner 9*, NASA SP-337. Washington D.C.: U.S. Government Printing Office, 1974.

7. New information on this controversy recently came to my attention—unfortunately, too late to be included in this edition, except as a footnote.

Only days after the Mariner 4 encounter, noted space pioneer, Eric Burgess (whom we shall meet again later, in another context), called attention to a suspicious set of linear "faults" cutting diagonally across a large crater in one television frame (#11). In a paper published in the *Annals of the New York Academy of Science* (Vol. 140, Dec. 16, 1966), he tentatively identified the feature as "evidence of large [Martian] rift valleys with widths of about 30 miles from escarpment to escarpment . . . ," adding "when the area of this photograph is compared with surface maps of Mars . . . the rift structure may be precisely superimposed on a fraction of [a Lowellian] 'canal.' "

Geological implications of such an identification with Lowell's planet-wide "canals" (which was in direct contradiction to official conclusions of the Mariner Team, that "no trace of these [canal] features was discernable") could have been profound—including possibilities for previous "mountain-building episodes" on Mars, if not outright volcanism and resultant epochs of denser atmospheric evolution, with all the attendant biological implications.

Burgess' cartographic attachment of this discovery to Lowell's infamous network, however, was apparently the "kiss of death"—at least insofar as certain Mariner scientists were concerned. According to geologist (then Mariner 4 Team Member) Bruce Murray: "An analysis of this type of feature must be based upon *original data* received from the spacecraft . . . " (italics added—*Los Angeles Times,* August 15, 1965). However, Arthur C. Clarke (another space pioneer, somewhat known in these areas . . . ) soon confirmed Burgess' discovery, writing: "I was looking at the [Mariner 4] photograph when the linear feature jumped out . . . It certainly is striking once you've spotted it . . . " (personal correspondence to Burgess).

Ironically, Burgess, several years before, published the first technical paper specifically outlining an unmanned mission to Mars (Burgess, E., "The Martian Probe," *Aeronautics,* November, 1952; NASA would later recognize this very term—"probe"—and who originated it, and apply it to not only the Mariner unmanned spacecraft, but every far-flung descendant—*Origins of NASA Names,* (NASA SP-4402). Despite this considerable "track record," and despite the fact that NASA management (in the person of Urner Liddel, then Assistant Director and Chief Scientist for Lunar and Planetary Programs, at NASA Headquarters in Washington DC) even grudgingly acknowledged Burgess' discovery (" . . . the evidence for a lineament had not come to my attention until you pointed it out"—official correspondence to Burgess, August 30, 1965), nothing "official" ever happened—

For six long years . . . until the "rediscovery" of the Mars that Burgess had inferred from his suspicious "lineament"—by Mariner 9.

The effect of this "delayed discovery" on planetary science was probably not critical (except insofar as properly crediting Burgess with an historic "find"—and its sweeping geological implications for the "new Mars!"); the *political* effects—from public disappointment, as a result of the erroneous conclusion that Mariner 4 had found only a "dead and cratered world" were definitely profound . . .

And the "canals" . . . ?

Perhaps the matter isn't as settled as we thought.

8. Leighton, R.B. and Murray, B.C., "Behavior of Carbon Dioxide and Other Volatiles on Mars," *Science* 153:136, 1966.

9. Murray, B.C., Ward, W.R. and Yeung, S.C., "Periodic Insolation Variations on Mars," *Science* 180:638, 1973.

10. *Op. cit.*

11. *Ibid.*

12. Personal communication; Ward, W.R., "Large-scale Obliquity Variations on Mars," *Science* 181:260, 1973; Ward, W.R. "Climatic Variations on Mars: Astronomical Theory of Insolation," *J. Geophys. Res.* 79:3375, 1974.

# IV

# TESTING FOR THE
# REALITY OF "MARTIANS"

"They made their way to the outer rim of
the dreaming dead city in the light of the racing
twin moons. Their shadows, under them, were
double shadows. They did not breathe, or seemed not to, perhaps, for several minutes . . ."

Ray Bradbury
*The Martian Chronicles*

My mind hovered between emotions, between an almost "floating" sense
of wonder—and outright disbelief.

A City . . . on the planet Mars?

Could it be real, or the product of an overworked imagination; for
there was no denying that I had *wanted* such a city to exist. Yet at the
same time I was strangely afraid. So many generations had wished for
evidence of life on Mars, so many had had their hopes shattered by the
"real Mars" of the spacecraft revolution.

Part of me cried out inside, "They *did* exist!"—an ultimate vindica-
tion of "the Martian Legend." Another, calmer voice reminded me repeatedly that since I'd embarked upon this "quest" with the deliberate
intention of discovering where "they" might have lived, of course, I
"found" them.

Which "reality" was real?

There is a misconception about Science, that the practice and the practi-
tioners must remain uninvolved emotionally—separated and aloof. In
reality, nothing could be further from the truth.

Science, when it's at its best, is carried out by passionately involved people. The stereotype of the "cold, unemotional scientist" just doesn't wash; take a look at any of the "greats"—Newton, Einstein, or even Lowell.

What separates these individuals from the more traditional public image of a "creative type" is their willingness to *test* their ideas and beliefs—against the "objective reality" of the Universe itself. The greatness of Newton and Einstein is beyond dispute, not because a small segment of society "liked" their work, but because other, far less creative individuals, could test the predictions of their work against some "standard" that is accessible to anyone—in the case of these two men, the movement of objects in response to forces that were repeatable within the greatest laboratory of all: the Universe.

Lowell, I would contend, deserves a place among the legion of "creative, true scientists" because he took his observations—that there were indeed line-like markings of the planet Mars—and he formulated a series of interlocking tests—the annual shrinkage of the polar icecaps, the then-measured thinness of the atmosphere, the (then primitive) theory of planetary evolution—and he proposed a straightforward hypothesis to explain his observations of the "line-like markings":

That they were made by Martians—by intelligence.

In Lowell's pursuit of the objective truth behind this, admittedly, extravagant hypothesis, he himself advanced the whole of astronomical research—by constructing the Lowell Observatory in an arid desert, where "seeing" conditions were superior to the more traditional locales for telescopes (which were then, by and large, in cities). If nothing else, this practice (which Lowell instituted in his relentless, yes, passionate, search for even more "proof" that there were "Martians") became the standard for the rest of the astronomical community; now, major observatories place their telescopes on the tops of mountains, in the most stable and arid air as it is possible to find on Earth—the direct benefit of Percival Lowell's pursuit of new technology to test his "truth" concerning Mars.

The fact that Lowell was wrong regarding the canals says far more about the limitations of the technology at his disposal than it does about Lowell "the scientist." For, restricted essentially to visual observations of the threadlike detail upon the planet, then believed to be "canals," Lowell lacked the ultimate, objective, test of their existence—let alone their reason for existing:

A photograph.

In the case of the Martian City, we had already passed beyond Lowell's first obstacle—proving there was something enigmatic on the planet. Viking had furnished exquisite photography of both the Face and

*The Monuments of Mars*

the City. Here, then, was something that could be measured objectively—no matter what our interpretation of the reason for its being there.

That ability to fall back on the tried and true "scientific method" (even if it is commonly misunderstood) was to prove an invaluable anchor to my own tendency to become too attached to a particular hypothesis or theory regarding these fascinating "artifacts." It would also be my ultimate support if someone ever asked "Why you—a science *journalist*?" For Science is the ultimate democracy; anyone can play—providing he or she plays by one simple rule: submitting the idea—no matter how far out—to the ultimate test: the ability of other investigators independently to arrive at the same "truth."

More because I couldn't truly believe that I had found "a Martian City" than to convince anyone else, I began devising tests for my own "truth."

The line that seemed most profitable was a continuation of the reasoning that had led to the discovery of the City itself—what was its purpose? This question had proved extraordinarily useful when I'd applied it to the Face, in that it forced one to think *culturally* about the reasoning behind the creation of such a monument, from which it was a short step to thinking about the other needs of such creators—where they lived. Now, I couldn't help but return to that first question, "Why go to all the trouble to create such a massive, stupendous piece of engineering—unless it has an overriding importance to the people who created it?"

I stared at the image in the Martian desert. It stared back—a distant, serene, almost god-like representation of a human likeness, on a world where it simply didn't belong.

I was reminded overwhelmingly of Egypt.

From my (then) limited recall, during Egypt's long and remarkable persistence as a culture, from its dim beginnings to its eventual disappearance as a creative force on the world stage at the end of the so-called Dynastic Period—a continuity of 3000 years—there was an era when its inhabitants erected gargantuan structures in veneration of the "god-king" of the culture—the pharaoh. When one thinks of Egypt, one simultaneously calls to mind this form of "massive engineering," not only those "artificial mountains"—the pyramids, but the direct attempt to immortalize in stone a human being—the pharaoh himself: to deify him by sheer scale.

Throughout Egypt's enormous span as a monument-building culture, the image of the pharaoh was recreated as heroic—more than life-like. During the middle of the so-called New Kingdom (1567 B.C. to 1085 B.C. in some chronologies) one of the pharaohs—Ramses II—erected a collection of statues, temples, and finally, a set of gargantuan representations of himself cut out of the "living rock" overlooking the Nile, which took the glorified representation of the ruler to its ultimate.

In a massive temple at a place called Abu Simbel, Ramses II had four 67-foot-high statues in his image carved in the sandstone cliffs, guardians worthy of Ramses II's image of himself—as a "god." A short distance north was an almost equally impressive temple dedicated to Queen Nofretari, with four more massive sandstone effigies depicting Ramses II, and two in the likeness of the Queen herself.

This was at the height of Egypt's power as a far-flung empire, ruler uncontested of the Mediterranean region. As one expert would write, "The age of Ramses, with its great temples, its colossal statues, its glowing accounts of the leader's exploits, marks the climax of the age of the pharaohs . . . an age that spanned several millennia, throughout which the pharaoh and his subjects, despite vicissitudes in fortune, were born on a faith in themselves, in their superiority over other nations, and in the pantheon of which the pharaoh was a member."[1]

In looking at the massive monument gazing upward from the reddish deserts of the planet Mars and, a few miles away, the collection of gargantuan "structures" apparently related to that monument, I imagined that the author could as well have been writing about whomever built the City and the Face. Its very scale made one think of other lines, those of the poet Shelley, inspired by one of Ramses' own imposing statues:

> On the Sand,
> Half sunk, a shattered visage lies . . .
> And on the pedestal these words appear:
> "My name is Ozymandias, king of kings:
> Look on my works, ye Mighty, and despair!"

The parallel between the larger-than-life culture which had grown up beside the Nile and what was represented on the Viking photographs an entire planetary orbit away, was eerie . . . and compelling.

While one part of my mind darted down a score of interlocking canyons ("Could there have been any possibility of actual Contact . . . between the people who built to last Eternity on Earth . . . and those who apparently built here . . . ?"), another, more disciplined, was warning that such an avenue of inquiry was extremely premature. Yet, the comparison was useful; notwithstanding the possibility these two cultures ever came in contact, the *form* was outwardly the same: massive works of engineering—including a likeness of some "god" or person—requiring an enormous investment of resources. The "driver" for such investment

*The Monuments of Mars*

in ancient Egypt had been simple: a religion, connecting the physical and imaginal worlds.

Was such a religion a primal "Rosetta Stone" to unlocking these extraordinary Martian puzzle pieces?

Then, I had a flash of insight, a flash which leapt the distance between Earth and Mars and took a giant step toward proving as well as explaining the existence of a City and a Face.

Alignments.

Egyptian temples—even the pyramids themselves—were *celestially aligned*, some to certain stars, such as Sirius, others to the rising sun. DiPietro and Molenaar had noted the "pyramid" found standing on its hill about ten miles southwest of the Face was pointed directly north—as the famed pyramids of Egypt.

It wasn't the orientation of this object which now captured my attention, but the more general purpose its compass direction served; for it was (if DiPietro and Molenaar had done their measurements correctly) a known reference for every other feature in the picture.

It took only a second or two—once I'd recalled this fact—to realize the obvious: the entire "main avenue orientation" of "the City/Face unit" was aimed not "east/west," as I'd previously unconsciously assumed, but northeast/southwest: *toward the sunrise/sunset.*

Was the City so specifically aligned that, on a certain Martian day, the sun rose directly from behind that monumental visage—if you were standing in the "City Square" I'd found? Was it an enormous Solstice marker!?

I began a series of calculations to find out.


The connection between religion and astronomy was, in the case of countless developing cultures back on Earth, fundamental. In repeated instances around the world, the practice of "sun worship" (as it used to be dismissed somewhat derogatorily by early ethnologists) was an elegant means of codifying—through geomantic symbolism—a knowledge that was absolutely essential to the survival of the culture: the length of the year and its seasons, as Earth spun inexorably around the sun. From the striking (and almost clichéd) remains at Stonehenge, to the brilliant example of Egypt itself—which, of all the developing societies in the Near East, knew the exact length of the sidereal year (the length of time it takes for the earth to complete one orbit of the sun)—to the still controversial Anasazi *noon* Solstice marker in the American Southwest, such celestial knowledge has been crucial to the economic, political, and social development of cultures the world over.

If I could mathematically establish a "solar connection" between the City and the Face, not only would it be a milestone on the road to proving they were the products of Intelligence—it would be a vital clue to determining at what stage in the evolution of that intelligence, and its associated culture, the City and the Face had been constructed.

But my own schizophrenia still stood in the way. While one part of me turned methodically to the task of measuring the precise angles of the relationship between the City and the Face, another part was observing all of this, amazed: how could I so calmly apply procedures from another world to whomever had lived here—if they *had* lived here at all!? Wasn't it stretching plausability totally out of shape to presume that "rules of cultural development" learned on Earth would apply here? Why should "Martians" (or whoever) create a "Stonehenge" in the desert?

Because (the first part of me, still methodically measuring and calculating, answered) certain things were universal. Apart from certain small colonies of organisms living several miles beneath the oceans, all life on Earth depended on the sun. It was part of the "paradigm"—the Search for Extraterrestrial Life—that if we found life elsewhere, it would be the same—it would also be linked to starlight as a source of energy and food.

The near universality of astronomical awareness and interest on the part of widely diversified cultures on Earth had little to do with such abstract things as "curiosity" or "being human." It had to do with the one common denominator of all pre-technological cultures: survival.

You couldn't live without a knowledge of the sun—how it moved throughout the year, and how its movements affected every other living cycle on the earth.

If I was right, if there was a solar alignment of the City and the Face, it would be an extraordinary "find" on several levels—not the least of which would be a confirmation of a "truth" I'd long advocated for so-called "primitive" cultures:

That every society, in the beginning, must develop an "astronomical connection"—an awareness of its basic link to movements of the planet of its birth—if it is to survive, evolve, and flourish.

That I was about to extend this basic principle of "archaeoastronomy" to another planet—Mars—and the first discovered extraterrestrial culture (if I was correct!) was mindboggling.

\*     \*     \*

In principle, the calculations were quite simple.

Let's start with some basics: living on a planetary globe. The globe is spinning, turning on an axis. At the same time it is revolving around the sun. There is a specified tilt between the planet's orbit (its path around the sun) and the planet's spin axis (the reference of the spinning motion).

For the earth, the particulars of this scenario are well-known: the angle between the spin axis and the orbit is 66.5 degrees (where one degree is 1/360th of a complete circle). If the reader has noticed this is *not* the familiar "23.5 degrees" quoted as the Earth's tilt, that is because that is the complementary angle—between a line perpendicular to the orbit (straight up and down) and the planet's axial tilt. Insofar as that angle—known technically as "the obliquity"—is used in the calculations (and because it's also well-known as "the Earth's tilt") I'll use this spin-axis deviation from the perpendicular to the planetary orbit in the rest of this discussion.

Now, to what happens to an observer on the surface of this globe—in this case, the Earth—as the globe goes around the sun:

The spinning motion about the planetary axis causes all objects external to the globe—stars, the Moon, and, of course, the sun—to appear to "rise in the east and set in the west." But this is only a general description; a celestial object can only appear to rise precisely due east and set precisely due west if it is on that imaginary line across the sky formed by the projection of the Earth's equator onto the "celestial sphere." This imaginary line, dividing the "northern" and "southern" regions of the heavens (from the perspective of someone sitting on a planet), is called (for somewhat obvious geocentric reasons) the "celestial equator."

Again, from the perspective of someone observing all this from the surface of the Earth, the daily spinning motion imparted to the heavens by the turning of the planet on its axis is compounded by additional motions imparted to some of those celestial objects by the annual motion of the Earth around the sun. Which is why it took thousands of years for people to unravel these interrelated movements of celestial objects . . . and why it is going to take a few paragraphs and a good deal of patience on both our parts for me to communicate them to you here!

During the course of one terrestrial "year" (which we now know is one complete orbit of the sun), the sun's motion in this complicated dance is far from simple: because we're looking at it from a moving object (the Earth) it *appears* to move against the background stars—the same way someone sitting on a chair will appear to move against background furniture in the room, if we move around them in a circle.

If the Earth were oriented "up and down" relative to its solar orbit, the apparent annual path of the sun would be along the projection of the Earth's equator, that "celestial equator" mentioned a moment ago. This would occur because these two planes—that formed by the extension of the terrestrial equator into space, and the plane of the Earth's orbit of the sun—would coincide. In this extreme case, where the planet had no obliquity, no tilt, the sun would truly rise precisely east and set due west—all the time.

The real situation is quite different.

Because we're on a tilted planet (with respect to its orbit around the sun), the sun can only rise or set exactly in the east or west when its annual motion *crosses* the celestial equator. And that happens only twice a year—in the Spring and Fall. The days that happens are called the Equinoxes (for "equal night"), as the length of the day and night are then precisely equal. In terms of the overall geometry, it is at these moments in the Earth's orbit around the sun that another imaginary plane—this one containing the Earth's spin-axis—lies at right angles to the sun. (That plane, oriented "inertially" in space, is, for our purposes here, unchanging with respect to the distant stars. Relative to the sun, however, it *appears* to rotate in a complete circle in the course of one year—presenting itself "broadside" to the sun twice during that annual orbit. This is the moment of Spring and Fall Equinoxes.)

If you were located far out in space, "above" the plane of the Earth's orbit of the sun, you could easily see this geometry in operation. Looking down on the northside of the orbit, you would see the Earth's north polar axis tilted off the vertical by that familiar 23.5 degrees. And you could "see" the plane containing that axis as it moved around the sun, sometimes presenting itself "sideways" to the center of the solar system, other times aimed directly toward and away from the sun itself.

Now, back on Earth.

When that plane—containing the tilted polar axis—is aimed toward the sun, one of two additional seasons can occur—"summer" or "winter," depending on which hemisphere you happen to be standing at that moment. The observable fact that "summer" and "winter" occur six months apart (half an orbit) is due to the built-in fixed geometry in stellar space of the plane containing the Earth's axis of rotation. The fact that they occur three months away from "spring" or "fall" (a quarter of an orbit) is due to the apparent rotation, relative to the sun, of this axial plane by 90 degrees (one quarter of a circle of 360 degrees), as we move 90 degrees around the sun.

[If you still don't understand this, don't be too discouraged; take a pencil, tilt it, hold it fixed in a single plane in space (relative to some distant piece of furniture)—and move it around a nearby table lamp, to represent the sun. Eventually, you'll get the hang of it.]

For a given hemisphere, when the axis is tilted toward the sun it is

*The Monuments of Mars*

"summer." Six months later in the same hemisphere, when the axial plane against sweeps across the sun and the axis is now tilted away, it is "winter." (At the same time, of course, in the other hemisphere, the counter season is occurring—because the tilt is just the opposite there at the same instant in the orbit.)

The extremity of this occurrence is called the "solstice" (for "sun stands still"—a reference to its momentary pause in the north/south motion of its apparent rising and setting, as you'll see below).

For ancient peoples (or, for that matter, for everyone unfamiliar with the reality of all this "orbit stuff") the result of these complicated interactions between the Earth's orbit and the planes containing its terrestrial references—the axis and the equator—was best perceived through the motion of the sun.

At some times of the annual cycle, the sun would indeed rise directly in the east, proceed across the sky during that day, and set precisely in the west. But that was rare—only taking place two days out of the entire year (if you really took the care to measure it). At other times, it rose and set *close* to these two cardinal compass points, but not exactly. And, as the year progressed, the discrepancy became decidedly apparent—the sun rising well north of east during the height of "summer," while rising far south of east as the season hardened into the so-called "dead of winter."

For the careful observer, the position of the rising and setting sun, along the horizon, became a telltale clue as to the season. And the seasons, for peoples dependent on animal migrations (if they were hunters) or on annual cycles of rainfall and the length of the day itself (if they were farmers), were of inestimable importance.

But there was more.

These were not modern twentieth-century "dispassionate observers." In a world full of uncertainties, from apparently random storms to doubt as to where your next meal was coming from, the idea that the seasonal cycle could be *relied upon* was also foreign; as primitive societies saw the sun rising farther and farther south each day, and the days themselves seemed to get shorter and shorter (to say nothing of colder!), who was to say that the "sun god" wouldn't keep on going—until it completely left the sky?

Thus, in order not only to mark the seasons (by the progression of the rising and setting points of the sun along the horizon), but also to know when to begin crucial ceremonies designed to get him to "hang around," early cultures erected a variety of monuments around this planet with the expressed purpose of warning when the sun was about to "disappear."

These familiar "solstice markers" took many forms, from massive "megalithic stones" erected in places which had no handy natural markers of their own—such as the Salisbury Plain of Stonehenge—to a variety of manmade structures oriented in exact geometrical relationships to key horizon markers—such as mesas or mountains. All, however, despite a vast variety of external differences of form, have been found to have a common purpose: to mark the "endpoints"—the solstices—of the annual progression of the sun's rising and setting on the horizon, as it "moved" north and south to the rhythm of the seasons.

Now let's consider Mars.

Everything I have just explained—the annual orbit of the planet around the sun, the tilt of its axis to that orbit, the progression of seasons on the surface of the globe (caused by the changing amount of solar energy hitting the surface as a result of that tilt)—applies to an observer—or an entire culture!—on the planet Mars. As does the apparent motion of the rising and setting sun along the Martian horizon.

While the specific numbers may be somewhat different (due, for instance, to the much wider orbit of Mars around the sun) the *principles* remain exactly the same—as does the geometry. By a curious coincidence, the tilt of the Martian axis is currently about 25 degrees (compared to 23.5 for Earth), creating an almost identical geometrical solution for any given latitude upon the planet, when compared to a similar solution worked out for Earth. (This coincidence is not important for a calculation of alignments of the Martian City and the Face, as the equations used apply to *any* given tilt or latitude. But it is curious . . . )

*         *         *

What, then, did we discover?

For any given tilt (obliquity) of the planet's axis to its orbit, the annual swing along the eastern horizon of the sun is equal to *twice* the obliquity—for a position on the planet's equator. In other words, if you're on Earth (and on the equator, in some lovely tropical paradise, or on a ship) the annual swing of the sunrise point north and south of the due east point will be a total of 47 degrees (2 x 23.5). On Mars a similar observer on the equator (where there is currently a noted absence of an ocean, let alone any liquid water . . . ) would mark a 50 degree swing of the sunrise north and south.

What happens if you are standing at some other position on the globe, at some other latitude for instance?

The equation which predicts this motion looks like this:

$$\text{Sin } D = \sin \delta / \cos \phi^{2}\cdot$$

Where D = max deviation (north or south) of the East/West line

$\delta$ = planetary obliquity (tilt)

$\phi$ = the latitude of the observer

Plugging in the appropriate numbers allows you to predict the swing of the sun north or south of the east/west direction at its maximum excursion—defined as the solstice.

For the most familiar example on Earth—the orientation of Stonehenge—the calculation works as follows:

The latitude of Stonehenge is 51.2 degrees (North, of course—in Britain). The obliquity of the Earth is 23.5 degrees. Plugging the appropriate sines and cosines into the equation by a pocket calculator, one gets the answer:

The annual swing of the sun to its most northerly rising point, as seen from Stonehenge, is 39.5 degrees.

When we look at an aerial map of this famous monument, there—at the end of a "main avenue" heading 39.5 degrees north of east out of the center of the stone circle—is a marker: the "heelstone" (so termed because of a small heel mark carved on the stone, by someone in the millennia its ageless vigil was attended . . . ).

When it's clear on the morning of the Summer Solstice (which doesn't happen very often in the present climate of Britain), the sun majestically, rigorously rises over this specific marker—fulfillment of some ancient architect's celestial vision.

Was there an architect of equal genius on the planet Mars?

The latitude of the City and the Face was easily available: it was on the computer printout of the National Space Science Data Center's versions of the Viking photograph. Each corner, and the exact center of the Viking image, had a readout noting its exact longitude and latitude on Mars.

After a bit of measurement (in terms of the distance of the center of the City from the center of the photo), I settled on 41.8 degrees N. latitude for the location of the "City Square"—the equivalent of the center of the Stonehenge circle back on Earth. Using the orientation of the pyramid published by DiPietro and Molenaar to establish "north" and "east," I proceeded to measure a line from the center of that peculiar circle of five "buildings" toward the Face itself. It measured 23.5 degrees—north of east, from the center of the "city square" to "the mouth." From the top to the bottom of the Face presented a range of angles, from 27.5 degrees for a line just skimming north of this strange object, to 21.5 degrees for a line which ran just south of the "chinline." This range of angles, in turn, presented something of a problem.

Unlike the almost pointlike heelstone, as viewed from the center of Stonehenge, the Face presented a relatively large viewing angle to a hypothetical observer standing in the center of "the square": about 7 degrees, or a full *fourteen times* the apparent width of our own full Moon as seen from Earth. While an impressive view, this large range of angles made for a difficult decision: which angle—if any—represented the Summer Solstice for this latitude on Mars? Or, viewed another way, which part of the "anatomy"—the forehead, the nose, the mouth, or chin—was the "significant" marker for the sunrise moment to be calibrated?

Leaving aside this sticky question for a moment (which was analogous to a similar argument at Stonehenge some years ago, over whether the sun would *rise* or stand over the heelstone when *fully risen*—as the sun rises at an angle to the horizon, and moves south as well as "up" when it is rising—at 51.2 degrees), I decided to complete the calculation, which would tell us if any of the angles were, indeed, significant.

They weren't.

The Summer Solstice point on the eastern horizon, as currently seen from the center of the City, occurred 34.9 degrees north of east—over 7 degrees north of the Face itself.

I repeated the entire sequence—and got the same result. Okay. If this was a Martian "Stonehenge," perhaps I'd chosen the wrong location in the City from which to view the sunrise; there were a variety of other objects scattered in and around the collection geometric structures. Too many. None of them had that simple elegance of the geometrically precise set of four "buildings"—with the circular fifth one set in the center. Nor, did any other location have that line-like mark in the ground—pointing directly at the Face itself. (There was another kind of marking, just west of the first, however. It formed a "V" in the ground, its apex pointed west, with one arm aimed just about due east, the other northeast at about 66 degrees. At the end of the apex there was a pyramid-like structure, located just south of the southernmost "building" of the City Square. Its immediate significance, if any, was not apparent.)

Perhaps my intuition—that the Face marked a Solstice alignment—was plain wrong; there could be a lot of other days in the "Martian calendar" which might be significant, before the sun reached its annual end-point in its "travel" northward, along the horizon, or after, when it was again headed south. Perhaps when the Earth was seen near the sun . . .

What if the reason the Face looked so human was because it was supposed to commemorate some "connection" with the one place in the solar system where "humans" live—Earth? And suppose the geometry inherent in the City/Face layout was designed to provide a visual reinforcement of that link . . .

Allowing an observer to see the Earth rise in the Martian dawn over the only "human" thing on Mars—the Face itself?

Or suggesting that the builder intended to draw a terrestrial humanoid link to future extra-Martian discoverers of his handiwork? One more reinforcing element of a code?

But how did that help resolve my problem: that the Summer Solstice currently takes place a full "head length" north of the Face?

It didn't.

The Earth would be seen rising in the Martian sky at any time it was physically west of the sun—half its orbit. (At other times, it would be east of the sun, and therefore appear as an "evening star" before the sunset.)

The two exceptions to this rule of thumb were these: when it was on the far side of its orbit (as seen from Mars), and thus too close to the overwhelming glare to be seen; and when it was passing "between" the sun and Mars, and was also lost in the sun's much brighter scattered light, in Mars' own atmosphere.

There was a time, I realized, when these *un*favorable viewing situations ended: when the constantly changing geometries of Earth and Mars, in their respective orbits of the sun, "conspired" to bring the Earth out of the glare—just before the Martian dawn.

Such an event would be the "heliacal rising"—the rising of a celestial object just before it is overwhelmed by the increasing sky glare of the dawn. (The term comes from "helios"—meaning "pertaining to the sun.")

My mind flashed back to the example of the Egyptians.

The rising of celestial objects—such as the star, Sirius—just before the sun, had been a crucial aspect of Egyptian religious, cultural, and *economic* existence. For, the annual flooding of the Nile took place coincident with this beautiful and endlessly repeatable phenomenon—the sparkling rising, just before the sun, of the brightest star visible in the entire sky—the heliacal rising of the brilliant Sirius.

The Egyptians immortalized this natural phenomenon in many temples, arranged geometrically so that they viewed the eastern dawn—and the annual ritual of the rising of the sky's brightest member, the "Dog Star"—before the appearance of the sun-god himself—Ra.

Of all the "magical" appearances of Earth—a brilliant "star" as seen from Mars—none would be so significant as when it first appeared . . . after having disappeared in the evening solar glare some weeks before.

Was the Face a marker of the heliacal rising of the Earth, before the dawn?

On Earth, there is another culture which observed precisely such a recurring planetary "reappearing act":

The Mayans.

The so-called "Dresden Codex," a fragment of a Mayan manuscript which survived the Spanish efforts to eliminate all traces of this most remarkable culture, seems to be an astronomical ephemeris. In it, the planet Venus has a prominent position: there are several tables devoted to its apparent motions near the sun, including the times of its disappearance in the west at sunset . . . and its subsequent reappearance in the east, just before dawn.

Dr. Anthony Aveni, perhaps the most noted authority on New World archaeoastronomy and the Mayans' extraordinarily persistent observations, once said, " . . . predictions of the heliacal rising of Venus seem to be the theme (of the table in the Dresden Codex)."[3]

Geometrically, the situations between the Earth and Venus, and between Earth and Mars—as viewed from Mars—are very much the same; a slower-moving exterior planet being overtaken by a faster-moving inner one, the latter reappearing a predictable number of "days" (planetary spins) in the same celestial location, relative to the sun. The biggest difference is in the fact that the Earth's orbit and that of Venus are almost exactly circular, while the Martian orbit is definitely not—giving rise to another observable phenomenon: the fact that at some years in this recurring cycle the Earth would appear very bright, indeed. At other times, it could be noticeably dimmer.

And the interval between these periods would be that familiar (see Chapter III) seventeen years—for exactly the same reason: Earth and Mars would then be closest, because Mars was closest to the sun when Earth passed it.

But unlike the heliacal rising of Sirius, which takes place at *exactly* repeatable intervals each year (because it is a very distant star—8.7 light years away—and not an irregularly moving planet), the heliacal rising of the Earth, as viewed from the City, would slowly progress and regress— occurring at different places on the horizon and at different seasons throughout the Martian years. Without a sophisticated computer program, to figure out the range of possible motions of this complicated dance— made more complicated than the case of Venus because Mars' orbit was elliptical—I didn't stand much chance of determining when key configurations of the Earth's rising before the dawn, above which points of the horizon—including the Face—would occur.

[Come to think of it, I don't recall any detailed astronomical calculations relating the phases and the appearance of the Earth as seen from Mars—for the simple reason there's no use for them.

[Aside from a few brief descriptions found in science fiction, there has never been the interest in the kind of intense, mathematically detailed analysis required to turn up such interesting cycles of our planet viewed from the Red Planet. Some of the more obscure motions of Venus only came to light, in fact, from the decipherment of the Mayans' record of the planet (and these exist, not because the Mayans were such whizzes at mathematical predictions, but because they had a long and detailed data-base—several *centuries* of patient observations).

[The only recent attempt to calculate where Earth would be—from Mars—occurred during the recent Viking missions to the planet. Some investigators (notably, Dr. James Pollack, of NASA-Ames) wished to use the Earth as a point-source reference for the transparency of the (now known) very dusty atmosphere of Mars. The intention was to generate an ephemeris (a mathematical prediction) for the Earth, aim the Lander's cameras at it, and scan a photograph or two. The appearance of the Earth—as a mere point of light—would reveal the degree to which a relatively dim object could be viewed through the current planetary atmosphere.

[Besides the fact that the experiment turned out to be a dud (because there is much too much dust suspended in the atmosphere of Mars—making the Earth totally invisible to the relatively insensitive Viking surface cameras), the "ephemeris" itself wasn't very useful, designed merely to last a few months of the "nominal Viking mission."*

[What would be useful would be a computer calculation of the position of the Earth, as seen from Mars, extending across several hundred—if not several thousand—years. But no one has had sufficient reason to wonder about the appearance of the Earth as seen from Mars . . .

[Until now.]

It was while wondering about these exciting possibilities for checking the "Earth's heliacal rising" theory, that I suddenly had another thought:

---

*It should be mentioned that the *primary* reason for generation of this JPL ephemeris was simply to enable NASA's DSN (Deep Space Network) to communicate directly with the Landers.

There was one point on the horizon that I didn't need a large computer to calculate, where—if it and the heliacal rising coincided—it would certainly be "sufficiently propitious" to rate a marker:

The Summer Solstice itself!

But the summer solstice point occurred significantly farther north than any portion of the Face, my internal watchdog reminded me. Yes, I said, but that's *now*. What about sometime in the past? What was the obliquity of Mars itself . . . *then*?

I had remembered a bit of trivia picked up somewhere, something about the entire Martian polar axis shifting . . .

A few moments' frantic searching through my library turned up a reference, by a friend, in fact, Hal Masursky—formerly of the Viking team itself. And there, in print (with diagram!) was the depiction of the polar axis shift for Mars and Earth in the course of several thousand years—and Mars' was significantly greater. In fact, the article declared, the polar axis of the Red Planet was capable of shifting by up to 10 degrees, plus or minus, its current "mean position" of 25 degrees.*

What would a dramatic change in tilt, from a minimum of 15 degrees to a maximum of 35 degrees, do to the solstice points on Mars? A moment's thought told me I had a way of finding out—by reversing the equation I'd used before!

A few punches on the buttons of my calculator, and I knew: for the sun—and Earth—to rise over the center of the Face, the tilt of Mars to its own orbit had to be slightly over 17 degrees—almost 10 degrees less than it was now. The actual range of angles, equivalent to the range of sightlines from the "city square" to the 7 degrees subtended by the Face, ranged from a tilt of 15.8 degrees (for a line that just grazed "the chin") to 20.1 degrees (for a line just passing north of the "crown" of the enigmatic "head").

But which was the correct angle?

I was intuitively drawn to the one coinciding with that rakish slash corresponding to "the mouth." Not only had I traced it (within the limits of measurement) in a direct line toward the "city square," it *looked* like a perfect riflesight—a long, narrow "canyon" aimed outward, toward the desert to the east of the mesa so resembling a human, toward the horizon over which—I had now calculated—the sun would rise in perfect alignment—

---

*This was my first effort to track down the significance of Mars' "tilting" motion. Later, I discovered Ward's original paper and a slightly different set of numbers (see Chapter III).

*The Monuments of Mars*

*If* the planet's spin axis was 17.3 degrees!

But it still could be coincidence, I found myself saying. There had to be another means of confirmation, some other piece of interlocking data, some other measurement . . .

Then I realized I had been staring at it: the Face itself!

Even DiPietro and Molenaar had commented on its remarkable bisymmetry, the perfect "match" of the left and right halves. If it was truly bisymmetrical, then it should be possible to express that as a *measurement* (something DiPietro and Molenaar hadn't published—except in the form of an almost quirky cartoon on page 36 of their own booklet).

There, in that booklet, above another cartoon representation of the Pyramid, was a small drawing of the Face—with a line depicting the center of the image. What fascinated me was the *tilt* of the line drawn through the center of the Face in the cartoon, for it was obvious that it was shifted by some angle to the north—*only DiPietro and Molenaar had never mentioned what that angle was!*

*If* the Face was a constructed object, and *if* its purpose was to mark the position of the Summer Solstice for someone standing in the City, and *if* the "sightline" was that perfectly straight "mouth" running northeast/southwest—and aimed directly toward the city square—then that sightline would have to be oriented at a 90 degree angle to the line marking the bisymmetrical centerline of the Face itself.

A moment's measurement on the blow-ups of the Face confirmed the 90 degree angle—between "the mouth" and a line perpendicular, from "the chin" up through the center of "the nose," right between "the eyes," and off into the desert.

Now for the critical measurement.

*If* any of this was really true, and not merely a figment of my overworked imagination, then the centerline itself had to lie at a *specific angle* off true north.

That angle (within the limit of measurement) should be 23.5 degrees.

With great care and trepidation, going back to the original computer printout on the NASA version of the original Viking image (for confirmation of true north), I proceeded to establish a grid system for the Martian landscape in frame 35A72—on a piece of clear plastic placed over the photograph itself. Once my set of reference lines were drawn—from which I could now establish the orientation of any "structure" in the picture—I drew the centerline of that amazing visage, the Face, on the plastic.

Then, very carefully, I measured its exact tilt—relative to the planetary meridian on the established grid.

It measured 23.5 degrees.

Sometime in the cavernous, yawning geologic past of this extraordinary planet—a world that had held center-stage on Man's attention as the one place in all the solar system where there might be other "Life"—if one of those hypothetical "Martians" had stood in the center of "my" City, in the city square—they would have seen the Earth rise brilliant in the Dawn. And, a few moments afterward, the sun itself would have "magically" appeared . . .

Rising directly out of the "mouth" of the god-like figure in the Martian desert—if it even was a "desert," then. For . . . the last time this alignment would have worked was half a *million* years ago.

In my own mind the mathematical odds, against these two "alignments" being the result of mere chance, were compelling. The "Martians" had moved a quantum leap closer to reality.

## Notes

1. Casson, L., Krieger, L., consult. ed., *Ancient Egypt*, Time-Life Books, 1965.
2. Gingerich, O., "The Basic Astronomy of Stonehenge," *Astronomy of the Ancients,* The MIT Press, 1979.
3. Aveni, A. and Urton, G., eds., *Ethnoastronomy and Archaeoastronomy in the American Tropics.* Vol. 385. 1982 Annals of the New York Academy of Sciences, New York: The New York Academy of Sciences.

# V

# A CITY VIEWED FROM ORBIT . . . EXPLORING RUINS FROM A THOUSAND MILES

> "They walked forward on a tiled avenue.
> They were all whispering now, for it was like
> entering a vast open library or a mausoleum in
> which the wind lived and over which the stars
> shone . . ."
>
> Ray Bradbury
> *The Martian Chronicles*

Staring at the tiny photograph, at the small collection of peculiar pyramidal structures huddled off to the southwest of the enigmatic Face—with its mouth aimed, like time's arrow, at its heart—I imagined what it must have been like to step into the past of once-great-cities with fabled names like "Troy" or "Ur," or what it would be like to discover the most legendary of them all—"Atlantis."

If I were correct, I was staring at the ruins of a culture which had vanished at a time when men on Earth were just learning to tame fire . . . if not long before.

The solstice alignment theory could not "date" these ruins with any greater precision than I have attempted to describe. They could be half a million years old, a million and a half, two and a half million—or even older. There was no way yet for truly knowing. The whole idea was already so fantastic; precision was hopeless.

Sitting there, armed with the $5 \times 5$ inch SPIT-processed image of 35A72, my vision returned: archaeologists, digging in the sands of lost worlds, entering cities whose inhabitants hovered, invisible, behind the long abandoned doors and windows . . .

I could feel it through this computer-generated image, taken a thousand miles above a world whose "inhabitants" were supposed to have evaporated into that mythic reality reserved for unicorns and dragons. "They" had once been there . . . someone, to whom the massive pyramids, arranged in their meticulous rectilinear array, had had meaning, soaring against a nonblue sky of a world that perhaps had once been home. In itself it was ludicrous. It wasn't part of any current "reality" that geologists and planetologists referred to when they spoke of Mars. But it was there: palpable and as real as the gold color of the couch that I was sitting on. I was gazing at the remains of something that had once been grand and beautiful, whose grandeur—across countless millennia and the vast emptiness between the worlds—still clung to the crumbling shapes casting their long shadows out across the sands of this strange world.

There was something . . . timeless . . . here.

*      *      *

The usual image of an archaeologist is of someone constantly digging in the ground (or, even better, overseeing a bunch of undergraduates doing it—usually in summer!).

With changes in technology, however, have come new approaches; I remember a visit to Chaco Canyon, New Mexico, site of a rather well-kept secret in prehistory: the first urban civilization in North America, about a thousand years ago. And as I walked up dusty trails and gazed at intricately-laid sandstone masonry that formed the massive pueblos of the settlements scattered down the Canyon, I heard our guide refer to what the Canyon looked like from about 500 miles above—on LAND-SAT photographs from space.*

"The archaeologists attached to the Chaco Canyon Project," he volunteered, "have discovered that they can use these satellite images to map features otherwise totally invisible from down here on the ground. Like the roads.

"The Anasazi Indians," he went on to explain, "left a puzzling series of 30-foot wide roadways through the Canyon, even though we know they *didn't* have wheeled vehicles, or even horses; those didn't come in until the Spanish brought them, in about 1540 A.D.—over 500 years after what you see here had been built.

"You can see some of the roads literally climbing up the cliffs; there

---

*A fascinating history of "aerial-archaeology," as it pre-dates images from space, can be found in O.G.S. Crawford's *Wessex from the Air,* Oxford: The Clarendan Press, 1928.

*The Monuments of Mars*

are hand-chipped steps leading from down here on the floor," he said, pointing, "to up there, on the rim.

"Then, they take off across the desert—or at least that's what many of our guys supposed; the wind and rain have pretty much washed out all traces leading from the Canyon itself.

"That's how things stood until somebody got the bright idea to look at some LANDSAT images of the Southwest. And there, on the ones taken with the right combination of filters, were the roads—standing out like they'd been made yesterday! You can see them leading out from the Canyon in all directions—to the west, where the timber was for the beams in the pueblos and that kinda stuff; to the north, toward Mesa Verde; and south, toward Mexico. The network just suddenly appeared, connecting up a lot of the tiny pueblos set out, alone, by their lonesome from the Canyon.

"But," he finished, proudly, "Chaco Canyon was the Hub, the central Authority, it looks like. Almost all the roads lead back here . . . and apparently start out from the ones known here in the Canyon.

"Without those satellite photographs, we wouldn't have found out."

Looking at the pocket-sized image of 35A72, at the rectangular outline of the City, I remembered this example of satellite-assisted archaeology from Chaco Canyon.

What I was going to attempt to do—using this processed Viking image of Cydonia—was no different in principle from what the Chaco Canyon Project had accomplished vis-a-vis the Anasazi, the "Ancient Ones" who lived and built a remarkable civilization all across America's Southwest—a thousand years ago. What was different was the scale.

I was dimly aware of at least two previous systematic efforts to identify patterns of intelligent activity on Mars through the use of satellite photography. Both had been conducted by Carl Sagan, based on a previous assessment of the visibility of *terrestrial* civilization on satellite images taken in the 1960s. On those—the rather primitive TIROS and NIMBUS television pictures of the Earth designed for weather forecast information—Sagan's ultimate analysis was this:

> . . . there have been several hundred thousand photographs taken and examined in the Tiros series. On some of these photographs, objects as small as 2000 feet are discernible. Yet in all of these photographs, only one shows any clear sign of life on Earth. This is a photograph taken by Tiros 2 of the forest near the Canadian logging town of Cochrane, Ontario, on 4 April, 1961. In the upper left part of the picture, several wide parallel stripes can be seen; and at right angles to them, another set. Swaths one mile across

have been cut through the Canadian forest by the loggers. The swaths were separated by about two miles. After the swaths had been made, snow fell, enhancing the contrast between the trees still standing and the treeless swaths. But even here, on this one-in-a-million photograph, do we have unambiguous signs of life, from the vantage point of Mars . . . ?

. . . with resolutions of a few tenths of a kilometer (on a study of the higher resolution photographs available from the Nimbus satellite), we have discovered a recently completed highway in Tennessee, perhaps a jet contrail in the Davis Straits, the wake of a ship in the Red Sea, but also a very straight feature off the northern coast of Morocco which had all the apparent signs of intelligent design, but was, in fact, a natural peninsula. At a few tenths of a kilometer resolution the signs of intelligent life on Earth can be detected but not unambiguously. Convincing photographic evidence of intelligent life on Earth requires resolution of 10 meters (30 feet) or better.[1]

This inability to confirm conclusively the presence of intelligence on Earth (no jokes, please!) through satellite photography, opens up some profound questions:

Just what does "intelligence" look like—when viewed from orbit? For, while it is true that the Chaco Canyon Project had access to better satellite images than Sagan's earlier studies, the key is *they knew what to look for.*

Which brings us to the two most recent analyses of satellite information pertaining to the possible presence of intelligence on Mars.

Shortly after the successful flood of spectacular Mariner 9 pictures of Mars, in late 1971, Sagan and a graduate student, Paul Fox, undertook a search of the first Martian satellite's imagery for evidence of intelligence.[2] The criteria for "unambiguous signs of life" were apparently the same as those used in the previous TIROS and NIMBUS studies: a comparison of the Martian and terrestrial features at similar resolutions, with our known artificial structures on Earth used as a baseline measurement of scale if not the *kind* of "artificial structures" expected in any hypothetical "Martian civilization."

In 1976, under a $50,000 grant from NASA, Sagan and another graduate student repeated the search—using Viking images as the means to look for evidence of intelligent design. Both studies, as far as were ever published, turned up nothing.

The reader might well ask, why should our evidence for life on Mars be accepted now, when previous searches for such evidence—conducted by a scientist of impressive background in the subject, if not sheer repu-

*The Monuments of Mars*

tation—have failed?

The answer, I believe, lies totally with the assumptions underlying these very independent studies—Sagan's and this author's.

Fundamentally, it comes down to the issue of scale.

It should be quite obvious from the preceding satellite investigations carried out by Sagan on the Earth, that his criteria for extraterrestrial artifacts were strongly biased by his terrestrial experience; his very line, "convincing photographic evidence of intelligent life on Earth requires resolution of 10 meters or better . . . " is an important clue. For, while admitting these criteria apply strictly to the Earth, phrases in the Mars analysis like, " . . . at a resolution (on Mars) where an equivalent (terrestrial) civilization would have been detected (on the Mariner and Viking imagery) . . . no evidence of intelligence appears . . . " make one at least wonder if the Martian studies fell prey to "creeping terrestrial chauvinism."

For nowhere is the presence of the Face addressed—let alone discussed.

The fact that we (after several years of "prodding") finally took this object seriously, and Sagan didn't, would seem to be the answer to the question: why have our two investigations reached such radically different conclusions concerning the presence of intelligence on Mars? For the sheer presence of such a monument—a megalithic work of art on a vast scale—should have been a clue to the scale of other potential artifacts . . .

*If* you accepted the presence of the Face as a work of art!

The fact is that the Sagan searches for signs of intelligence on Mars were profoundly affected by the "obvious" anthropomorphism inherent in any such acceptance—he rejected it flatly out of hand—and, ironically, by a more subtle terrestrial influence than the recognition of a "face": the anthropomorphic assumption that we "know" what the signs of intelligence will look like from space, because we've seen our own world from orbit—and thus know the scale.

Hence, the combined presence of such a "terrestrial" monument (the Face), albeit on a huge scale, and the concomitant absence of familiar "farms" or "highway networks," apparently conspired to make the presence of intelligence on Mars invisible . . . at least to Sagan.

Otherwise, these words would have a different author.

\*           \*           \*

Unlike the images of Chaco Canyon, the one thing I was certain we would *not* find on the Viking images taken of Cydonia were roads . . .

For countless hundreds of millennia the winds of Mars had been at work, winds clocked at over 300 miles per hour by the Viking Landers themselves. Such winds, I knew, lifting their violent world-wide duststorms high above the planet, would have drifted over any surface transportation networks long ago . . . if the "Martians" ever built such roads to start with. The same reasoning applied to other "well-known" features of intelligence with which we were familiar; even several million years under current Mars conditions would probably have obliterated most evidence for smaller "structures" on the planet, leaving only those objects constructed on a truly massive scale . . .

Like the Face, the massive "pyramid," and, of course, the strange rectilinear collection I now increasingly thought of as the City.

It wasn't truly like a city—at least any modern city such a reference conjures up. There were no obvious spires, like Manhattan; no grid-like streets marking out "downtown." So why was I thinking of it as the City . . . ?

Because (my internal watchdog prompted) it *feels* like a place where "Martians" should have gathered.

The layout—a rectangular complex of larger and smaller "structures," apparently aligned lengthwise towards the rising sun—was eerily reflected in the similar layout of a countless succession of terrestrial complexes—"ceremonial centers" stretching from the banks of the Nile to the jungles of the Yucatan.

It wasn't a "city," my mind suddenly intuited, as much as some kind of "sacred center"—perhaps dedicated to the "god" represented by the Face itself.

A series of images cascaded before me: silent processions of robed and hooded "Martians" gathering among sacred stones on the Solstice . . . all attention turned toward the heavy-browed god-like "being" in the desert . . . watching . . . as Martian dawn turned the ink-black shadows to a sun-tipped crimson glory . . .

It took real effort to shake this vivid picture, to see once again the small black and white photograph on the desk before me, the varying shades of gray that represented whatever the strange "complex" truly was—be it a pile of rock or something else . . .

I had always believed that measurement and prediction separated "science" from that other stuff. So. What to measure? The consistent irony of another Sagan comment, "Intelligent life on Earth first reveals itself through the geometrical regularity of its constructions,"[3] floated through my mind.

The area in question seemed to consist of approximately seven major pyramidal forms—in various states of preservation. Among these were at least another half dozen smaller objects, some also having pyramidal forms, but at the same time consisting of apparent "domes," "cones," isolated "walls," and a suggestion of buried rectilinear markings to the southwest of the main complex. A rough measurement of the average spacing of these objects from one another in the City indicated that the range varied, from about half a mile to three quarters. This contrasted sharply with the much wider spacing of the randomly scattered mountains to the southeast, which averaged two and a half miles. These latter objects were also much larger than the forms within the City, consisting of large, irregular blocks measuring about two miles across, and with no particular orientation.

The City was thus qualitatively and quantitatively *different* from the surrounding countryside. Its dimensions were rectangular, contained within an area approximately $6.4 \times 11$ kilometers. The northeast "boundary" was marked by the straight "wall" of "the Fort," at 11.2 kilometers from the midline of the Face. The southwest boundary seemed naturally marked by the last large "pyramid" located in the complex, a deteriorated form situated on an elevated "platform."

The City Square described before seemed located in almost the exact center of this complex.

On a hunch, I placed a compass point on the small round "building" in the center of the Square. The compass inscribed a precise circle around the entire city complex, with the radius just touching the outside edges of the "ruins" northeast and southwest. In terms of the overall axis of the rectangle, the Square lay exactly one quarter the distance from the "southwest edge of town" to the Face.

Returning to my original intention, to compare the apparent parallel edges and walls in the complex with the central axis of the Face, I carefully measured the alignment of the pyramidal object marking the "southwest edge of town." I chose its "base platform," as it alone had a clearly defined edge.

It was parallel to the central midline of the Face, to within less than a degree.

I then measured the alignment of two other major "walls" within the City—one defining the edge of a Mexican-looking pyramidal object, with an apparent flat roof (!), the other a very Egyptian-looking pyramid located just east of the "Mexican Pyramid"—and both edges were again parallel to the central axis of the Face, at least within the errors of my straight-edge and the small scale of the photograph. The same was true for the northeast boundary of the complex: the wall of the "Fort."

Working with my homemade grid, I proceeded to measure features in the City arranged at right angles to this parallel alignment to the Face. These were abundant.

Again, starting at the "southwest end of town," I found a linear, ramp-like feature apparently ascending from ground level to the level of the base platform of the previously mentioned pyramid. It was at right angles to the mile-long edge previously measured. Furthermore, it appeared in the exact center of the span.

This symmetry pointed up another, more significant, I felt. In the center of the City, north of the City Square, lay the best preserved image of a classical pyramid, including an apparent pointed apex casting a long pointed shadow out across the Martian afternoon. Two sides (of an apparent four-sided object) appeared sunlit: the southwest and northeast. And in the center of each—in the *exact* center—lay two much smaller objects. The one on the northeast side resembled a "broken cone." The southeast object appeared like a much smaller version of the pyramid it nestled up against, though set at a definite angle to its southwestern edge.

The main pyramid was a vast object, at least a mile on each side. The smaller objects appeared about a thousand feet across—about the scale of the Great Pyramid at Gizeh, in Egypt. Against the sheer enormity of the larger pyramid, however, each appeared as tiny objects, mere attendants to the centerpiece they framed—

For "frame it" they did.

A line run perpendicular to the main pyramid's southwestern edge, through the apex of the smaller pyramid, ran parallel to the overall axis of the City's rectangular configuration. In addition, it passed very close to the apex of the large pyramid, so close, in fact, that I realized that it would pass directly through it . . . *if* the 12 degree tilt of the Viking photograph off the vertical was allowed for! Continuing the line northeast produced a *most* interesting discovery; it intersected another "object" located between the "main pyramid" and "the Fort." But the real discovery was this:

If the previous line was divided in half, and another line run northwest—perpendicular to the first line—it intersected the "broken cone" lying in the center of the main pyramid's northwest side! In the opposite direction (southeast) this line intersected both a peculiar dome lying about a mile from the apex of the large main pyramid, before passing through the highest point of a badly eroded, rounded object also measuring approximately a mile across. If the latter was once a pyramid, it was now merely a shapeless hump casting a whale-shaped shadow toward the southeast. But the fact that a set of perpendicular lines included it in the expanding pattern I was coming to discover in this single Viking image, spoke reams for a status it once shared with its less eroded neighbors . . .

As part of a deliberate design.

Of this I was becoming more and more convinced, with every line and angle that I measured.

Looking at the small image of 35A72 furnished by DiPietro, I realized I was rapidly approaching the limits of what could be done with such a small-scale print; to accurately refine the previous measurements, to "fine-tune" the angles in search of subtleties beyond the limits of a hand-ruled grid and a magnifying glass, would require photographic blow-ups of the image. In addition, in looking once again at the mosaics DiPietro and Molenaar had assembled in their booklet, I realized there existed, in addition to 35A72, a *second* sun angle image of the City—only Vince hadn't furnished it with the sheaf of material he'd sent. Taken on orbit 70, it was 70A11—a companion to 70A13.

The thought of being able to compare the alignments and details of the City at a higher sun angle, to see into the shadows of the pyramidal objects and to trace the outlines further, was enough to send me to the telephone. For, even if DiPietro hadn't sent a SPIT-processed version of the higher sun-angle image, I knew of a nearby source of the next best thing: a NASA version of the photograph.

I called Mike Carr, an old friend located a few miles away at the United States Geological Survey, in Menlo Park, California.

Mike was a geologist, actually a new breed of the same cat, an "astrogeologist." He and other members of the profession had led the pioneering studies of the surface of Mars, analyzing the countless Mariner and Viking images to produce the amazingly detailed reconstructions of Mars' history—as gleaned from studies of craters, volcanos, ancient river channels, and other Martian surface features. Not only would Mike probably direct me to whomever I should talk to regarding getting a hard-copy print of 70A11, he could also tell me a a lot about the current ideas regarding the geology of the Cydonia region.

Mike was also the former head of the Viking Orbiter Imaging Team. What he didn't know regarding Mars and Viking wasn't worth knowing.

When he came on the line, however, I got cold feet. I mean, how do you tell a friend—and one so "officially" connected with such a vast survey of information—that you—a "mere" journalist—have discovered a city and an implied civilization . . . on "his" planet?!

The answer is: you don't.

Mumbling something about "wanting to check the regional erosion features" in this section of the planet, I asked Mike some guarded questions regarding the past history of the Cydonia area. I also asked where I could get copies of orthographic versions of 70A11 and 70A13. To my surprise, Mike offered to send me the file copies out of the USGS library, provided I didn't need more than "a couple of prints." The budget for new prints "was awfully tight" he said, so he could only have a few replaced.

How far the billion dollar Viking Mission had fallen! The situation reminded me of the sad reality: Congress had displayed an annoying habit over the years of approving lavish planetary missions, such as Viking or Voyager, with price-tags of several hundred million dollars. But, they would begrudge NASA a *few million dollars more* to analyze the data properly—as evidenced by the fact that the USGS (a NASA experimenter group) didn't have the necessary funds to furnish researchers or (in this case) a member of the press with more than "a couple of prints."

Before I hung up, still not divulging the reason I needed the information, I casually (I hoped) asked Mike who I should talk to regarding the changing obliquities of Mars. (Remember, it was the changing obliquity of the Martian pole that moved the horizon position of the Solstice sunrise—allowing my provisional estimate for the last inhabited epoch for the City.)

"Oh," Mike replied, "you want to talk to Bill Ward at JPL. He's the outstanding obliquity expert of the solar system." His grin was obvious without seeing it, even over the telephone.

"By the way," he finished, before he had to hang up, "you're not looking at that damn face, are you?"

And the line clicked off.

I sometimes think that the hardest part of research is waiting for the mail. During the couple of days it took for the new prints to reach me, I busied myself with measuring, and then remeasuring, the details I've described on 35A72. And checking them on the mosaics printed in the DiPietro/Molenaar monograph.

As part of the latter, I took the booklet, opened it to page 37, and proceeded to draw the alignments and angles in fine colored pencil, extending them in every direction as far as the mosaic reached. Perhaps, I thought, there were other significant lineups . . . No one could live in a region, inhabit it for centuries or thousands of years (another terrestrial oversimplification?), and not leave clues. Perhaps not more "faces," but almost certainly something else . . .

That's how I found "the cliff."

At first I didn't see it because it was perched on top of one of the most ubiquitous features of the Martian surface (or any surface in the solar system, for that matter): a crater. This crater was one of literally millions littering the Martian landscape, a roughly two-mile impact scar complete with a roughly circular "apron" of debris surrounding it. The

*The Monuments of Mars*

debris pattern was an "ejecta blanket." Peculiar to Mars, this type of pattern resembled nothing so much as a splash around a stone dropped in mud. And the explanation, from Carr and others, was just that: the debris "splash" created around craters on Mars apparently came from the impact of asteroids or comets into *very* soggy ground, or even penetrating to a layer of liquid water underneath the surface. The resultant crater was clearly identifiable: a circular basin surrounded by a "rim" of material, around which stretched in all directions for at least a crater diameter a radial pattern. The result looked simply like a splash of frozen mud.

Not a reason in the world to give such a crater even a single glance . . . unless the Solstice sunrise line drawn from the City through the Face happened to *cross* this particular crater, located 14 miles *beyond* the Face itself.

In the DiPietro/Molenaar booklet "the cliff" was a very tiny, linear object, seemingly perched on the ejecta pattern about a mile in front of the crater—as measured from the direction of the Face. What was intriguing was its presence—a definite object which started and stopped well within the pattern, and at *right angles* to the outward flow of material which made up the ejecta blanket itself. It also seemed exactly parallel to the Face.

This orientation created something of a stir within me as I measured it, for the tiny "cliff" was also at right angles to the Solstice sunrise line, extending from the innards of the City through the slit-like mouth of the Face. Almost as if someone had placed it there, as a convenient place from which to watch the other celestial spectacular visible along that line, but in the opposite direction . . .

The Winter Solstice sun, *setting* behind the Face!

But why a *cliff*? Why not a single marker, so an observer looking into the setting sun on the evening of the Winter Solstice could mark the sunset behind the slit-like mouth of the Face? Unless the cliff was designed to mark the continuous southward drift of the sun as it *approached* the solstice . . .

My mind again conjured up visions of robed priests, parading along the length of this approximately 3.2 kilometer-long cliff each sunset, marking the blazing image as it sank into the west far beyond the upturned gaze of "god," lying a few miles to the southwest. I had intimations of a kind of Martian "sun dagger," like the ancient Anasazi solstice marker found in Chaco Canyon: a line-like image of the sun as it was "focused" through the "mouth"—only 21 kilometers away . . .

But the mosaic was too tiny to check the details of this hypothetical reconstruction, such as the location (if any) of "seasonal stations" located at intervals along the cliff. If they were there, they were totally invisible against the dot-like pattern of the reproduction in the booklet. Still . . .

On another hunch, I placed a straight-edge on the single page mosaic, stretching from the City Square across the Face to the northwesternmost extension of the cliff. The distance (as near as I could make it out on the tiny reproduction) was something over 32 kilometers. But the interesting part was this: the line passed directly through the *eyes* of the upturned Face . . . and ended at the northern end of the peculiar cliff. Another line, begun in the same starting place and drawn directly past the chin of the enigmatic visage, ended *exactly* at the southeasternmost extension of the cliff—

Almost as if the cliff were designed to be a backdrop to the Face . . . when viewed from this precise location in the City?

But why . . . ?

Then all at once it came.

The cliff was part of the deliberate design of this entire Complex—City, Face, and cliff. The reason was so simple: to see the Face as a proper silhouette against the rising sun on the morning of the Solstice required that it be the *only* silhouette on the horizon. The crater out there some 22.4 kilometers beyond the Face would have been a problem, as the literal *curvature of Mars* brought the hilly crater rim down to the sightline from the City—

Messing up the pièce de résistance on the horizon!

The cliff was simply an enormous *artificially flat horizon*, exactly like the mound of earthworks which completely circles Stonehenge—and for that exact reason.

So the sun and Earth could rise over the sharp edge of the world, behind the Face, without interference from the true topography of that horizon.

Suddenly, it all *fit*—the upturned silhouette, the sightlines, the convenient cliff stretching along the horizon just far enough (and no farther) to provide a flat knife-edge for the all-important sunrise . . . That this could all be sheer coincidence was asking more of my credulity than to believe that someone had carefully designed it. I felt as though a science fiction tale were not only writing itself but coming true inside me. "Monoliths" indeed!

A day or so later, Mike's pictures came—70A13, 70A11, and good ol' 35A72. Eagerly, avidly, I ripped open the envelope and spread them out on the desk, with Vince's version of 35A72 beside them as a comparison.

There was the City! The higher sun angle print from the USGS Library files clearly showed it, looking even more rectilinear than on the

*The Monuments of Mars*

late afternoon version—35A72. But whereas the deep, long shadows of the latter hid most of the details on the southeast sides, this print of 70A11 showed a lot—including a surprise—

The main pyramid within the City, the one neatly "framed" by the three much smaller pyramidal objects placed at exactly right angles to each other and the apex of the large pyramid, was different. It wasn't square. In fact, the steep sun angle allowed me to see that the two other sides (I had inferred from 35A72) were missing! At this lighting, the "main pyramid" looked three-sided.

Or (I took a closer look) as if the two others had been destroyed . . .

There was an apparent selective degradation of all the "structures" in the City from the southeast; the "Fort," the "main pyramid," the "Mexican pyramid," and many other objects in the complex were definitely more eroded on that side than on the opposite. The reason for this directional erosion wasn't clear but probably had something to do with the regional winds, etc. I made a mental note to look that up at some point.

With the availability of 70A11—taken at 30 degree sun, as opposed to 35A72's 10 degrees—I also had the potential for a stereo pair. For, while the different sun angles would make for some confusion in looking at the two photographs through a stereo viewer, the fact that they were taken from different orbits (35 days apart), and thus over a baseline of at least several tens of miles, meant that one could actually use the two images to see the topographic relief within the City! (A similar technique had been used by Vince and Greg to demonstrate the three-dimensional nature of the Face at the AAS meeting of astronomers in Baltimore, in 1980.)

The two images—DiPietro's SPIT-processed version of 35A72 and the USGS 70A11—were *almost* the same scale. With a bit of squinting, my eyes at last were able to fuse both images under the viewer into one—

And I was hovering over an ancient ruin on a sun-bleached plain.

The intimation of looking down on something that had once had people in it, had been a center for commerce, art, communication—all the things that people do in pursuing their own lives—was overwhelming. I could trace the broken edges of the pyramids, feel as though I was floating down half-mile wide avenues between monstrous "buildings," looking at the decay of a place which once might have bustled with activity.

The City seemed higher to the southwest than to the northeast—towards the Face. That fit. If the complex was arranged so that it had a view of the spectacular image lying out in front of it, the lower elevation of the structures towards the northeast would have afforded an unimpeded view from almost anywhere within the complex.

*A City Viewed from Orbit . . .*                                                                 77

The exaggerated elevations in the stereo perspective also served to highlight something else: the City was perched at the edge of a flat plain. In fact, it didn't take much imagination to trace the outlines of an ancient shoreline—perhaps a lake or even a protected inlet from the ocean (!)— around a perimeter which stretched from the northwest to the southeast. In the stereo view, the City was located at the base of a steep mountain to the southeast. In fact, a permafrost "creep" could be seen flowing down the mountain (which resembled another "face"—perhaps a bit satanic) towards the small rectilinear array of objects, looking as if it were trying to push them off the plateau and into the waters of the dried-up lake or bay. A couple of "walls" could be seen between the main City complex and this advancing wall of water-laden material. In my imagination I could even see this as part of the design: climatic conditions had changed and the builders had to confront the worsening conditions of the planet—which included soil characteristics similar to those in Alaska or elsewhere in the Arctic here on Earth.

Ironically, the object I had been thinking of as the Fort was situated just about where a real fort might have been designed to sit: on the last point of land before the "open water." And the Face . . .

With a start, I realized that when any of this might have been inhabited, *real* water might have flowed in this dried-up basin, where now there were only features looking suspiciously like permafrost cracks. And that meant . . . that the Face might have been viewed across the water, with the reflections of the Martian dawn mirroring that image looking *down* as well as up . . .

The artistry of such a possible design didn't come now as any great surprise—considering.

For days I hovered above the City, hungrily devouring the details, the multitude of pyramids—large and small—and the mathematical layout of the complex. But some objects didn't conform to any here on Earth. Besides the ruins of four-sided structures, the "Martians" seemed to have created their own unique form: a *trapezoidal* pyramid.

Consisting of a rectangular outline, when viewed from above, the complex shape was created by two inward slanting ends, and two parallel slanted sides which met at a linear "ridgepole"—as opposed to the pointed traditional Egyptian apex. There appeared to be several of these "trapezoidal pyramids" scattered through the City, including at least one member of the City Square itself. The magnifying glass, and the raised relief afforded by the stereo viewing of the two images, allowed me to discern many details as the days passed, details that had initially escaped me. Not for the first time was an observer of the planet Mars becoming more

*The Monuments of Mars*

open to discovery, as the eye/brain combination got used to the "newness" of the detail it was perceiving.

I was thinking specifically of Percival Lowell, who claimed that the reason why most observers never saw "canals" was because their eyes were simply not accustomed to the richness of the details on the planet. I couldn't help but think that the comparison was fitting . . .

For who was going to believe what I was seeing? And I had pictures, unlike Lowell.

Appearing as if only a few kilometers beneath my "space ship," the City looked worn and ancient, a tired ruin sagging back into the legendary "sands of Mars." One of my childhood heroes, Arthur C. Clarke, had written a novel immortalizing (at least for me) that phrase. It concerned a journalist who journeyed to the Red Planet . . . and ultimately discovered its semi-intelligent inhabitants.

Years later, while advising CBS News on going to the Moon (when we as a nation were actually doing those things), I became friends with Arthur—a man whose fiction and non-fiction were the reason, in part, why I was there at all.

I learned later that *The Sands of Mars* was his first novel . . .

And now here I was, gazing down on the ruins of a lost civilization on the Red Planet, thinking, "Of all the people who are *not* going to believe me, the first is Arthur Clarke."

As I roamed the ruins from a thousand miles above, tracing the crumbling outlines in the sand, I couldn't help but speculate about the "who" and "why" of what I was exploring. Who had created all of this? When? And, perhaps most important: Why?

The current Martian environment was totally inhospitable to the level of cultural development indicated by the existence of these ruins. The City/Face geometry implied a fixation on the sun, and its seminal importance to a fledgling agriculture. If I was right, if the level of technology was comparable—"Egyptian"—they shouldn't have existed—let alone originated—on the "current" Mars.

And that implied that when they did exist, Mars was another world.

Mariner 9, in 1971, had opened up our eyes to the possibilities that previous Martian "epochs" had been different—very different—from the current "cold, glaciated world" that we had come to know as "Mars" (see Chapter III). Carl Sagan was among the first, in the wake of the successful Mariner 9 mission, to propose a cyclic climatology for the planet. To explain the apparently layered terrain at the polar caps as the byproducts of such a cyclic climate, Sagan proposed that every 50,000 years a "Martian spring" would come—created by a favorable combination of planetary orbit changes and the changing tilt of the rotational axis to that orbit. These effects, he said, could create a global "thaw," particularly at the polar caps, which would liberate huge quantities of frozen carbon dioxide, which in turn would trap more solar heat (through greenhouse effects) thus raising the global temperatures above the freezing point of water . . .

And "spring" would have arrived.[4]

As I gazed down at the small collection of strange shapes in "my" City, I speculated what it would be like to live on a planet with these vast extremes. What would be the cultural response to an environment which turned from bad to worse-than-worse . . . every 50,000 years?

For, while Viking failed to find deep carbon dioxide at the poles (as had been hinted at by Mariner 9), the pictures coming from that spacecraft and then from Viking were more than proof that *something* had been different in the past. The evidence for running water, and subsequent erosion, were compelling—if not in 50,000-year-cycles, then perhaps at longer intervals in Martian history.

Which raised again the question: what would it be like to live on a world where literally the "stuff of life"—the air—could freeze?

Perhaps, I thought, one might build suspended-animation chambers . . . to wait for Spring.

Beneath the bright reflection of the late afternoon Martian sun off the sharply symmetrical outlines of the main pyramid within the City, I imagined (within the pyramid) rows of these waiting chambers, each containing—

A sleeping Martian nurtured, through some kind of dormancy, into the inevitable spring.

Was it even possible that the entire Egyptian legend associating "immortality," and the pyramids' connection with the "afterlife" shared origins with this exotic City of pyramids?

Was this the source of one of the greatest civilizations the world—our world—has ever known? Were the legends true? Was there indeed a mile-wide pyramid that one could enter . . . to wake up to a new life after—how many—million years?

On Mars.

It was a speculation that could never have an answer—unless we went there and entered those same pyramids . . . and looked.

But, as I kept exploring, measuring, the question haunted me. Why pyramids? It was on my mind as I strove to understand an equally mysterious feature of the City, located directly northeast of the main pyramid of my exotic speculations.

*The Monuments of Mars*

The object I had termed the Fort.

It didn't make sense, even to my now educated eye.

Parts of it seemed very simple, like the linear wall that defined the northeastern edge of the City itself. This feature was straight as the proverbial arrow for more than a mile—northwest/southeast. And parallel to it was a second linear feature, which gave this wall its appreciable width. On it one could see some strange substructures, with their own curious angularities and apparent right angles.

At the southeastern end of this linear feature was another—but at a sharp 75 degree angle to the southwest. It too seemed to have two parallel "inner" and "outer" edges, which were the same width as the previous wall. The inner "courtyard" created by these features (and by the third, major section of the Fort itself) was dark and mysterious—even at the 30-degree sun angle of 70A11. Its existence was one of the primary reasons why I had coined the term "The Fort" for this strange geometric object.

The complicated part was to the northwest of the preceding features. For the main puzzle of this object seemed to be its shape; it was long and fat, like a beached whale. Oriented northeast/southwest, a long, fish-shaped profile—it was this object in particular—which had me mystified.

The northeast end was pointed, and seemed to lie atop the previously described platform like the apex of a fallen spaceship. I'm serious; that's what it looked like. This impression continued right down to the other end. There, the object seemed to split—into two parallel and highly symmetrical "tailfins," like the kind they used to put on those 1950s rocket ships to Mars!

And in the center of these "tailfins," looking for all the world like the exposed rocket motor, was a peculiar "crater"—with about one half abruptly cut off.

And there was no apparent reason why.

The more I stared at this peculiar geometry (it obviously couldn't be a *real* spaceship!), the more I was confused. The "truncation" which seemed abruptly to terminate the crater at the end of this linear shape also seemed to determine the cutoff for the "tailfins" as well.

Whatever was causing this sudden termination of the substantial relief in this region, across a linear extent of at least two miles, was apparently almost invisible—except as a faint scattering of light *between* the "tailfins" and the "rocket motor." There was only one logical conclusion: this linear "something" was a physical material, somehow veiling the underlying terrain.

That this was the case was supported by observations of the region on the higher sun angle print, 70A11. On this image you could see the "fins" and the "rocket motor" extending further to the southwest—as if the underlying topography were getting more light at the higher lighting angle. The conclusion seemed inescapable: something was blocking the light between the "surface" topography of the Fort and the Viking cameras—a thousand miles above. The question: What?

I tried several plausible geological explanations, such as a sheet of ice, but had to discard them one by one as simply too contrived. (For instance, what would make the ice sheet perfectly straight for over 3 kilometers, when it had to lie across substantial geological relief underneath? Or, even simpler: why didn't it melt? At these latitudes exposed ice would have a very short life. And finally, if it was ice, why was its reflectivity so low—lower than the terrain it was covering up? And don't say, "It was dusty ice." Dust would have caused it to melt even faster— it was *summer* in the northern hemisphere when these images were taken!)

In the end, the answer came simply. I had been staring it "in the face" for probably two weeks, as I went over the small images again and again with the magnifying glass.

Late at night, on a Saturday, in fact, and about two o'clock in the morning, I was taking a last look at the City before turning in. As I bent over the small print—the SPIT-processed version of 35A72—I noticed a small smudge between the Fort and the main pyramid that I hadn't caught before, just southwest of the abrupt and maddening cutoff of the "rocket motor" aspect of the Fort. I focused the magnifying glass . . .

And realized I was staring at a tiny pattern of honeycomb-like cells.

Several seconds passed as I slowly comprehended what I was seeing. Moving the magnifying glass down a bit I traced the "honeycomb" across the two-mile region which had nagged me for so long . . . and saw the shadow of the adjoining main pyramid falling across this tiny region, clearly outlining a raised, three-dimensional relief.

And suddenly, it all fell into place.

"Oh my God," I said softly, in the night with no one to hear me except the clock and the cat.

I was gazing at another mile-square pyramid. The shadow of the adjacent pyramid confirmed that; it hid the lower levels and allowed sunlight to reach a topmost "deck" visible southwest of the Fort. But the amazing, the almost unbelievable thing was the "honeycomb pattern": this pyramid was apparently without a covering . . . and was filled with tiny cell-sized "rooms"—

Exactly as I had predicted only a few nights before!

I sat there in the night and realized with a chill that my entire "suspended animation story" could be real!

*The Monuments of Mars*

This wasn't Earth, where every structure built until this moment had existed for all time, in the minds of each of us. This was another planet . . . and I was staring at a something which couldn't be easily dismissed as mere "geology" or "wind erosion." It looked for all the world like a multi-leveled structure, with numerous decks descending to the Martian surface from that shining upper level, still in sunlight. And it definitely looked artificial!

Its presence in a stroke explained the mysterious "veiling" which created the 3-kilometer knife-sharp contact across the nearby regions of the Fort. Some of the "decking" was built over the adjacent levels of the Fort, preventing sunlight from reaching these lower levels—except at higher sun angles. And, as if to cap this theory, I could literally trace a portion of the honeycomb descending from a higher level into the strange "rocket motor" crater. There appeared to be at least one level of honeycomb within the crater itself (although to term this feature now a "crater" was a serious misstatement). And, just as easily, the presence of a honeycomb-like structure, strung *between* the "tailfins" and the "rocket motor" of the Fort, explained the faint shimmer of sunlight which I could barely detect above the light levels in the shadows; the honeycomb built over this entire area was just bright enough to scatter some sunlight from its "mesh."

The primary effect of this "open girder framework" (as I interpreted the structure) was to obscure other levels of the structure underneath—much as a screen dims and diffuses the scene outside a window. Imagine a multi-layered window screen, with successive "floors" arranged in a vertical building of substantial relief—yet without a "roof" or covering.

That's what this structure looked like—with one side caved in!

I sat there, stunned by the implications of the honeycomb, a thousand questions swirling through my mind. Was it real? Or, was it an "artifact" of the special computer process Vince and Greg had applied to 35A72?

If it was merely an effect of the computer, why was it in the *one location* on the almost 3000 square-kilometer photograph that made sense—both architecturally as well as optically? For, one had to explain the presence of the peculiar "truncated terrain" at the southwest end of the Fort, terrain seen on both the processed and unprocessed versions of this image. Then, one had to explain why more of the "underlying structure" was visible at the higher sun angle.

A multi-layered, veiling superstructure—a honeycomb—precisely fit these optical requirements.

In terms of architecture, both the geometry of the honeycomb as well as its precise location—in the one square mile with the best view of the enigmatic Face—argued that it was a conscious construction. Its outlines, dimensions, even its orientation, seemed smoothly integrated into the surrounding geometry, both of the Fort and the adjoining main pyramid.

My "gut"—and all these factors—said that it was real.

But that threw open the doorway to a completely different interpretation of the City, the Face, and whomever had designed this extraordinary complex . . .

For the presence of a vast, mile-wide open framework structure, bared like bleaching bones against the reddened Martian sky, called into immediate question all earlier intuitions about the architects who had mapped and assembled this. This was not an "Egyptian level" culture, it was something far more—

Something . . . that thought nothing of throwing up mile-wide pyramids against the sky, constructed out of materials vastly superior to rock and stone, materials that had stood on this dry shore for an eternity—for at least half a million years . . .

And left a mile-wide effigy to us—or something we once were.

Even now in the frigid Martian night or just before the dawn, one could stand in the center of the all-too-ancient ruins, the stark skeleton of the "honeycomb" off to the left, its shattered remnants towering against the sky, and see that Face out beyond—softly silhouetted in the dark. And sometimes, like a cool green star, a brilliant point of light would seem to hang above it . . . the next world inward toward the sun.

Earth.

For how long, for how many million trips around the sun of both these worlds, had this been happening . . . waiting . . . for us to finally come?

I suddenly shivered, not so much from fear as with anticipation. The long wait was over—for both of us. The question was: what were we going to do about it, when we returned to explore—

I remembered a line from Harlan Ellison,

"A City on the Edge of Forever . . ."[5]

While the Face—the "Guardian of Forever"—looked on.

## Notes

1. Sagan, C. and Shklovskii, I.S., *Intelligent Life in the Universe*, New York: Dell, 1967.
2. Personal communication.

3. Sagan, C., *Cosmos*, New York: Random House, 1980.

4. Sagan, C., *The Cosmic Connection: An Extraterrestrial Perspective*, New York: Doubleday, 1973. *Ibid.*

5. Taken from Harlan's Hugo award-winning script, for my favorite *Star Trek* episode of all time: "City on the Edge of Forever."

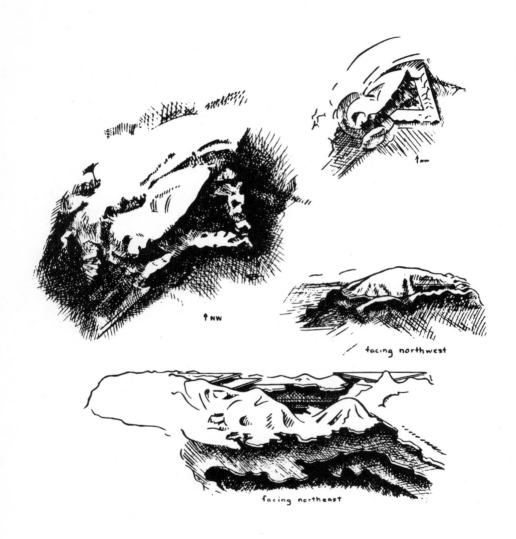

The Fort: upper left, rendering of Fort as seen from composite NASA photos; upper right, simplified aerial sketch of structure of Fort; lower left, side elevation of Fort facing Northeast, with section of Honeycomb in foreground; lower right, side elevation of Fort facing Northwest, with section of Honeycomb on the left. Drawing by Kynthia Lynn.

# VI

# INITIAL SPECULATIONS

> "Its makers had prepared it for many things, and this was one of them. It recognized what was climbing up toward it from the warm heart of the Solar System.
> "If it had been alive, it would have felt excitement . . ."
>
> Arthur C. Clarke
> *2001: A Space Odyssey*

Again and again, as I measured—and remeasured—the collection of pyramid-like objects in the City, a part of my subconscious kept insisting, "This is all delusion. From everything we know, it's just impossible for Mars to have supported life—let alone to have developed it. And Intelligence? Come on, Hoagland. Of all the nutty ideas you've come up with, this has got to be *the* nuttiest!!"

But the data remained—a series of "anomalies," each of which was unusual enough in itself to warrant some kind of explanation beyond "weird geology." More than that, however, it was the way these anomalies all "stuck together"—forming a truly inexplicable "integrated anomaly" on the Martian landscape—which demanded a complex in-depth explanation.

When I had started, there had only been a "face" and a strange "pyramid-like object" several miles away. Now, measured by my own hand, I had added a collection of equally mysterious objects—an entire precisely arranged "city" of "pyramids"—to the growing number of anomalies peculiar to this region of the planet; but more than that, I believed I may have discovered the *purpose* of this extraordinary collection of objects: to focus attention on the most extraordinary member of them all—

A representation of a human face—on Mars.

Time and again I was brought back to that central figure . . . and the problem: it shouldn't be there.

Despite my reservations, each new measurement merely added to the evidence that there was a conscious design in both its proportions and its orientation on the Martian surface. For instance, measuring its central axis, I became aware that I could lay the meridian line (the north/ south line) across the Face in such a manner that, beginning at the southwest corner of the mouth the line extended due north *right between the eyes*. For such a thing to happen, the Face had to be perfectly proportioned (the mouth couldn't be too long, or the space between the eyes too high above the mouth) *and* the central axis had to be tilted to the meridian by a precise amount.

In other words, to expect this degree of precision from several separate elements *by chance* was asking for a lot—a remarkable fortuity, or a virtual miracle.

Then there was the relationship of this remarkable representation to the other "anomalous objects" at Cydonia: the peculiar pyramid of DiPietro and Molenaar, and the even more significant cliff lined up in a specific formation with the Face. On the small print of 70A13 provided by DiPietro (the "second confirming image" of the Face), I could measure the alignment of the pyramid they had discovered on the frame. One of the "buttresses" seemed aimed directly at the Face. Another "coincidence?" Or an architect?

As for the objects in the city, the first one (the one I'd mentally noted as the Fort) also had a key alignment towards the Face: the southeast wall which formed the southern boundary of its interior keep. The second major pyramid behind the Fort (moving southwest) also had a feature which aligned with the Face: its entire northwestern edge formed a linear "contact" on the surface which continued back towards the Face, through the Fort itself (!), until it disappeared at the edge of the northeastern wall which formed the other boundary to the keep. Incidentally, that same northeastern wall-like feature of the Fort in turn seemed aimed directly towards the pyramid southeast of the City—the one with those exotic "buttresses." And so on, throughout the City.

Any geological or meteorological explanation for this complex, and its apparent myriad interlocking geometric and mathematical relationships, would have to be pretty darn good. In fact, mentally reviewing the coincidence piled upon coincidence each new measurement revealed,

*The Monuments of Mars*

I was slowly and against my original credulity being drawn towards an almost inescapable conclusion. What had started for me as a kind of "Einsteinian thought experiment" ("What kinds of questions should one ask, if these objects were by remote chance the work of an extraterrestrial intelligence?") had produced such copious supporting data, that the opposing viewpoint—that these objects were the result of strictly natural forces—was becoming extremely difficult to continue to maintain.

What was the simpler explanation for this set of facts? That a fiendishly clever geologic environment had "conspired" to create several familiar "terrestrial" objects of differing morphology, yet all in the same location on the planet Mars, and to array them in both striking geometry and mathematical precision . . . or . . . that a straightforward—if awesome—architectural design was the ultimate explanation for this data?

As implausible as it had first appeared, the Intelligence Hypothesis seemed more and more to be the *simpler* explanation.

Part of asking the right questions in Science is to know how to speculate. A hypothesis, no matter how tentative or "far out," gives some kind of order to raw facts; one can then seek new facts. Ninety-nine percent of this scientific speculation never sees the light of day for one reason or another. Perhaps it's because scientists believe that uninhibited free association will be misinterpreted; that the sometimes truly wild bits of speculation will be mistaken for "scientific truth," when in fact they are merely part of the road to such an understanding. The result is that, for whatever reason, the truly exciting "road to discovery" almost never gets reported: the play of ideas that occurs between colleagues over coffee, the notes that get jotted on the backs of envelopes (really!), or (more often) in restaurants—where such impromptu conversations are jotted on napkins and left behind (to the bewilderment of more than a few waitresses).

By the time a "discovery" gets to the stage of a scientific paper or (worse) a book, it's been "all prettied up." Lost are the random bits of inspiration, the trial balloons, the blind alleys that force one to go back to the original data . . . to rediscover something somehow missed.

Thus, when I decided to write a book about Mars, one of the key elements I wanted to include was an exposure of the *process*—how I (and others) arrived at a particular interpretation of the material. For it is these snippets of speculation, mixed with the raw facts, which have carried the on-going "Mars Investigation" to new heights. Some of these speculations may seem absurd. That is not the point. The key factor on which they should be judged—on which the entire investigation into the extraordinary objects on Mars should be judged—is the same one that is used to judge all other scientific inquiries:

Is this testable? Does it predict new facts—which in turn can themselves be tested by the process?

For this is how our entire scientific picture of reality is constructed—and an inquiry into the possibility of an ancient "lost" civilization on the planet Mars should be no different.

*      *      *

Even with all my residual discoveries, still the most baffling aspect of this investigation, the one thing which really seemed to defy a reasoned analysis, was the central construction of the entire Cydonia Complex.

The Face.

It shouldn't be there.

That simple "fact" had apparently prevented any serious inquiry into this material (including mine!) for seven years. It was at the root of Viking Project Scientist Gerry Soffen's off-hand comment that "a picture taken by Viking of the area a few hours later showed it wasn't there . . . it was merely a trick of light and shadow . . . " The idea that such a Face could be real was so absurd that, without even checking the actual Viking data set, Soffen had confidently asserted what any reasonable scientist would "know" to be true: any confirming images would, of course, prove that it was only a peculiar-shaped mesa sculpted by sun, wind, and natural geology. It *couldn't* be real; therefore, it wasn't real.

Which, of course, isn't Science at all.

The true scientists in the beginning of this saga were Vince and Greg, who refused to believe even the word of the Viking Project Scientist. They insisted on checking the actual picture file . . . and came up with the vital confirming data that—whatever the true origin of this remarkable object—it was "real": a truly bisymmetrical representation of a human (or humanoid) . . . on Mars.

The refusal of the scientific community to accept the sheer reality of this feature has persisted. Long after DiPietro and Molenaar had discovered the second, higher sun-angle image of the Face, "official NASA spokespersons" continued to maintain "there is no second confirming image . . . the object only looked like a 'face' at one lighting angle . . . " Some have even maintained that the second image (taken 35 days after the first, but at a difference of two hours earlier in the Martian "sol" or day) shows *no difference* from the first. They insist that the shadowed right-hand side of this object is still shadowed in the second image (70A13)—despite the fact that any casual inspection of the image allows one to *see* the "missing" right-hand side—with its inexplicable bisymmetrical aspect blatantly evident. Further, a moment's calculation of the rate at which the sun moves across the Martian sky reveals that in two hours (because the Martian rotation rate is almost identical to Earth's) the sun will have moved almost 30 degrees—creating substantial differences in lighting angle.

The situation reminded me of the various reported observations of the most famous stellar explosion in recent astronomical history—the Supernova of 1054 A.D.

Bright enough to be viewed around the world in broad daylight for 23 days, and reported in the annals of astronomical phenomena from China, through Korea, even to rock art depictions in the "primitive" American Southwest, the only place on Earth lacking *any* recorded observations of this exploding star was Mediaeval Europe—which was filled with monks otherwise chronicling almost everything in sight. The only rational explanation for this otherwise inexplicable circumstance (as the supernova continued to blaze in the nighttime skies for almost two *years*) was the simple fact that the Church refused to acknowledge the existence of any marring in Heaven; the monks literally could not believe their eyes . . .

So they didn't "see" the star.

Something very similar seemed to be occurring with the Face on Mars.

Behind this apparent inability to accept the presence of a "human" face on another planetary surface was perhaps a much deeper "paradigm": that human beings—even "humanoid beings," such as apes and other simians—have *only* evolved and *only* live on one planet in the solar system, nay—in the entire Universe!

Earth.

Perhaps the best expression of this viewpoint was voiced in an article in *Science* magazine in 1966, by noted Harvard biologist George Gaylord Simpson: "The Nonprevalence of Humanoids." In this article Simpson expressed the opinion that human intelligence—let alone human morphology—is uniquely the product of the environment we call "the Earth." As Carl Sagan paraphrased it in the discussions at the first Soviet-American Conference on Communication with Extraterrestrial Intelligence, held in Soviet Armenia in September, 1971:

"The principal argument of this paper is that there are a large number of individually unlikely steps which are required for the evolution of man, and the chance of the random recurrence of this sequence of steps is so small as to make, I would say, the existence of mankind (elsewhere) impossible." [1]

At this point another conference participant, John Platt, University of Michigan, pointed out that Simpson had been specifically talking about man and not the likelihood of extraterrestrial intelligence itself. To which Sagan answered, " . . . that is precisely what I think the criticism of Simpson's view is—that there may be many, many other pathways to an organism which is functionally equivalent to a human being *but which looks nothing like a human being* (italics added)." [2]

This, then, reflects the body of scientific opinion on the matter: there may be extraterrestrial intelligence, but because of the literally trillions of evolutionary steps which have been required to produce the human race, the odds in favor of that intelligence physically resembling us in any way are essentially *zero*!

Now, in the face of this opinion, the discovery of a mile-wide "human" head on Mars was, shall we say, just a bit unsettling. For, if it was real, then some aspect of Simpson's (and by inference, a central premise of Sagan's) carefully constructed logic simply wasn't . . . or so any narrow interpretation of this chauvinistic viewpoint might initially construe it. Incidentally, as an appendix to the previously cited conference, Simpson was invited to amplify on his remarks made seven years before. Said he in part:

"Knowledge of Venus and Mars has increased considerably in the last eight (sic) years. It confirms the virtual impossibility of carbon-based life on Venus and greatly reduces the chances of such life on Mars. The chances of intelligent life (even *remotely* human) on either planet, or any other body in our solar system, are evidently as near nil as possible." [3]

As were the chances, evidently, that in this intellectual environment Vince and Greg could receive anything like an open-minded hearing.

There is an opposing minority viewpoint within the scientific community, sometimes termed "the school of parallel evolution," which holds a position at the other extreme from Simpson (and to a lesser degree, Sagan). Isaac Asimov, perhaps the world's most prolific science writer—with over 350 books to his credit at last count (this morning!)—has argued in favor of this viewpoint. It is his contention that, even as Sagan argues for multiple pathways to the "holy grail" of Intelligence, there are also good reasons for multiple pathways to the humanoid "package' in which that Intelligence might be housed: a bilateral form, with short nerve paths between the eyes and brain, and manipulable hands in front of the body (where the eyes can see what they are manipulating!) are perhaps elements of design employed in widely varying planetary environments.

In other words, Asimov contends, evolution will not likely produce intelligent "spiders" or "clams," but will wind up with a creature that more often than not looks something like us. Robert Bier, in an article in a 1964 edition of *American Scientist*, "Humanoids on other Planets," made a similar case.

Which is why determining the reality or falsity of the Face should be highly relevant to this discussion. If the first bonafide "artifact" of a civilization on another planet were a mile-wide representation of a humanoid of some kind, that information had to be of key import to the dialogue described above—or so one would think.

In point of fact, that potential artifact was conscientiously ignored.

My own prejudices in this matter ironically happen to lie with Simpson/Sagan. Despite occasional cases to the contrary (the examples of the apparently independently evolved eye—mammalian and octopus—which possess essentially *identical* structures; or, the streamlined shapes of fish, reptilian ancestral forms of dinosaurs, and modern mammalian dolphins—which seem influenced to a high degree by the medium in which these separate species evolved), my "take" on the evolutionary history of this one planet is one of remarkable diversity—not convergent development.

There are far more examples of unique morphology, coloration, and behavior among the current (estimated) several *million* separate species on the earth, than those of independent mimicry or outright "parallel evolution." In fact, the overwhelming number of species (of an estimated half a *billion* that are now extinct!), which demonstrate shared characteristics do so because of direct genetic inheritance, although often along widely diverging pathways. Or, to quote Simpson, "It is improbable that convergence ever produces literal identity in structure and certainly *no such case has ever been demonstrated.*" (italics added). [4]

To find a humanoid example on another planet, to me, was as startling and inexplicable as if NASA had photographed a dinosaur out there. That chance could independently invent a look-alike on Mars—when it had *never* come close to doing so here on Earth—seemed to me absurd.

If the Face looked like something humanoid, then chances were overwhelming that—again, if it was "real" (i.e. artificial)—it somehow referred to the one planet where humanoids exist.

Earth.

The operative word underlying this particular prejudice, of course, is "chance." If factors other than strictly random selection were ultimately to be shown to determine evolutionary endpoints, then the preceeding argument would, of course, be false. That, in turn, would only become known if we encountered a true example of life (based on DNA) which had evolved independently from the environment of Earth. And, again, the discovery of the Face on Mars seemed uniquely relevant to this inquiry into a possible universal mechanism of planetary evolution, or intelligence.

Mars, as a planet with a little over one third the surface gravity of Earth, with a radically different geological and climatological history, to say nothing of the absence of a magnetic field to shield the surface from effects of ionizing radiation coming in from space, seemed the *last place* one would look for such an extraordinary example of "parallel evolution."

To repeat: if the Face looked like us, then chances were it *was* us. The problem was to figure out what it was doing on the fourth planet outward from the sun . . . when all the rest of us were on the third planet!

In an earlier chapter (II), I presented the three immediate explanations which came to mind. Reiterating them in the context of the previous discussion seems appropriate. They were:

1) The figure had been created by indigenous "Martians"; i.e., implying a remarkable instance of parallel evolution on the neighboring planet and an "Egyptian" level culture (later modified to include progression to a "high tech" civilization).

2) The Face was the product of a designer from beyond the solar system—a so-called "cosmic greeting card"; i.e., implying an advanced culture capable of travel between stars.

3) A previously unknown technical civilization had arisen on Earth, gone to Mars, and left an image of themselves—us!—for us to find.

In keeping with my own prejudices a la Simpson/Sagan, I was, as I described, initially drawn to number 3—a theory which did least violence to what we "know" of our own history. That there could have been a previous technically inclined civilization on this planet—which we've somehow "misplaced" in the intervening several thousand years—at first didn't strike me as anywhere near as challenging to current concepts, as would, say, the discovery of a completely independent mirror-image of ourselves—on a planet nothing like the one that we've evolved upon. After all, archaeologists were all the time rediscovering "lost" civilizations on this planet. Why not one which developed just a bit beyond the others—including an ability to construct spaceships?

Why not indeed.

Over time, the more I looked into this "least far out" possibility (and none of the choices presented above was exactly conservative), the more disenchanted I became. For, not only was direct evidence of such a civilization missing . . .

So was any evidence of its impact on the planet—*this* planet!

A technical civilization such as ours, which lately has developed an ability to construct spaceships, along the way to that singular accomplishment, also develops the ability to destroy the environment of Earth—as it busily mines, pours concrete over, and pollutes the resources of an entire planet to support an increasing minority in the style to which they've become accustomed . . .

Which now includes an ability to take unmanned jaunts around the solar system.

That a previous technical civilization could have invented the myriad supporting technologies required of "an ability to construct spaceships," and not have left a trace on the environment of Earth—no stripmines, no depleted oil reserves, no "Love Canals" of toxic wastes—was quite improbable. Considering the trash left behind by previous *untechnical* cultures, the existence of a much more complicated and (of necessity) far-flung technical society, able to leave behind no environmental evidence of their existence, seemed beyond serious consideration.

Having effectively (in my own mind) disposed of this terrestrial explanation for the Face, and having also formalized the logic which essentially excluded humanoid life from having originated on the planet Mars, I was left with only one remaining option:

Number 2. That someone from elsewhere in some distant eon past, had carved it on the rusted sands as "a Message to Mankind."

By a curious twist of fate, I had a personal familiarity with a somewhat less expensive message . . . the famed "Pioneer 10 Plaque"—aboard the first terrestrial artifact, Pioneer 10, at this very moment speedily exiting the solar system on its one-way journey to the stars . . .

*      *      *

The Plaque had sprung into being during another mission, the historic ceremonies of Mariner 9's insertion into the first orbit of Mars, in November of 1971.

I had been at JPL as the events of that memorable afternoon unfolded: a human-built artifact successfully slowed by a small on-board rocket motor, into a grand looping trajectory around the fabled planet of Man's Dreams. At the moment the spacecraft's velocity dropped below the "magic velocity"—that which, even if the motor should at that instant fail, the spacecraft itself could not escape again from the gravity of Mars—the Auditorium, filled to the walls with hundreds of newsmen, cameramen, scientists, engineers and visitors, erupted in applause and whoops of exultation.

We'd done it! We had reached across space, outwards towards the Red Planet of Lowell, of Burroughs and Wells, of Bradbury . . . and we had placed a splinter of plastic and aluminum around it as a tiny man-made moon—a moon carrying a camera—which would soon (although we didn't know it yet) breathe new life into the "Martian Legend."

And in the midst of this rejoicing, for a spacecraft which had finally become a permanent captive of another world, *that* world, there was one man who wasn't rejoicing—which intrigued me.

"What's wrong," I asked, as we stood at the rear of the Auditorium and watched the grinning faces, the successful climax of almost ten years of effort come to fruition on a sunny November afternoon.

"Well," he said quietly, almost dejectedly, "I have a spacecraft going to Jupiter, called Pioneer, and nobody cares . . . "

His name was Pete Waller, and he worked for another NASA center, NASA-Ames—up the Coast near San Francisco, at a place called Mountain View. For years there had been a friendly rivalry between these West Coast facilities of NASA—Ames and JPL—for attention as *the* unmanned exploration center of the solar system. Waller worked for Ames as a Public Affairs officer, and he informed me he was arranging a little "party" for members of the press the following Tuesday—a Special Briefing on Pioneer F by the principal scientists who had experiments aboard the mission.

Now picture this: I had flown three thousand miles to be here, three thousand miles to see live images returned from *Mars* by the first artifact we as a species had ever placed in orbit around it, and here was this guy trying to interest me in attending a briefing with a bunch of scientists on a mission which wasn't leaving Earth for months, which wouldn't arrive at Jupiter for over two *years*. Yet . . .

The following Tuesday, as a raging dust storm prevented Mariner's *two* cameras from returning images filled with more than varying shades of grey ("bedsheets" followed by "enhanced bedsheets"—in the words of one disgusted reporter), I found myself driving south on the San Diego Freeway, heading toward a small community on the coast with the very California name "Redondo Beach," where the aerospace concern which put Pioneer together—TRW—was hosting the morning's briefing.

After the three-hour ritual (which was as long and as tedious as I had feared) Fate stepped in . . . well, actually, Pete Waller—

"This morning," he announced cheerfully (at the close of the scientists' seemingly interminable explanations of "magnetospheres"—Jupiter has one), "we've a little surprise. We've arranged for you to see the actual Pioneer spacecraft, before it's shipped to the Cape (Canaveral) in a few days. So, if you'll get on the busses outside . . . ''

Well, things were looking up!

Getting a close-up look at an actual flight spacecraft, even for members of the press (who can usually get in *anywhere*) was about as rare as seeing the swallows *leave* Capistrano. Suddenly, a peculiar thrill went through me as the busses caravanned their way toward the rear of the sprawling TRW complex, towards one dome-shaped vacuum chamber in particular which stuck above the landscape like a gigantic white salad bowl some alien had dropped upsidedown, and I began to understand where *this* spacecraft was going . . .

Unlike any previous artifact we had tossed into the bottomless "ocean" of deep space, this object was going to be different. For, no matter what happened to its instruments during its encounter with Jupiter's vast radiation belts (the "magnetopshere" to end all magnetospheres!), Pioneer itself, in two short years upon encountering Jupiter's gravitational field, would be literally tossed out of the solar system—forever.

By a quirk of celestial mechanics, Jupiter was going to add enough velocity to this tiny ship to eject it *permanently* from the gravitational confines of the sun . . . to wander for eternity (or the next best thing to it) through the infinite reaches of the stars . . .

Pioneer was going to become our first—if inadvertent—emissary to the Galaxy.

Musing on this transcendent, yet apparently unnoticed, aspect of the Pioneer mission, I entered the towering "thermal vacuum tank" in the milling crowd of newsmen . . . to see what an "emissary" looked like.

The chamber was actually two chambers—an inner, real vacuum tank, and an outer shell, painted white (for cooling purposes in the hot California sun?). Inside the shell, the organizers of this impromptu tour gathered us together at the foot of an ascending set of open metal stairs— the kind like they have in high school gyms. In groups of twos and threes, they let us ascend to a work platform high above the concrete floor, to sets of thick quartz windows that allowed us to gaze inside the real vacuum tank . . . at Pioneer.

Through the frost rimming the edges of the windows (testimony to the liquid-nitrogen level temperatures being maintained inside the chamber, to test various subsystems in simulation of the frigid depths of space at Jupiter's distance from the sun) an object glistened in the bright white work lights—Pioneer! The various layers of gold-plated mylar and bare aluminum making up its booms and electronics compartments reflected back the incandescent lamps like a miniature galaxy trapped inside the chamber, like the real stars that it would one day join . . . forever.

Its magnetometer boom and nuclear generators were tucked up against it in the confining tank, unlike the true way they would be extended once the satellite was lofted into space. Pioneer was some kind of electronic preying mantis, stuffed inside a cosmic bottle, waiting to be freed . . .

Then, it was someone else's turn.

At the bottom of the stairway on the opposite side of the inner vacuum chamber, those who had already "had their look" were gathered, waiting for the rest "to get it over with," so they could get back on the busses and get some lunch—at TRW's expense. Among these was an old friend, Eric Burgess.* Unlike many others milling around at the foot of the stairs, however, Eric didn't seem to be impatient to get to the TRW cafeteria. Rather, he seemed lost in thought . . .

Catching my eye as I reached the foot of the stairs, he gestured up toward the towering chamber behind me, "Do you realize that . . . thing . . . in there is going to escape the solar system?"

I nodded.

"It ought to carry a message—"

*That* was the unvoiced thought, which had been nagging at me ever since I'd gazed in through those icy, thick quartz windows.

But Eric (cofounder, along with Arthur C. Clarke, of the British Interplanetary Society) was missing one thing; in terms of whomever might find it in the long dark light-years ahead, *Pioneer itself was the message.*

Compared to the whispered dots and dashes of radio energy which had long been envisioned as *the* form of communication across the interstellar deeps, discovery of Pioneer—as a derelict somewhere in space—would be a veritable symphony of information. From painstaking analysis of the spacecraft—with its electronics, its metals, even its microbes—hypothetical trans-galactic investigators could learn in days items of information that *years* of sending messages across the night by radio could not supply.

Both Eric and I knew that, once beyond the familiar planets of our solar system, the chances of *anyone* ever seeing Pioneer again were . . . nil. A line from *Star Trek* crossed my mind, spoken by Uhura, "But, it's a big Galaxy, Mr. Scott!"

Yet into that vast Galaxy we as a species were about to catapult one tiny sliver of ourselves. Of that we could be certain. On how many worlds across the interstellar night had like events also come to pass? And what would those who stayed at home think—while their "emissary" cruised

---

*See Chapter III, ref. 7.

*The Monuments of Mars*

unseen and forgotten in between the stars . . . ?

If we meant any of the fine words which had been uttered in recent years, about "life being plentiful . . . " and the Galaxy being home to "myriad habitable planetary systems . . . " what was about to happen here must have been repeated . . . would be repeated again and again . . . millions of times . . .

But for us it would happen only once . . . and that was now.

I looked at the impatient crowd of newsmen waiting for their colleagues to finish giving a cursory once-over to the imprisoned Pioneer. How many of them realized—or could communicate to their readers—the significance of this . . . evolutionary . . . moment?

"Eric," I said, "You're absolutely right; it's got to carry a message. But—

"*The message will be to ourselves!*"

And we both began to plan to make it happen.

*      *      *

Looking at the small image of 35A72, at the Face staring back from the tiny photograph, I remembered every detail of that afternoon—vivid and sharp as if it had happened only moments before. What if the ultimate explanation for the Face on Mars was that simple:

A Message to Mankind?

Our tentative thoughts about Pioneer's message—Eric's and mine—evolved swiftly that distant afternoon, five years before Viking would take a photograph of a curious face-shaped mesa on Mars, twelve years before my own haunting contemplations of its meaning . . . Within hours we had identified the only person on the planet—Carl Sagan—with even a ghost of a chance of getting NASA in the allotted time, a mere couple of months, to include on Pioneer an object with no immediate scientific value—a "Message."

We had then raced back up the San Diego Freeway, to JPL, where by a fortunate coincidence Carl was scheduled to deliver to a group of newsmen a discourse on "the scientific benefits of the Martian dust-storm . . . " Cornered behind two steaming cups of coffee in JPL's tiny "spacecraft museum" of past relics of our explorations of the solar system, Carl had listened politely to our vision of Pioneer's impending journey. Then, a familiar grin had spread across his face and he said simply,

. "What a nice idea."

The rest, one might say, is Galactic History.[5]

None of us associated with this project "with a timescale somewhat longer than usually contemplated in NASA's budget cycle," (so Carl once phrased it) had any illusions regarding the chances that "someone out there" would one day find "the Plaque." Space was too vast and Pioneer's trajectory not likely to take it within light years of any star—let alone a potentially inhabited planet in any star system (and it would certainly not survive entry into an atmosphere and a fall to the surface of any such planet—and the chances of this happening anyway were literally millions to one!). No, Pioneer was carrying its message for one main reason: to inform the culture which had launched it here on Earth of this moment's unique meaning.

But what if this ambitious project—backed by miniscule resources— had been "the real thing"?

Instead of a drifting speck lost in an immensity of time and space as a "platform" for such a message, what if we had been able to target a specific audience—say, a known intelligent species developing on a specific planet in a charted solar system? Would we not—given the "appropriate" resources—design a message both unmistakable in content—

And in a place *impossible* not to find?

On a planet in their very solar system?!

Planets are inherently difficult to overlook. They are also logically the first targets for reconnaisance by any developing species . . . as they reach out to explore their neighboring planetary system. If one really wanted to communicate with such an evolving intelligence, what better way than a *physical storehouse* of information (which could contain, for an indefinite time, literally billions more "bits" of information than a constantly expanding—and weakening—wavetrain of radio emission)?

And, to call attention to its presence, why not place an unmistakable "marker" as a Guardian . . . or a Sentinel . . . in *their* image, in the hope that someday they would come . . . ?

Of the three explanations for the Face I could most readily imagine, this one—the Cosmic Calling Card—seemed best to fit the data. It "explained" the presence of a "human" face on a planet where there shouldn't be one, and in a way that violated neither our imagined understandings of our own origins and evolutions nor our limited knowledge of our neighbor world. In short, it explained everything . . .

And nothing.

Left unanswered by this "easy" explanation were a host of related questions: ranging from "when" to "why" and, finally, "who" might have left such a message on the Martian landscape. But most troubling, to me, was the absence in this hypothesis—that the Face was a message of some kind—of a ready explanation for the objects at Cydonia which had focused my serious attention on this problem in the first place.

The pyramids.

Why—if the intent was to simply send a message—construct a vast and megalithic *complex* on a "hostile" planet? Then, why orient it in a very specific manner towards the Face? Why carefully position a cliff 14 miles behind that Face—as viewed from the center of the pyramids—which spanned the precise optical angle as the Face itself . . .

Unless the entire pattern was part of a complex specifically designed for an *intrinsic* use?

Perhaps some kind of message to potential future visitors had been intended by this complex—the same way "the Plaque" now exiting the solar system on Pioneer had been, at best, an afterthought, a trivial aspect of its essential purpose:

Exploring Jupiter and its environs.

What, then, could possibly have been the *prime purpose* for constructing such a complex of massive, interrelated objects . . .

Habitation?

But why on Mars—a desolate, almost airless, frigid, barren world—when the blue-green "jewel" of the entire solar system—Earth—was so invitingly at hand?

The questions, the images, the paradoxes tumbled across my mind. The pieces all were apparently there, only they wouldn't fit into any ready explanation of the Mars I knew . . . which meant one of two things:

Either the Intelligence Hypothesis was wrong, or something that we "knew" about the planet was.

Which meant it was time to review all the original data—*all* the data we've acquired on this planet, from Mariner 4 to Viking.

My mind returned to another image in my childhood . . . most astonishingly prophetic of the Face . . .

The Sentinel.

Arthur Clarke's famed short story (which had later been transmogrified into the classic, *2001*) was the prototypical "encounter" paradigm: Mankind, reaching upward like a great vine into space, would one day find "an object." In Clarke's original story, it had been placed atop a sunlit pinnacle on our own Moon—to be encountered when Man developed spaceflight and reached the surface of his nearest steppingstone. (In *2001* it had been an enigmatic "monolith"—originally buried on the Moon, only to be uncovered in the crater Tycho by a human expedition pursuing a peculiar "magnetic anomaly" . . . When discovered, the monolith emits a piercing signal toward the outer solar system, where it is relayed by a "big brother" version orbiting Jupiter . . . to the unknown builders far beyond . . .)

This "cosmic trip-wire" scenario was adopted in the years following Arthur's original short story (1953) by some members of the astronomical community interested in the subject of contacting extraterrestrial intelligence, long before the appearance of the movie. Dr. Ronald Bracewell, of Stanford University, a well-known radio astronomer (whose 1950s technique for making radio "images" of the sun has now evolved into the highly complex machines in every major hospital for making x-ray "cat scan images" of patients) was the first seriously to propose a somewhat complementary scenario:

In Bracewell's concept,[6] one day we might discover a "robot probe" orbiting the sun—similar in kind to, but obviously much more complex than, our own "Mariners" or "Pioneers." The probe's purpose, Bracewell argued, would be to "lie in wait" in a target solar system until it heard radio signals, presumably "leakage" from fledgling radio and television transmissions developed for internal purposes on Earth (if our solar system was one of the choices for inspection). When it received such signals (for instance, Marconi's initial dots and dashes) it would answer—echo—those that it received. Presumably, according to Bracewell's model, we'd get curious—and eventually send a *deliberate* message at the probe . . . at which point it would begin a complex question and answer program . . . and relay "success" to its creators far beyond the solar system at "communications central," where (presumably) a vast fleet of probes "seeded" into target solar systems with likely planets were being monitored.

The obvious advantages of Bracewell's scheme were that such a probe could contain within its own internal memory a storehouse of information on both its home planetary system, and of its creators—which (again, presumably) it would impart to us if we discovered it and began communicating. The obvious disadvantage of the concept was the fact that any complex, almost "intelligent" machine would probably have a finite life; it could wait in orbit of the sun only for so long, before a meteor or even an energetic cosmic ray wiped out some vital part or memory—leaving but a silent derelict.

There was a tendency, in this Electronic Age, to imprint our "high tech" approach to everything we saw. But perhaps the older cultures—for instance, the Mayans or Egyptians—had been more in tune with the pace of "cosmic thinking" when they constructed their "messages to eternity" in stone. Electronic systems will inevitably fail, but "solid state messages" in granite—even in the corrosive environment of Earth—will last across the ages . . .

Was the Face an interstellar visitor's attempted combination of these

*The Monuments of Mars*

strategies—a treasure-trove of information, perhaps stored on literal "stone tablets" so as to outlast the weathering of ages, even on Mars—with an attention-getting "sentinel" waiting patiently, to relay to its creators somewhere in the interstellar night that someone had finally reached it?

But, again, what about those massive, latticed pyramids?

[All at once, I flashed on a detail that the original Sentinel in Arthur's story had been a . . . *miniature pyramid*. It wasn't an answer, but . . . ]

Bracewell's ideas, on the advantages of physical emissaries with interstellar messages that lay in waiting, had been published in 1960 . . . and in 1968 had come *2001*, to be followed, in 1976, by a photograph of the most unlikely "sentinel" of all . . .

A line from Clarke's foreword to the novel came unbidden:

"The truth, as always, will be far stranger . . . "

Not only had Arthur correctly envisioned what was to come, he had once again strategically arranged to have the "final" say; but even he had not been bold enough to propose that the mile-high "monolith" that Bowman landed on—

Would look like Bowman.

## Notes

1. Sagan, C., ed., *Communication with Extraterrestrial Intelligence (CETI)*, Cambridge: MIT Press, 1973.

2. *ibid.*

3. *op. cit.*

4. Le Gros Clark, W., *The Fossil Evidence for Human Evolution*, Chicago, 1964, *op. cit.*

5. Sagan, C., Sagan, L.S., and Drake, F., "A Message From Earth," *Science* 172:881, 1972.

6. Bracewell, R.N., "Communications from Superior Galactic Communities," *Nature*, 186:670, 1960.

# VII

# CHALLENGING ASSUMPTIONS

"Not only is the Universe queerer than we imagine
"It is queerer than we *can* Imagine."

J. B. S. Haldane

If the Intelligence Hypothesis was true, then obviously something concerning our current understanding of the Martian planetary history—and perhaps our own—was not.

That seemed the place to start.

Determining a planet's history begins with gauging a planet's age—and then extrapolating the age of various events throughout that history. For terrestrial scientific unravelling of past events, several methods have developed—beginning with relative dating via layered strata of sedimentary rocks and their imprisoned fossils, and culminating with absolute radiometric isotope dating applied to volcanic rocks. The simple principle behind these methods can be stated thus:

The youngest strata or rocks usually lie on top.

The preceding statement is called "the Stratigraphic Principle," and is fairly easy to understand; as mud is deposited in a stream, or shallows at the edge of an ocean by erosion off the land (during a rainstorm, for instance), the build-up of mud will form horizontal layers. This layering reflects, of course, the relative ages of each storm which caused the mud to be washed to the stream or off the continent into the ocean. Thus, the layers sequentially "date" the sequence of erosion; the oldest events are represented by the strata of mud at the bottom, the youngest by those at the top. Compressed and dried, these layers become rock.

If it weren't for other disruptive geologic processes—such as earthquakes, or the slow tilting of some parts of Earth's landscape—this simple rule "the youngest stuff lies on top" would allow easy and quick *relative* dating of much of the Earth's surface. But there are other forces at work, besides erosion, which complicate this simple picture.

Volcanos, for instance, can erupt *through* a previously layered strata of sedimentary rocks—through a crack upward toward the surface (a crack created, for example, by an earthquake). "Recent" lava becomes interjected between layers of older sediments—messing up the simple relative dating method. Or, after a period of slow deposition of mud (which will eventually turn into rock over millions of years), climate can change in the region of a one-time stream; the stream will dry up—leaving the layers of now sedimentary rocks exposed. There will be new rainstorms, and these recently deposited layers are *eroded*, to mingle with uncounted debris washed down from higher ground on its way to a new place of deposition. Thus, future geologists seeking to date the sequence of strata at the old stream site, if they don't realize the period of layering was followed by a period of removal of some of the layers, will arrive at an erroneously *younger* relative age for the sequence of rocks. To complicate matters, the ancient climate could have changed back again after the "erosive" period, and new layers of mud could have been deposited at the site of the old stream once again—eventually to turn into new layers of sedimentary rocks on top of the old layers that had been partially eroded.

In fact, such periods of deposition and erosion are tediously familiar to experienced geologists. They are called "unconformities"—meaning, the layers don't conform to one another across large areas. These unconformities are a key means of telling when the ancient climate in a particular region changed. Other geological events—such as volcanic eruptions and impact cratering—can also produce unconformities. But these are usually local anomalies.

On Earth, the presence of life has created another dating technique (as well as, of course, the "daters"). Fossils of dead organisms, trapped within deposits of mud or lava, were discovered to have global correlations; discrete biota seemed to occur in layers of rock laid down atop one another in *particular sequences.*

These sequences became the basis for the "epochs," "periods" and "eras" of the geological timescale in use today. The successive changes in terrestrial organisms—from the first recognizable shellfish, called "trilobites," to *Homo sapiens*—are thus faithfully reflected in their fossils, lodged in ascending layers (when all other geologic processes were accounted for). A particular species (represented by its fossils) appearing in a rock formation could be used as a tool for estimating the relative age of that strata—anywhere in the world; you would *never* find muds containing trilobites, for instance, deposited above rocks containing fossilized bones of much younger dinosaurs . . .

Thus was born the science of palaeontology, and its geological application—palaeogeology.[1]

To get a feel for the absolute time progression required for depositing these sedimentary layers, turning them into rock, and interleaving them with successive lava flows (to say nothing of uplifting them, tilting them, cracking them, and eroding portions of this "record") is another matter.

So far, as I have explained, the dating mechanisms are all relative. There is no way to determine if the activity seen in the "geological record" (as the rock formations are generally referred to) or its associated sequence of palaeontology all occurred in the Biblical "6000 years . . . ", a statement made infamous by Ireland's Archbishop Ussher in 1664 (i.e., that the Earth was created at 9:00 A.M., October 26, 4004 B.C.!), or if they had taken longer—even a *lot* longer. There was, however, a suspicion among geologists that the time-frame for deposition of the observed *miles-deep* rock formations had to be at least a bit longer than the Archbishop's emphatic assessment. The problem was proving it scientifically—"by the numbers."

For instance, one could estimate the amount of mud washed off the land in any particular storm, measure the frequency of these events currently, then use that *rate* to estimate the amount of time required to have left deposits of the measured thickness—with due allowance for gross errors caused by the erosive unconformities. Such a technique, first developed by a Scottish gentleman farmer named James Hutton and published in 1785, has been called "the Principle of Uniformitarianism." It basically says that "nature behaves in a uniform fashion throughout time, so that by studying the present one can infer the behavior of past processes."

Hutton was the first to understand the cyclic nature of this entire process—of upheaval, erosion, subsidence, and sedimentation—and the role of these in creating the unconformities which "messed up" an otherwise logical tool for estimating rough ages of the observed rock formations making up the surface of the earth. It is not without good reason that Hutton is "the father of modern geology."[2]

Age estimates based on Hutton's approach made it abundantly evident that the rocks of the Earth—and thus the Earth itself—had to be a *lot* older than the good Archbishop's estimate. By 1830, using only the technique described above, the general scientific "age of the Earth" was approaching 100 *million* years.

If the strata laid down as sediments were this old, then the fossils trapped within them must be at least as old—which made life on Earth suddenly a much vaster and grander drama than anyone had dreamed before.

So the stage was set for the 1859 publication of Charles Darwin's *On the Origin of Species by Means of Natural Selection*, and the beginning of the revolution in our own understanding of the truly ancient history of both the earth . . . and the life which lived on it.

While the geologists and biologists were laying the foundations for this revolution, the physicists were approaching the problem of the age of the earth from a completely different angle.

Hermann von Helmholtz, a nineteenth-century leader in the field of thermodynamics (the science of heat flow) became fascinated by an apparent paradox: the Earth was older than the sun!

From the accumulating geological evidence, staggering layerings of rock—such as the exposed levels of the American Grand Canyon—were estimated to be at least 75 million years old. Yet, according to calculations made by one Immanuel Kant, a leading nineteenth-century philosopher of science, the sun could be no more than 1000 years old—*if* its light was generated by processes familiar to the nineteenth century—Combustion!

Von Helmholtz came up with an ingenius alternative (since it was obvious that the apparent contradiction between the age of the sun and the age of the Earth—as derived from well-founded geological processes—had to be merely that—apparent). He proposed that another well-known physical principle—gravitational potential energy—was what powered the sun. A slight shrinkage of the sun's outer layers each year, something like a couple of feet (in a diameter of almost a million miles for the entire sun!) would be enough, he calculated.

The geologists heaved a sigh of relief . . . until the actual numbers of the calculation were revealed; for, according to von Helmholtz, even a generous estimate of the total potential energy available to power the sun's heat and light would give it "only" about 40 million years—shy from the geologically accepted age of the Canyon by almost a factor of 2.

The Earth was *still* older than the sun!

(In fact, the situation was much worse, for the famous physicist actually felt that 20 million years was a much better figure for the sun.)

Now you might ask precisely why this seemed to be such an embarrassment. After all, if the Earth was older than the sun, or even the same age, why did it matter? The answer was as simple as it was puzzling: if the Earth was older than the sun, where had the energy come from to maintain *liquid water* on the Earth during all those dark millions of years . . . ?

For, liquid water—as rain—was essential to drive the entire cyclic geologic process; without evaporation of ocean water, without condensation of that water into storms, there would be no erosion of the continents, no sediments flowing to the sea and the continental shallows— and thus no layered deposits to turn into strata of sedimentary rocks. In short, without sunlight, the entire terrestrial geological process (and biology!) unravelled by collaboration of these two sciences simply wouldnt't work—yet it had!

So, what was the answer to this paradox?

Von Helmholtz had published his calculations in 1854. In 1899 the entire problem came to a head: a true giant in nineteenth-century physics— Britain's Lord Kelvin—gave an address to the prestigious American Association for the Advancement of Science. Kelvin had taken up the reins from von Helmholtz, and confidently announced that the Earth could be no older than the results of *his* extensive expansion on von Helmholtz's original calculations:

20 million years.[3]

Then, in a rejoinder that was almost prescient, one of the audience— T. C. Chamberlin, head of the Department of Geology of the, then, brand-new University of Chicago—speculated that perhaps everything wasn't known to physicists regarding matter . . . including the possibility that there might be undiscovered sources of energy (unknown to them then) within particles of matter which would eliminate the need for either "burning" or "contraction" as the source of the sun's energy . . .[4]

This debate occurred four years *after* the discovery of radioactivity, in 1895 by the French physicist, Henri Becquerel. The fascinating thing is that, none of the participants in this "age of the sun versus the Earth" debate seemed to be aware of this discovery—nor the immense implications of a source of energy contained in decaying atoms to the problem of the source of energy of the sun, despite the fact that both the scientific journals *and* the popular press were veritably "popping" (to quote Dr. Frank Press, now head of the American Academy of Science) with stories of the news.

In 1905 the critical conceptual breakthrough occurred, which was to bring all these separate threads together—and resolve the paradox.

Ernest Rutherford, the physicist, was to change our modern world forever—by discovering the now-familiar building block, the proton, which determines the separate nature of each element. Rutherford proposed that radioactive minerals could be used to date rocks—

By acting like microscopic atomic "clocks" trapped within solidifying lava!

Only eighteen years elapsed between Becquerel's original discovery and the publication of the classic work, *The Age of the Earth*, by Arthur Holmes, a young geologist who had yet to receive his doctoral degree. Holmes plotted radioactive dates opposite the stratigraphic time scale by determining the age relations of the sediments (which were dated by fossils) and interpolating these with the radioactive ages of the volcanic intrusions (dated from the cooling and solidification of the lava "dikes"). The result was little short of remarkable: in a field which had literally just been invented, Holmes's radioactive calibration of the stratigraphic time scale has remained, with minor variations, the same since his first measurements—over a half century ago.

And the sweep of geologic and biological time he revealed was little short of awesome.

The earliest sedimentary rocks which contained fossils, including those "trilobites" mentioned before, were dated at about 600 million years before the present (BP). Yet, before the dramatic appearance of such complicated life forms, Earth's geological history appeared to stretch backward *billions* of years—as revealed by the tell-tale decay of imprisoned radioactive elements in lavas that had crystallized eons ago.

For technical reasons, radioactive age determinations get worse (in terms of precision) the farther back in time one tries to push them (which only stands to reason: the dating relies on a *decreasing* amount of an already very scarce radioactive element, compared to normal minerals making up the rocks; the older the rocks, the more the radioactive material will have disappeared). Thus, age estimates near the beginning of the Earth's history—some 4.5 billion years ago—are typically "off" by as much as 100 million years either way. This is equivalent to almost half the "recent" span of life on Earth—and on a planet where getting datable samples is no problem (a foreshadowing of how this applies to Mars—which we haven't forgotten!).

Finally, with the discovery of radioactivity, the entire cosmic sequence of nuclear reactions was soon unravelled—from the decay of trace elements within the Earth (which provided the heat that "fueled" volcanos and the creation of the very lava flows that the same radioactivity allowed geologists to date), to another kind of nuclear reaction—

Fusion.

In the 1930s astrophysicists, such as Hans Bethe, would finally solve the source of the sun's energy: the fusion of light elements into heavier ones—with the subsequent release of a minor fraction of the mass difference as raw energy.[5] Calculations revealed that with this source of

110

energy—which was fully *one million* times as energetic, gram for gram or pound for pound as any chemical reaction—the sun would have roughly a million times the expected lifetime, compared to a sun actually burning something. Thus, the original estimate of a thousand years, for a sun which "burned," was replaced by an age a million times as great—a billion years. With allowances for the actual numbers, the age of the sun was eventually calculated at about 10 billion years—of which half had already elapsed, since the formation of the Earth, the planets and their satellites [For the latter two classes of objects—planets and satellites—meteorites were used for dating; it was assumed from the radioactivity of these fallen shards of primordial stone that the approximate age of the entire collection of objects orbiting the sun was some 4.65 billion years.]

Thus the "embarrassment" of an Earth older than its sun disappeared immediately; replaced by a consistent picture which envisions it—and all the planets orbiting our star—as having originated in the same "primordial nebula" eons ago.

Thus, Mars is as old as Earth. Of this we can be very sure . . . which is the last declaration of certainty we can utter with regard to the age of geological events on the Red Planet.

In terms of precision, our best estimates for Mars are probably no better than our worst estimates for the stratigraphic time scale on Earth—prior to the application of radioactive dating. Yet, the "reasonableness" or "unreasonableness" of the Intelligence Hypothesis depends on our current understanding of the dating of geological events on the Martian landscape.

For, we have no direct samples of the Martian surface actually to date—despite the Viking landings at two points on that surface (the equipment required for such a test is too heavy).

Instead, we must rely on an indirect method of dating of the surface—the counting of the relative numbers of impact craters on different geological "units" (regions)—and attempt to correlate this "cratering curve" with a similar curve derived for our own Moon: the only extraterrestrial body for which we do have direct samples (thanks to the Apollo astronauts)[6].

Estimates of the relative ages of various parts of Mars, made by different investigators using the same Mariner or Viking images (to count the craters) differ by as much as *half a billion* years—the amount of time roughly which encompasses the entire history of multi-celled life on Earth![7] Further, these estimates depend critically on assumptions for the impact craters on Mars compared to those produced on our own Moon—the "standard." Mars' proximity to the asteroid belt has raised the possibility in the minds of some investigators that the impact flux would therefore be higher at Mars—creating the appearance of a falsely "older" surface for the same area measured for cratering.[8] Finally, there is a major problem in calibrating such a cratering curve, which is completely absent on the Moon:

Erosion.

Evident in even the first crude Mariner 4 images of Mars taken that week in 1965, the shapes of Martian craters above a certain size differ dramatically from those photographed on any other body in the solar system. They are relatively flat—even for depressions 20, 40, 100 miles across; as if something had "filled them in" and rounded off their rims, completely different from similar-sized "craggy" spectacular craters (such as Copernicus or Tycho) on the Moon.

The apparent cause of this pervasive difference is erosion—caused by a denser Martian atmosphere sometime in the past. *When* such an atmosphere existed, and how much erosion it has caused (and how many ancient craters it obliterated) is *the* geologic question which now confounds all attempts to date the Martian surface. A correlative question is how much erosion the *current* atmosphere of Mars exacts—for that leads to the gradual disappearance of small craters, those below a mile or so across.

Thus, in place of the methodical progression of terrestrial sedimentary layers and the evolutionary sequence of fossils, which allow independent correlations of rock formations—and thus dating—across an entire planet (corroborated by the absolute radiometric dating of volcanic rocks back almost to the beginning), the situation on Mars is radically different—

A current dating mechanism totally dependent on one feature—craters—and the relative distributions of these landforms across the surface of that planet. Critically lacking for Mars then, is the certainty of any *absolute* calibration of this "cratering curve," from Martian samples returned to terrestrial laboratories—for the simple reason that there haven't been any.

Despite these vast uncertainties, several inferences can be made about the time-scale of major geologic events on Mars:

1) Given that every other solid planetary body in the solar system has major portions of its surface "saturated" with large (above 20 km) craters—Mars has roughly one entire hemisphere (the southern) covered with such craters.

2) Given that the central assumption for the creation of these "shoulder-to-shoulder" craters is that there was a period of intense bom-

bardment very early in the history of the entire solar system—Mars' heavily cratered terrain most probably dates from this "heavy bombardment period" as well.

3) Given that the presence of these ancient cratered terrains on the Moon, Mercury, and the satellites of the outer planets (Jupiter, Saturn and Uranus) bespeaks preservation in the airless vacuum of these objects' surfaces—the preservation of similar terrain on Mars argues forcefully for negligible erosion, most likely from the lack of dense atmosphere for most of the lifetime of the planet.

4) Given that the *other* hemisphere of Mars (the northern) matches similar terrains on the Moon and Mercury, in terms of far less cratering—this "two-faced" distribution of craters argues (as it does for the Moon and Mercury) that there was a dramatic drop-off in the rate of cratering for Mars, consistent with a solar system-wide termination of "heavy bombardment" 4 billion years ago.

The physical explanation for this solar system-wide "heavy bombardment" is simple: as the planets accreted out of the spinning, disc-shaped nebula of dust and gas swirling around the sun, the last stages of this process—specifically, the formation of Uranus and Neptune far in the outer solar system—caused millions of icy chunks of "planetismals" to be diverted into orbits which took them into the inner solar system . . .

Where they proceeded to collide with the solid surfaces of Mercury, the Moon, Earth and Mars . . . until the supply was almost totally depleted—at which point the bombardment "suddenly" ceased ("suddenly" to an astrophysicist means anything under a hundred million years!).

The "two-faced" situations of many of the inner planets (Mercury, the Moon, Mars)—one hemisphere intensely cratered and the other relatively unscarred—are assumed by astrogeologists to underscore the fact that during this *external* modification of their surfaces by cratering, *internal* geological processes (mainly the eruption and surface flows of lava) were also occurring. After the heavy bombardment ceased, this internal activity continued—on at least one half of the surface of these planets, thus wiping out the population of large craters created during the heavy bombardment.[9]

Subsequently, across the ensuing billions of years of the solar system, the only activity on these worlds has been an occasional collision. With no further internal processes (except on Earth) to wipe out craters—old or new—the surfaces of the airless bodies of the system essentially preserve a record of the violent events "in the Beginning."

With the complicated exception of Mars . . .

For, in addition to these two major solar system-wide "terrain types"
—intensely cratered and relatively uncratered—Mars possesses a host of
other features which defy easy explanation—at least in terms of the "nothing much has happened since" scenario. Further, attempting to *date* these
features is very difficult, as the only mechanism available at present is
the "cratering curve" provided by "recent" craters on the relatively un-
saturated portions of the Moon . . .

Where varying estimates as to the current rate of actual new impact
craters on Mars, versus the Moon, become critical—making the difference
between a feature being "several hundred million years old" or "several
billion . . . "

Which is why NASA would practically sell its soul for a returned,
*datable* sample from the surface of Mars! Actually, a whole range of
samples, gathered at specifically selected sites designed to back-up the
cratering studies, is what would be most helpful—for even a few samples,
chosen wisely, could provide essentially planet-wide calibration of the
cratering on Mars, and thus ages for the myriad perplexing features *not*
present on the other inner planets of the solar system.[10]

So what has all this got to do with faces, cities, pyramids . . . and
life on Mars?

Some of the features peculiar to the Martian surface are precisely
those which would be required of any "scenario" for the origin and evolution of indigenous life—such as liquid water. There are two sets of geologic
features—the "outflow" channels and the "network" channels—which
strongly indicate to some investigators that liquid water . . . at one time
in the past history of Mars . . . flowed across its surface (see Chapter III).[11]

The problem is precisely when . . . and for how long.

The "network" channels are pretty much what the name brings to
mind: the dendritic pattern of eroded gullies, valleys and arroyos familiar
to anyone who has gazed out of the windows of a plane flying above
the American southwest. They are found throughout the equatorial regions
of the planet, but seem essentially concentrated in the "ancient cratered
highlands"—leaving some investigators with the strong impression that
such a relationship is more than strictly accidental.[12]

The problem of the origin of these channels is highly controversial.
The "rainfall" enthusiasts believe they represent a familiar pattern of
erosion caused by running water, in the wake of Martian rainstorms . . .
which implies a Martian atmosphere—and temperatures—thick enough
and warm enough for it to rain![13]

And that, of course, implies a planet much more like the
Earth—sometime.

As one might expect (if there is a controversy) the other side has a different set of ideas regarding the origin of this peculiar network. They believe that underground melting (perhaps of permafrost), a process called "sapping," causes an erosion of the landscape at the ends of stubby valleys. The network channels in this model have no need of denser atmosphere or warmer temperatures in order to form; all they need is a ground layer saturated with water (mainly as ice) which melts during the current Martian summers, eroding the surrounding "soil" and causing the regression of the landscape—forming stubby canyons.[14]

The fact that these channels have been seen overwhelmingly in the ancient cratered terrain and rarely in any other, supports a very simple assumption:

That they formed at the same time as the earliest craters!

If the network channels (which some investigators call by the much more provocative term "runoff channels") indeed formed during an earlier "warm, wet epoch," then their connection with the oldest terrains on Mars seems to define clearly the time and duration of that epoch: at best, during the first half billion years or so of Mars—the same time as the estimated duration of the heavy bombardment period. That the span of conditions didn't last much longer can be inferred by the essential *absence* of these types of tributary channels from any younger areas of Mars—again, as dated by the (relatively) imprecise post-bombardment cratering statistics.

(One might at this point ask the simple question: why is it necessary to attempt to date the channels by dating the *terrain* beneath them? Cannot one date the channels directly, by counting craters on them? And the answer is that such a technique, sound in principle, fails by simple virtue of the lack of sufficient *area* of the channels on which to do the counting! The statistics are too small.)

Now, the reason I have taken you, dear reader, through this detail is to demonstrate how easily *assumptions* can shade conclusions regarding supposed "objective" data.

For, a completely viable alternative to the theory that the run-off channels, found almost exclusively in the ancient cratered highlands, must therefore be as old could be as follows:

The channels appear primarily on the cratered terrain because that is *the most easily eroded* landscape on the planet—having been bombarded and shattered by the effects of the heavy asteroid impacts time and again in those first half billion years of Martian history!

In this scenario, the less channelled terrain could be simply more resistant to erosion (being formed, for instance, of "new" lava flows).

Thus, rainfall and erosion on Mars could preferentially appear in selected areas, not because they formed concurrently, but because the ground was simply softer!

If the second scenario is correct, it extends the time significantly during which supposed rainfall might have fallen from the Martian skies. In other words, it extends by several hundred million years—if not longer—the duration of the earliest "warm, wet epoch" for Mars. And that, in turn, extends the "window" for the kind of event of interest to us here:

The origin of life on Mars.

Then there are the "outflow" channels, which have provoked at least as much controversy (see Chapter III). First photographed by Mariner 9, these are completely different geomorphologically from the fine networks associated with the runoff valleys, and appear to have been formed in *awesome floods*—estimated at "ten thousand times the peak water discharge of the biggest recorded flood of the Mississippi!"[15] Some outflow channels are over 200 kilometers wide, stretch across the Martian landscape for over 2000 kilometers, and appear to have scoured everything in their path to depths of several hundred meters—as a veritable avalanche of water roared in fury across the Martian plains leading northward, down from the edges of the cratered highlands.

It was the original intent of the Viking scientists to land Viking Lander 1 on the (presumed) debris flow at the mouth of one of the biggest of these outflow channels. Only the detailed images from orbit, which revealed boulders littering the projected landing site the size of houses (!), caused a sudden change in plans—and a delay in the landing for over a month, as the Viking Orbiter frantically took pictures in search of a safe landing site. The vivid pictures returned by Viking in that last-minute search, with teardrop-shaped islands and other landforms obviously sculpted by hydraulic forces more savage than any recently on Earth, presented clear evidence that water—and a *lot* of water!—had, at some time, shaped the Martian surface.

Which left hanging in the air the small matter of "when."

Unlike the complex network channels, one thing seemed immediately obvious for the awesome outflow versions: they were almost certainly younger than both the ancient cratered highlands *and* the relatively uncratered "plains;" they cut across both terrain-types with abandon ("dissected" is the appropriate technical description). In some pictures, ancient craters could be seen with their rims breached (broken) by what appeared a turbulent current of raging water. Scour marks and all the signs of massive deep erosion—where the water pressures had lifted bedrock units

*The Monuments of Mars*

thousands of feet in length—were mute testimony to events of unimagined violence . . . which swept northward and literally "down" (as the Martian elevations fall by several kilometers to the north of the ancient cratered terrain), leaving in their wake a tortured landscape.

The nearest thing on Earth to these "outflow channels" were the catastrophic Pleistocene "Missoula floods," created by the breakage of an ice-dam in an ancient lake (Missoula) in the last terrestrial glacial period (some 10,000 years ago). Called now the "Channelled Scablands," these rare geologic features in eastern Washington State are silent evidence of the unchained force of water—when in flood. Much—if not most—of what astrogeologists believe regarding the nature of the Martian outflow channels has come from comparing the extraordinary features left in eastern Washington to the even more extraordinary features left on Mars.[16]

The ages of the channels, then, has been partially derived from the estimated ages of the terrain they have dissected—again, as determined from counting numbers, sizes, and the densities of impact craters on the underlying land. But in addition, provisional direct estimates of their ages seemed attainable—by counting the craters on their floors; unlike the runoff channels, the outflow versions (due to the area covered—which is the nature of catastrophic floods) was sufficient for suitable statistical analysis.

When this was done (almost immediately after the first Viking images became available), a puzzling thing became apparent: the ages of the outflow channels spanned a range—from very old to relatively young, from 3.5 to 0.5 billion years—

In stark contrast to the interpretation of the runoff channels: that— after several hundred million years "in the beginning"—Mars had "died."

Which returns us to the central question: life on Mars?

Disregarding Harvard biologist George Gaylord Simpson's premise, that "any close approximation of *Homo sapiens* elsewhere in the accessible universe is effectively ruled out . . . ," the question regarding the origin and evolution of *any* kind of life on Mars has so far not even been addressed. And it is in that context that the ages of key features on the planet—particularly, when water could have been liquid—becomes of paramount importance.

For water is the key ingredient in current scientific scenarios for the origin—and subsequent evolution—of all terrestrial life; its presence on Mars *as a liquid* can be considered, then, as *critical* to the appearance of indigenous Martian organisms as it was for the appearance of our own one-celled ancestors . . . several billion years ago.

Contemporary scientific thinking regarding the origins of living systems—on Earth and other "terrestrial planets" [17]—involves the collection in one place, usually termed "the primordial ocean," of the kinds of chemicals today found in living cells. These "amino acids," formed from the (presumably) random combination of pre-organic chemical precursors—hydrogen, oxygen, carbon and nitrogen—through the input of energy (lightning, volcanic heat, solar ultraviolet light, etc.), are the building blocks of proteins—the essential foundation of life as we know it here on Earth.

The time required for this synthesis to happen here is currently unknown—but some authorities have estimated that it took place in as little as a hundred million years, once the Earth had cooled from its intense process of accretion. What is known is that the first one-celled organisms to be identified in the geologic record—so-called "blue-green algae" (now termed "cyanobacteria")—are found in rocks dated by radiometric methods at 3.5 *billion* years.

Since the appearance of an organism as complicated as these blue-green algae is something not exactly expected as the *first* "invention" of a random chemical evolutionary process, biologists assume that what we are seeing is a later stage in the development of life; blue-green algae are already pretty sophisticated creatures, capable of using sunlight to combine carbon dioxide and water molecules—liberating oxygen as a by-product of the process . . . which (it is assumed) is how the Earth's current atmosphere came to consist of almost 21% free oxygen. Something living—presumably these photosynthetic cyanobacteria evolving billions of years ago—had been busily at work dumping their "waste by-product" into the primordial atmosphere . . . resulting in the vastly changed and oxidizing atmosphere of the present.

The bottom line of all this reasoning is this: since the oldest organisms currently observed in very ancient rocks were already quite sophisticated, and since they are found in formations already almost as old as Earth itself (there is "only" another 800 million years to go until you come up against the origin of the entire planet!), some *other* combination of chemical/biological evolution must have preceded blue-green algae—

Leading to the not inconceivable conclusion that the fabled "origin of life" occurred almost as soon as the initial oceans formed . . . and the "primordial soup" of pre-organic chemicals rained down—4 billion years ago! [18]

None of this could have taken place, it is assumed, without the vital presence of this mediating agent—

Liquid water.

Now, transpose the preceding scenario to Mars. The first question any exobiologist would ask (and has) concerning the origin of life on Mars, is, "Was there liquid water?" From a variety of evidence presented here we can now answer such a question: "Yes, there was—and *lots* of it!"

The second question would then follow, "Were the chemical ingredients of life, listed in the preceding scenario as it applied to Earth, available?" And, thanks to the Viking Landers—which performed a thorough soil and atmospheric analysis, we can also answer, "Yes, everything—including nitrogen."

We know that Mars orbited the same star as Earth itself, thus (presumably) this source of energy was readily available to drive the ancient Martian counterpart of storms, of evaporation/condensation cycles, and erosion. Thus, from geology to energy source all the initial conditions which apparently produced the origin and evolution of simple one-celled life on Earth were duplicated on the planet Mars . . . except perhaps for one—

Time enough for it to happen.

One reason geologists are comfortable with an "ancient" age for all the runoff channels is that this conforms to a basic preconception regarding all the planets: in their "youth" they were much more active geologically than later on. The reason for this sweeping generalization (which is definitely wrong in some particular examples—Io, Jupiter's inner large moon, for instance) is that its basis depends on an almost exclusively *internal* source of activity in planetary youth—

Radioactive heat.

Remember those traces of radioactive rocks that became so invaluable for dating, mentioned earlier? Imagine the situation "in the beginning," after Earth (and all the other rocky planets and their moons) had initially formed . . .

Radioactive levels which we can barely measure now with sensitive instruments (because of the nature of radioactive stuff—to go away!) were at much higher levels; the energy released from these decaying elements, surrounded as they were by trillions of tons of planetary rocks, was trapped by the confining mass of the planet itself . . . resulting in a tendency for things to melt! It was simply this small admixture of radioactive elements in the otherwise "normal" minerals making up the forming planets and their satellites which gave rise to internal sources of magma, volcanos, and eventually eruptive lava flows to modify the surfaces of all such planets—including Earth and Mars.

And over time, as these radioactive elements originally entrapped decayed, the levels of resulting heat also decayed . . .

Leaving some planetary objects in the solar system "dead"—such as our own Moon.

The Apollo astronauts, in bringing back samples from the lunar surface, brought home the proof that planets die. In the case of the Moon, the radioactive dating of the lunar samples confirmed suspicions raised by crater counts conducted on the "highlands" and the darker "mare": that, after the great lava flows which produced the "seas" that make up the dark patterning of features on the Moon, nothing much—save an occasional collision with a small asteroid fragment or meteoroid—occurred . . .

For several billion years.

A direct by-product of this reasoning led to equally depressing predictions regarding the origin and evolution of a planetary atmosphere. All planetary atmospheres (it is believed) are "outgassed" from the interiors of planets; any measurements conducted in the vicinity of volcanos—such as Mt. St. Helens—give ample proof that, in addition to rock pumice, (and in the somewhat different Hawaiian-type volcano, abundant lava), a profusion of *gases* is belched forth. The cumulative effect of countless such eruptions from the interior is the film of gases comprising a planet's "primitive atmosphere."

If the quantity of internal planetary energy declined with age, and with it the number of eruptions, so too would the quantity of "outgassing" produced in these eruptions . . . resulting in a gradually declining atmosphere . . . Why?

Because an atmosphere not constantly replenished would inevitably be lost—by escape of the lighter elements directly into space, and the combination of many of the heavier ones with the surface planetary rocks. Without a means to recycle those atmospheric constituents combined with the surface crust, or to replace those that simply floated off—a planet with declining volcanic output would be expected to have a constantly declining atmospheric density and pressure . . .

Exactly like the Mars we'd finally come to know.

So extrapolations of the lunar "cratering curve," backed by observation of a distinct similarity of the two "planets"—the Moon and Mars—created a perfect template for interpreting the "ancient" runoff channels as just that: ancient. The interpretation was totally consistent with the growing body of evidence regarding the other geology of Mars—that it lacked the key "recycling" mechanism of terrestrial sediments and volcanic rocks: plate tectonics ("continental drift"). And that it had somehow preserved major portions of its ancient crust *against* the forces of erosion across billions of years . . . indicating that whatever "warm, wet—and dense—atmosphere" it may have originally had—

*The Monuments of Mars*

Had long since disappeared—leaving the thin, cold taunting vestige that currently whipped up global duststorms across a basically senescent planet.

Which neatly wrapped up everything . . . except for those nagging enigmas: the awesome outflow channels. Where—on a dead and glaciated world—had all that *liquid* water come from . . . 3.5 billion years after (according to craters counted on the flooded floors) the other channels—the runoff versions on the ancient cratered highlands—testified Mars died?!

And where was it now?

Theories to account for this discrepancy ranged from "fossil aquifers" to "sudden, catastrophic melting of regions filled with permafrost"—to denial that the outflow channels were, in fact, ever created by raging floods at all (see Chapter III)!

Lurking behind this continuing controversy (at times as eventful as the purported floods themselves!) was an unspoken implication . . .

If Mars' imminent demise had been, in the words of Mark Twain "somewhat exaggerated," then perhaps the chances against life originating . . . and subsequently evolving . . . on Mars had been also . . .

Which brings us back to "faces," "pyramids," and "ruins"—and the probabilities that all of those were created by *indigenous* inhabitants.

For even granting the remote possibility that, against all odds, life had managed to originate in the (relatively) brief time "in the beginning" when Mars possessed a "warm, wet atmosphere," and granting some later renewal and warming from volcanism and periodic climate variations (created, for example, by William Ward's calculated obliquity variations—and their effects on long-term Martian climate), this still left a key factor unanswered in the "indigenous life scenario."

The *rate* of evolution.

Evolution depends, so say the biologists, on many factors—including environmental inputs (such as radiation), "harshness" or "benignness" (which determines species adaptability, competition, etc.), and one more . . .

Temperature.

On Earth, the effects of this single parameter can readily be seen in terms of the diversity of species—and their rates of metabolism—from the equator to the poles.

At the equator, life on Earth explodes in a display of color, diversity, and population. From insects to plantlife, the tropical rainforests of the Amazon and the jungles of New Guinea flaunt the greatest range of species, sizes, ingenuity of competition, and sheer numbers—of anywhere on Earth.

By contrast, the frigid polar regions of the planet harbor only a few (and well-entrenched) species—and the numbers which exist in each "environmental niche" are also few . . . by reason of the limited supplies of food to support any great numbers in a population.

Not by accident, then, is the rate of evolutionary change greatest in the equatorial region of the Earth—where the basic metabolic *rate* is also highest. For, underlying all the marvelously complex and interrelated arguments regarding ecological niches and "competition for territory as a function of population size . . . " is the basic *rate* of chemical reactions taking place within each cell of each member of each population of each species. That is, above all, the fundamental clock of evolution—the rate of life on Earth.

And that fundamental clock is based, in the final analysis, on the simplest of environmental factors . . .

Temperature.

Raise or lower the average temperature in which a cell exists by 10° Centigrade, and you speed up or slow down internal chemistry by a factor of *two*.[19] There are, of course, other mitigating factors—such as the development of self-regulatory systems within advanced species, like the mammals, to maintain even body temperatures. But in most life forms—such as insects, reptiles, even plants—where external temperatures govern internal metabolic reactions, the pace of life is set by forces totally outside the organisms themselves . . .

By the simple fact of their distance from the sun.

As we have seen, the current Martian temperatures range from mildly "arctic" to downright "cryogenic!" The Viking Landers faithfully recorded temperatures for several years following their touchdowns; the resulting weather reports consistently saw numbers several tens of degrees below Zero as the average daytime temperatures. The nights plunged to over 150 below! And in the dead of winter (which on Mars at present may be a bit redundant) the polar temperatures fall low enough for the very "air" to freeze, and solid carbon dioxide snow to cover the otherwise rusted, arid ground. Under these conditions, the words of Elton John ring very true indeed:

> Mars ain't the kinda place to raise your kids.
> In fact, it's cold as Hell . . .

But what about the so-called "Martian spring?" What of when the rivers ran, when rain actually fell from deep-blue skies, and something might have *lived* . . . ?

If we calculate (as several planetary scientists—including Carl Sagan—have done)[20] the minimum air pressures required to wrap the Martian surface in a warming "greenhouse" blanket, even at its greater distance from the sun, we discover some very interesting things. Such increased air pressures can, in fact, warm up the planet—enough for water to become liquid (at least at the equator), and—more important—to *stay* as liquid water for significant periods of time (see Chapter III). But the resulting average temperatures across the planet are hardly better than a summer day in our terrestrial *polar regions*. Which leads to the following set of arguments and conclusions . . .

Let us say that, somehow, life indeed arose on Mars—at approximately the same time as it did on Earth. And furthermore, let us say that it followed the same initial course of evolution—from simple cells which fed on preorganic molecules, to later more complex varieties which developed the ability to utilize the light and heat from the sun.

On Earth, according to the immutable record in the oldest rocks, such developments took *at least* 800 million years. And Earth, as we all know, is a good deal closer to that sun—at least 1.5 times. Its average temperature is several *tens of degrees* warmer than the average calculated for the planet Mars during its "warm, wet epoch," in the Beginning of the solar system. Which means, according to our rule of thumb for chemical reactions proceeding in those early one-celled creatures, that the *rate* of metabolism for primitive bacteria on Mars had to have been slower . . . a *lot* slower.

Now. After some 3.3 billion years of subsequent evolution here on Earth, life suddenly "exploded" in a wonderful diversity of form—after billions of preceding years of basically the same simple organisms living on the planet:

Blue-green algae.

All the marvelous developments we think of when we imagine evolution—the appearance of multi-celled varieties of living things, from trilobites, to sharks, to dinosaurs, and then to us—took place in less than half a billion years. But it took over 3 billion years before that, to set the stage for it to happen.

And that "stage setting" took place here on Earth—a tropical paradise compared to even Mars when it was "spring."

If we follow the preceding reasoning, and compare simple planetary temperatures as the basic "clock" of evolution, then Mars—even if its "spring" was still extant—would be far behind the planet Earth in evolution.

How far behind?

For the most charitable reading of the calculated temperatures, Mars would still be *several billion years* away from "our" appearance—and by "us" I mean the evolutionary appearance of Intelligence. And that, *only* if the average temperatures on Mars remained above the freezing point of water—as they have consistently for several billion years on Earth.

And they have not.

Long periods of cryogenic temperatures (when the atmosphere itself condenses on the planet in the form of solid "snow") are not conducive to a rapid form of evolution. Even positing some kind of alien organism (after all, this is Mars, right?) which can sleep through the intervening eons until another Martian "spring,"[21] the *rate* of evolution in the "on"/ "off" mode must be incredibly slow—even by the glacial standards set by Earth's own record; almost 3.5 billion years of dominance by microbes . . . then, finally, some real action!

In short, the "Martians" if they ever did exist were—at most—a very hardy strain of *microorganisms*!

Hardly the kind of planetary "neighbors" with a penchant for pyramids and faces . . .

Which meant that I was back to square one. Or was I?

It was an almost automatic (if erroneous) assumption that "the Martians" were from Mars. I had fallen into it myself, when initially looking at the Face and noting a striking resemblance to the massive monumental architecture of the Egyptians; my first thoughts had been, "This looks eerily like the beginnings of an indigenous civilization, complete with astronomical alignments . . . " One could easily accept (after the initial shock of finding *anyone* on Mars!) the notion that we were looking at some kind of extraordinary parallel evolution (Simpson notwithstanding). "Faces," "pyramids," the whole "megalithic architecture," it all fit in with our own history and known examples of "beginning pre-technical civilizations . . . "

Except that . . . these were customarily on Earth.

And this was *Mars*.

The more you delved into the hard, known facts about the planet, easy explanations like "parallel evolution" simply wouldn't work—as the preceding pages, hopefully, have demonstrated. Which leaves us with exactly what?

"The Martians," by the simplest application of the basic law of physics—themodynamics—couldn't have evolved beyond the stage of one-celled organisms . . . even in the entire lifetime of the solar system! Yet, if the Face and Pyramids were real, then someone—whom we would undoubtedly call "the Martians"—had created on that planet a complex

*The Monuments of Mars*

of extraordinary majesty and wonder. Perhaps with a technology and scope that dwarfed the whole history of life on Earth.

And then used it . . . for some time.

Who and why?

And now . . . from where?

Suddenly, an entire galaxy of possibilities opened up before me. No longer constrained by the unconscious need to "fit" the presence of a set of artifacts to a timescale of indigenous Martian evolution (because one simply couldn't!), I realized that even the provisional "solstice dating" of the Face and Pyramids at "half a million years" was merely an attempt to make the data fit with some known and understandable reference point—in this case, the fact that on Earth circa "half a million years ago" our distant ancestor—*Homo erectus*—had been taming fire.

But, who was to say that the ineffable "monuments of Mars" were that *young*—now that there was no way to imagine any conceivable Martian origin or evolution for whomever made them, not parallel "Egyptoid" Martians, and certainly not the suspended animation creatures of the honeycomb?

The alignment data, from which I derived my provisional estimate of age of the objects at Cydonia, also worked with impeccable geometrical precision—every million years—back through a near unending regression of obliquity cycles . . . almost to the origin of the entire planet.

The fact that these calculations couldn't be extended before that point was the consequence of an event in ancient Martian history termed the "Tharsis Uplift"—which had rearranged countless gigatons of planetary "stuff" within the mantle, and thus had changed the very Martian spin . . . and its response to outside forces (such as tides from Jupiter) which drove the obliquity variations. But subsequent to that "event," nothing had occurred to disturb this inexorable rhythm of planetary "nodding" (so Ward's calculations indicated) for some 2.5 billion years . . .

Or, the resulting summer solstice sunrises occurring at Cydonia in their million-year appearances . . . over the most inexplicable of "monuments."

If the astronomy apparent in this key alignment provided no constraint to the basic age of the objects at Cydonia, then what of other methods? The only other method open was the cratering . . .

The area of the single Viking frame encompassing the objects—35A72—was slightly less than a thousand miles in area—a thousand times *smaller* than the areas used for "crater counting" in the literature. Nonetheless, I decided to apply this tried and true technique to this single Viking frame, for it would give me some rough estimate of the numbers of craters in this region of the planet—and thus a rough idea of the underlying landscape's age . . .

And moments later I was presented with a paradox: there were "too many" craters.

The curves published in the Martian literature by previous investigators gave approximate calibrations of a landscape's age, using crater counts, in "numbers of craters in excess of a kilometer in size . . . per million square kilometers . . . per billion years." [22] Judging by these curves, the 50 by 50-kilometer region of Cydonia presented by the frame should contain roughly one kilometer-sized crater—if it was of roughly average age, compared to ages for similar regions of so-called "fretted terrain" listed in the papers.

But the actual numbers of craters in the picture, exceeding a kilometer in size, was *three to five times* the numbers in the published curves. This could be explained by "clumping" (the numbers were a statistical fluke, caused by a flock of secondary craters falling on this particular region of the landscape from another, larger impact some distance away), or they could be real . . .

In which case, this region of Cydonia was among the oldest on the planet—over 4 *billion* years in age!

The main problem with sampling such a relatively tiny area (as in one Viking frame) was this statistical uncertainty. Yet, something about the objects in the picture made me think of ways that such ages for the landscape could be real.

With the magnifying glass I examined carefully each of the "anomalous objects" in the image—the pyramids within the City, the pyramid that DiPietro and Molenaar had found, and, of course, the Face itself.

Not truly until this moment did I notice that there was a smattering of tiny "pits" on many of them—between the "eyes" of the Face, for instance—and that there were some considerably larger features such as the apparent thousand-foot crater on the shaded side of DiPietro and Molenaar's remarkable "buttressed pyramid" southeast of the City. The more I looked, the more it seemed unreasonable that these evidences of impact could have appeared on such small areas—as represented by the Face and pyramids—in "only" half a million years.

Could, in fact, the Complex be older? Yet, how could I square that with the apparent "freshness" of the Face and some of the pyramids themselves?

Then I remembered something that Mike Carr had written on this entire area of Mars, that "there appear to be features which have been 'stripped'—as if a covering of some kind had been previously deposited, then eroded away . . . ''[23]

Could the entire City have been *buried* . . . then exhumed by the natural forces of erosion?! But if so, when?—in the last few million years . . . or longer? Carr's rough estimate for when the "stripping" could have taken place, across the northern plains of Mars, including this region of Cydonia was impressive.

Not for "several billion years . . . ''

Could this fascinating complex at Cydonia be, not millions of years old, but literally *billions*!?

We had already seen, from at least two totally independent directions, why the very presence of the Face was "impossible." So, why not one more "impossibility"—if the evidence supported it.

I think it was that very moment when I finally realized what I had stumbled into; for I faced, not "just" a discovery of potentially immense significance—the first hard evidence that "we are not alone"—I was confronted with the real possibility that this evidence would ultimately overthrow everything we thought we "knew" . . .

About ourselves, about our fragmentary history of "humanoids" on this one planet, about even the essential mystery of life's "sudden" evolutionary leap—from the Pre-Cambrian one-celled creatures of our long, dark evolutionary night, into the shattering series of "inventions" in the last half-billion years that presumably had led to our appearance.

Were all these interlocking problems ultimately traceable to this one collection of exquisite and inexplicable objects?

For, if "the Martians" hadn't come from Earth . . . or Mars . . . then there was just one place left they could have come from . . .

From beyond the solar system . . . and bearing a humanoid image either in their "genes" or minds.

In which case, a scenario opened that was not unlike the very essence of Arthur's own grand vision in *2001*.

The "City on the Edge of Forever" as a truly ancient place of habitation . . . and an experiment. And the Face—possibly carved on another planet a billion years before its image would appear on Earth—merely a foreshadowing of what was to come . . . by Their presence on a small and rusted world.

In the last three hundred years the human race has lived through several wrenching shifts in its "world view" regarding its position in the Universe. From the Copernican "dethronement" of the Earth's position, to that of just one planet orbiting the sun with several others, to the equally amazing "dethronement" of the sun—to the status of a relatively minor star orbiting on the outskirts of an immense "star city" of a hundred billion suns—to the realization that even this "galaxy of stars" is merely one of billions . . .

Lost in an incomprehensible immensity of Space and Time.

Now, I realized, I might be gazing on the last moments . . . before the end of the Last Assumption: that in all that immensity, we—the human species—were still unique.

Was the Face but a gentle way of telling us that we are far older than we know . . . and an invitation to come and join "the family" . . . ?

Spinning dizzily through the echoing corridors of these boundless speculations, I clung resolutely to the one remaining piece of evidence of which I could be certain—

Regardless of the place the "Martians" came from, there was one place where we could find the truth—if any—underlying all these yawning questions . . .

Return to Mars ourselves . . . as soon as possible.


## Notes

1. Geikie, Sir Archibald, *The Founders of Geology*, Reprint, New York: Dover Publications, 1962. A classic.
2. McIntyre, D.B. "James Hutton and the Philosophy of Geology." In *The Fabric of Geology*, Reading, MA: Addison-Wesley Publishing Co., 1963; McIntyre, D.B., Craig, G.Y., and Waterstone, C.D., *The Hutton Lost Drawings*, San Francisco: Freeman, Cooper & Co., 1981.
3. Hallam, A., *Great Geological Controversies, Oxford: Oxford University Press, 1983*.
4. *Press, F. and Siever, R., Earth*, 3rd ed. San Francisco: W.H. Freeman & Co., 1982.
5. Noyes, R.W., *The Sun, Our Star*, Cambridge: Harvard University Press, 1982.
6. Soderblom, L.A. et. al., "Martian Planetwide Crater Distributions: Implications for Geologic History and Surface Processes," *Icarus* 22:239, 1974.
7. Neukum, G. and Wise, D.U., "Mars: A Standard Crater Curve and Possible New Time Scale," *Science* 194:1381, 1976.
8. Hartmann, W.K., "Cratering in the Solar System," *Scientific American* 236:84, 1977.
9. Hoagland, R. and Bova, B., "The Origin of the Solar System." In *Close-Up: New Worlds*, New York: St. Martin's Press, 1977.
10. Beatty, K., O'Leary, B. and Chaikin, A., eds. *The New Solar System*, Cambridge, MA: Cambridge University Press and Sky Publishing Corp., 1981.
11. Sagan, C., Toon, O. and Gierasch, P., "Climatic Change on Mars,"

*Science* 181:1045, 1973

12. Weihaupt, J.G., "Possible Origin and Probable Discharges of Meandering Channels on the Planet Mars," *J. Geophys. Res.* 79:2073, 1974.

13. Masursky, H., "An Overview of Geological Results from Mariner 9," *J. Geophys. Res.* 78:4009, 1973.

14. Sharp, R.P., "Mars: Troughed Terrain," *J. Geophys. Res.* 78:4063, 1973.

15. Carr, M.H., *The Surface of Mars*, New Haven: Yale University Press, 1981.

16. Baker, V., and Milton, D. "Erosion by Catastrophic Floods on Mars and Earth," *Icarus* 23:27, 1974.

17. Day, W., *Genesis on Planet Earth*, New Haven: Yale University Press, 1984.

18. Weiner, J., *Planet Earth*, New York: Bantam Books. Copyright © 1986 by Metropolitan Pittsburgh Public Broadcasting, Inc. and J. Weiner.

19. Francis Crick, in communication with Tom Rautenberg.

20. Sagan, C., "Reducing Greenhouses and the Temperature History of Earth and Mars," *Nature* 269:224, 1977.

21. Sagan, C., "The Long Winter Model of Martian Biology: A Speculation," *Icarus* 15:511, 1971.

22. Carr, M.H., *The Surface of Mars*, New Haven: Yale University Press, 1981. *op. cit.*

23. *op. cit.*

# SECTION II
# INVESTIGATION AND REACTION

# VIII

# INVOLVING A "SECOND OPINION"

> "Ask, and you shall have answers
> Seek, and you shall find
> Knock, and it shall be opened . . ."
>
> Biblical Quote

It's simple enough to say, "Return to Mars." It's another thing to do it.

In July, 1983, when DiPietro provided the initial Viking pictures he and Molenaar had processed—including the SPIT-version of 35A72—he also had a request: my help in securing confirmation of their work, and the promotion of a "sincere effort" to land men and materials on Mars "for the sole purpose of exploration."

There was a certain irony to this. In my opinion, I was probably the *last* person the planetary community was going to believe in urging an "unbiased hearing" for DiPietro and Molenaar; there was the small matter that, as a "mere" journalist, I wasn't even "a member of the club," and thus had even a slimmer chance of gaining a fair hearing in the scientific community than had already been evidenced in Colorado, where Vince and Greg—contracted employees of NASA—hadn't even been allowed to speak!

A return to Mars could be thirty years away, well into the next century—as neither the U.S. nor the Soviet Union had any funded plans (in the summer of 1983) to return to the Red Planet before then with a manned expedition. And, tentative discussions of a *possible* U.S. unmanned mission "sometime in the 1990s" specifically *excluded* mention of a camera!

While I pondered this, another part of me knew exactly what was needed, in terms of further scientific research of what was on those photographs . . .

A full, three dimensional reconstruction of the City—with state-of-the-art astronomical alignments; a thorough geomorphological analysis of the surrounding countryside (for statistical analysis of the probabilities underlying the "reality" of this discovery: from the chances of locating another "city" grouping, to an exact count of those all-important impact craters). I also dreamed of what the Face would look like—

In a carefully-drawn computer-accurate recreation . . . from the surface.

With the cliff behind it—that eerily too-well-placed "artificial horizon" . . . the solstice sun coming up behind them both!!

But then, the realities would loom . . .

The reality was that, given the situation within NASA—where even bringing up the subject of "the face on Mars" (as when Mike Carr asked, pointedly, if *that's* why I was interested in Cydonia) brought suspicion, if not outright mirth—the odds on getting any serious attention *there* seemed very low indeed. Yet, that was precisely what was needed—the full research resources of the Nation's space agency, if not its full attention—

If this was what it now appeared to be—the discovery of the first artifacts left by an alien presence (or . . . something) on another planet.

The trick was how?

My first need was for better data . . . well, at least a better *form* of the data that I had; all of the measurements that I had carried out, so far, had been conducted on the original 5-inch print of 35A72 supplied by DiPietro . . . and a couple of 8x10s that Mike Carr had sent from the USGS. It was obvious that the first requirement for any serious effort to get a "serious investigation" would be better prints—simply to allow others to see the contents of the data; squinting at tiny images through a stereo-viewer on my desk, as I had been doing for almost two weeks, didn't seem the way to show the evidence to full advantage.

Another thing that desperately needed checking, was "the honeycomb." If it was real, it was magnificent confirmation of the totally artificial nature of the City, and the purposes for which it may have been created . . .

If it was real.

But was it? Only a corroborating image could tell us, which is why I'd been so eager to get the prints from Carr—as on 70A11 (taken a few seconds before the famed "confirmatory image" of the Face, 70A13) the Viking Orbiter had snapped an equally vital *second* photo of the City, at that same significantly higher sun angle. But, after examining the USGS version of this photo (which seemed to have a number of defects), and the region of "the honeycomb," it was clear to me that the confirmation of this *essential* feature would be found only on a SPIT-processed version of this higher (70A11) sun angle—

Which only DiPietro and Molenaar could furnish.

During our conversation, however, I realized something which at first I had trouble coming to believe: DiPietro had *no idea* what was on 70A11; in fact, according to his best recollection, he and Molenaar *hadn't even processed 70A11*!

Reason: it didn't contain an image of the Face.

Furthermore, he seemed curiously reluctant to furnish the negative of 35A72—a reluctance that seemed to arise more from suspicion as to why I needed it than from any effort at concealment. Throughout, he seemed highly skeptical that I was serious regarding the need for an "in-depth investigation," and bitter for the years of criticism and outright accusations he and Greg had suffered.

I felt strangely fortunate that I had even a single, tiny copy of 35A72—and began to wonder why DiPietro had sent it in the first place. He obviously didn't believe that I could do anything about the problem, in fact he kept repeating, "You can't fight NASA!"

Having decided that the best way to initiate *some* kind of investigation would be to chronicle the facts to date, during the next few weeks I wrote furiously, and was able to send off to New York (and to my friend, Russ Galen, also my agent—at Scott Meredith Literary Agency) four chapters of my proposed manuscript. Russ promptly sent copies to "the top 20" or so publishing houses in New York—to reactions that seemed as bewildering (at first!) as disappointing; one editor responded, "although we have been known to be 'feisty and imaginative,' we're playing it safe and are looking for manuscripts *that are based on fact* (italics added) . . . "

Russ' only comment: "I'd say . . . that this person didn't really understand what she was reading, you know?"

Another woman certainly didn't; she thought we had already *been* to Mars with astronauts, "who would surely have reported a Face!"

Some people (though not necessarily in the New York publishing community!) *did* understand the extraordinary implications—and offered their assistance. One of these was anthropologist Randy Pozos, who offered to write a commentary for the manuscript: a critique of my earlier anthropological and archaeological analysis of the proposed "Martian ruins" (see Chapter II). Further, he proposed that he try to arrange a seminar, possibly at Berkeley (University of California), to discuss the initial findings of my work—and to plan for more in-depth analysis.

Thus was born a viable route to a real "Independent Mars Investigation."

I had my eye on a northern California "think tank," SRI International, as a possible ally in the Mars investigation—by reason of one of their most well-known physicists, Lambert Dolphin, who some years before had led a team investigating some of the pyramids in Egypt, using the latest "high technology." I reasoned that, if the Martian pyramids were real and resembled in their placement and design the ones in Egypt (which they did, to an amazing degree!), then there could be no better-qualified scientists to incorporate within a proposed team to investigate these "Martian anomalies" than those who were already deeply familiar with these terrestrial "analogs."

About a year before, during my search for answers to "the Thing in the Ring" (see Chapter II)—the radio enigma discovered by the unmanned Voyager spacecraft's radio astronomer, Jim Warwick, in the vicinity of Saturn—I had successfully "borrowed" SRI's giant 150-foot radio telescope—in an attempt to confirm from Earth what the Voyager spacecraft had detected orbiting in the middle of the rings.[1] And, over those long months of working with SRI's talented physicists "on Saturn" in the Radiophysics Lab, I had many times passed through a "corridor museum"—filled with spectacular color photographs taken during the Lab's Egyptian expeditions (in 1974). But, for some reason, during all those months, I had never met the leader of those expeditions: Lambert Dolphin.

Now, years later, at the Institute for the Study of Consciousness, in Berkeley, I did meet Paul Shay—Vice-President for Corporate Affairs for SRI. Shay listened quite attentively as I explained (without even the virtue of a photograph—I didn't carry any with me then!) what I thought I might have found lying on the Martian surface. After I had finished, he looked thoughtful (as well he might; I learned later that among his many accomplishments he had been a Rhodes scholar and an intelligence officer in Europe, after World War II). Cautiously, he began to explore some possibilities, including an introduction to "a well-known physicist at SRI . . . who recently completed a series of investigations of the Egyptian pyramids . . . Lambert Dolphin."

The next day, after Shay had notified him of the impending meeting, I decided to do what I should have done years before: I called Dolphin myself. I related my story, of passing through his "Egyptian Corridor" many times during our electronic Saturn observations, and his response was most gratifying. He said he was "honored to hear from Paul Shay about your extraordinary discoveries," and "look forward eagerly to seeing the actual photographs." Having waited impatiently for months (it was now October) for a chance to demonstrate to an appropriate expert what I believed lay on the Viking photographs, I impulsively suggested that I bring a few of them to his office, at the Lab in Menlo Park—about an hour's ride south of San Francisco. Dolphin instantly agreed.

During the long preceding "Martian Summer" other events had moved strangely in a parallel direction, events which eventually would come together to create the context for the "Independent Mars Investigation."

A good friend of Randy Pozos, Ren Breck, a former ABC producer, had been called upon suddenly to step in and manage a computer-conferencing company—InfoMedia, Inc.—in Silicon Valley. Randy, a "co-conspirator" and associate in many projects for over ten years, saw the potentials for a fruitful partnership between the needs of InfoMedia—to attract new customers—and the needs of our still latent Mars investigation. Why not, he suggested to Ren, set up a "demonstration" Mars Investigation Computer Conference—to illustrate how a scientific inquiry with scant money could still pool top talent from a wide geographical area, over a scientific problem of paramount importance.?

Immediately following my conversation with an enthusiastic Lambert Dolphin, I called Ren Breck.

"How would you like to accompany me on a 'Mars recon trip' to SRI?" I asked.

Two days later, as we sat in Lambert's office, I realized that here was a man who simply *had* to be part of any Mars investigation; his office overwhelmingly affirmed it. It wasn't very spacious, but was filled—almost to overflowing—with works reflecting a truly eclectic curiosity. The bookshelves contained some of the most esoteric works one could imagine—in a *physicist's* office; the spines of numberless volumes tantalized with titles like *The Collected Sayings of Buddha*, mingled with *The Handbook of Chemistry and Physics*, and *Biblical Archaeology of the Middle East*. Around the office were a variety of artifacts, including a small statuette of the Egyptian goddess Serket (the one with her arms protectively outstretched, guarding Tutankhamen's tomb) beside a cardboard replica of the Great Pyramid.

I knew we were in the right place.

While we waited for Paul Shay to join us, I spread the Viking pictures on a desk (after taking some pains to clear a space among the overflowing stacks of papers, reports, and aerial maps already taking up almost every available level surface). Dr. Dolphin bent over the Mars images, intent, as I quickly "walked him through" the tangled history—

Of NASA's mission "to search for life on Mars"—a mission that resulted in these pictures—and their years of subsequent neglect by the very agency which took them! Of DiPietro and Molenaar's first-rate efforts on the Face, and their subsequent rejection by the NASA community of planetary scientists. And then of my own efforts and discoveries . . .

I pointed out the proximity of, not one or two puzzling features, but a score or more of obviously related "structures" clustered in one tiny geographical location on this part of the Red Planet—and the mathematics I had found which seemed inextricably to connect them—

With the Face.

Then, I pointed out the "honeycomb," as the most blatantly "artificial" structure in the entire complex; how its shadow proved it was a three-dimensional feature (thus, not easily explainable as a "computer artifact") and how its curiously strategic placement strongly argued for its artificial nature.

Later, Dolphin would write about that meeting.

> Dick Hoagland briefed me thoroughly on the early work of DiPietro and Molenaar showing a clearly recognizable face on Mars and also a large pyramid, partially damaged on two faces. In addition Dick showed me carefully made enlargements of mountainous regions to the west of the face in which there are additional features *not appearing to be natural* (italics added) which DiPietro and Molenaar had apparently not noted before . . .
>
> The face by itself or even in combination with the (D&M) pyramid, impressive as they are, are not as startling as the 'city' region . . . which is obviously worthy of much more careful study . . . the city region clearly is something Hoagland's careful study has brought to light . . . (and) adds much more impetus to the earlier findings.

When Shay joined us, our discussions turned to how this "much more careful study" might evolve.

I reiterated my objectives of 1) securing the necessary data tapes from NASA and verifying all DiPietro and Molenaar's original work, and 2) securing the tapes of the *additional* frames represented in the regional mosaic I'd made from the pictures Mike Carr had furnished. I also outlined the need for extensive three-dimensional analyses of the City (the "table-top models" mentioned previously), a verification of the reality of "the honeycomb" (via processing the high sun angle frame—70A11— also containing the City), and a search for additional corroborating "honeycombs" and "structures" in those other frames—35A70, 71, 73, and 74.

A computer-facility equipped to image-process these NASA data tapes was essential for this work; image-processing of the data tapes had to take place before any *analyses* of what was on the images could even begin. This required either funding (to hire such facilities) . . . or, the willingness of a research institution (such as SRI) to "loan" us the necessary computer facilities "after hours." The latter approach, of course, would only work if the personnel needed to run those facilities—such as the image-processing experts and computer programmers—could be convinced to lend us their time as well.

Ren presented his willingness to organize a demonstration "Mars" computer conference—using InfoMedia's computer to link a team of scientists, like Dolphin, all around the country, even overseas. Such a conference—a modern management tool of inestimable value—would allow precise recording of all "conversations" over both the data, as well as the administrative details of setting up and maintaining the investigation. Each "entry" would be dated and timed, as well as private "notes" between participants. Each discussion could be instantly recalled and printed out; and older data compared to more recent developments or theories.

The value of such a tool to an on-going scientific investigation, particularly one with limited resources, dealing as it would be with controversial material, was impossible to overestimate. Ren was offering the only practical means for us to conduct an investigation of the Mars material—if some funding could be made available to support the costs of the telephone "connect time" (which would be expensive and not capable of being "written off"). Alternatively, if SRI undertook to use Ren's computer-conferencing facilities for other projects (such as a current job in Indonesia, requiring world-wide management and communications back to SRI in Menlo Park), some of those profits could be used to defray the costs of the "Mars" computer conference . . .

And so it went; discussion of all the ways that Dolphin's "much more careful study" of this data could be funded . . . or even "bootlegged." We considered everything—from assembling a "consortium of research firms" in Silicon Valley, to getting SRI itself to directly fund the study. Each had advantages and pitfalls; each involved finding someone willing to take the initiative and "sponsor" this most extraordinary of investigations—

A look at what could be *the first real evidence of extraterrestrial intelligence.*

Finally, Shay and Dolphin volunteered to take my proposal to their own internal management.

A few days later, Paul Shay and I flew to Los Angeles to see Merton Davies, a cartographic expert at the famed RAND Corporation—another "think tank," located in Southern California. Mert and I had known each other many years, a friendship which began during my days as a reporter covering the various missions at JPL. Mert is the chief "map maker of the solar system," a title I unofficially bestowed on him because of the fact that it was Merton's "control points," derived from key photographs taken by unmanned planetary spacecraft, which allowed the creation (by the USGS) of accurate maps of the surfaces of all the "terrestrial bodies"—planets and satellites—we've "visited" during the past twenty or so years.

What I wanted from Merton was twofold: additional confirmation that I wasn't crazy (!), that the objects at Cydonia really were as "weird" and extraordinary as I thought (from a man who'd seen a *lot* of planetary real estate over the years),and a set of accurate coordinates for the City and Face.

According to our preliminary conversation on the phone, the "datablocks" on the images I obtained from Mike Carr (and those that DiPietro and Molenaar obtained several years before from JPL) were in error—how much so would only be determinable *after* Merton had inputted the images to his planet-wide computerized "Control Net of Mars" (just completed in 1982).[2] Accurate coordinates and orientations of the features of Cydonia were a "must" for any detailed analysis of the "anomalous objects," or confirmation of the Intelligence Hypothesis—particularly, for researching "archaeoastronomical alignments."

In other words, you had to know where "north" was!

Merton greeted us warmly when we arrived at RAND, a tall, friendly giant of a man (it always amused me to see Mert standing literally "head and shoulders" above his colleagues at scientific gatherings or press conferences). He led us to a small conference room where, once again, I laid out the pictures for inspection (another ritual begun . . . ). And, as Paul Shay watched this time, I ran through the same litany I'd done for Lambert Dolphin . . .

When I finished, Mert sat puffing on his pipe, saying nothing. The silence was, as the cliché says, deafening.

God, I thought, he thinks I've gone off the deep end . . .

Puff . . . puff.

Finally, he lowered the pipe with one hand and simply said,

"Well, I've always thought evolution—as a theory—left something to be desired . . . "

Afterwards, Shay was to confess that that simple statement, coming from a man with Mert's background and familiarity with the planetary explorations of our last twenty years, "blew my mind."

Mert, after carefully defining what he considered his area of expertise and what he might contribute, agreed to supply the "Independent Mars Investigation" (whatever form it eventually took) with the necessary

*The Monuments of Mars*

cartographic data for Cydonia.

Another crucial step "on the road to Mars" had just been taken.

The next meeting was a few days later back in Menlo Park—literally around the corner from the labs of SRI. In attendance were Lambert Dolphin, Randy Pozos, and myself. Our objective: to secure access to the NASA data tapes—without which there could be no "Mars investigation." Our "target": Mike Carr, at the United States Geological Survey (USGS) in Menlo Park—

The former head of the Viking Orbiter Imaging Team, himself.

I considered Mike a friend. Over the years, ever since his favorable comment regarding my (now abandoned) theory to explain Voyager's discovery of "braided rings" at Saturn (that they were the "vapor-trail loci" of tiny icy satellites, orbiting around each other much like the "asteroid satellites" proposed for other regions of the solar system), Mike had been one of the few planetary scientists (Mert was another) who expressed a respect for my penchant for "theorizing." In fact, Mike's first comment when we met, the evening that the braided rings were first discovered, impressed me with his honesty and willingness to accept a valid theory—even if it came from a source outside the planetary community, from a "mere" reporter.

The Mike Carr we met this afternoon was very different. He merely glanced at the large ($16 \times 20$) print of the City and the Face that I spread out on the conference table in his office. And he barely listened to Dolphin and Pozos, trying to explain why they thought the images, containing these peculiar objects, "needed looking into." He politely—but quite firmly—refused any role in the "Independent Mars Investigation"—even though a man of his intimate association with the Viking Mission (and Orbiter technology), to say nothing of his Martian geological expertise, could have saved us from "reinventing the wheel" a hundred times or more. The fact that he listened at all, in retrospect, must have been only some vestige of our former friendship; at the end of our meeting he authorized what we had come for: access to the data tapes themselves, without which we could not have proceeded.

He suggested we contact the tape librarian, a woman named Linda Sower, at the main USGS offices in Flagstaff, Arizona (as there were no tapes in Menlo Park—just prints).

Shay's idea now was to get SRI to fund the entire Investigation out of something called the "President's Fund"—a discretionary source under direct control of the president Dr. William Miller (used primarily to open up potential new areas of research for later SRI development).

Miller was a personal friend of the President of the United States' own Science Advisor, Dr. George Keyworth; he was former Head of the "Computation Group" at the famed Stanford Linear Accelerator Center (SLAC), one of the best high-energy research laboratories in the world; and he'd been Vice-President and Provost of Stanford University, before joining SRI—as its first chief executive in many years with a "hands on" research background in the physical sciences.

During the Briefing, with Lambert Dolphin, Randy Pozos, and Paul Shay in attendance, Miller listened silently as I ran through the, by now, familiar data. Then, Lambert made a pitch for minimal in-house funding—$50,000—based on the extreme scientific potential—if the results were positive; and Shay made a case for the favorable public image such an investigation would create for SRI, as well as the favorable position it would place the corporation in—with regard to later governmental funding of a return to Mars.

Miller, after we had finished, asked a lot of tough and pointed questions, such as the "'favorable impact' if it all proves to be just a 'pile of rocks!'" To which Randy supplied some elegant answers, including pointing out "the value to the social sciences (one of SRI's fields of research) of the *methodology* developed in the course of such an investigation; just how do you answer the 'loaded' question: does this set of objects truly represent artificial design on another world?"

Compared to SRI's annual budget (something around $300 *million*), what we were asking for—as an investment in a possible "discovery of the Century"—was *nothing*.

We finished.

Finally, Miller leaned back . . . looking directly at me, and asked:

"Well, how do you want it: in cash or travelers checks?"

A few days later, based on Randy's report to Ren, and Paul Shay's promise to Bill Miller—to "send a formal proposal for use of monies from the President's Fund for an SRI investigation of the Viking images"—Ren officially set up the Mars Computer Conference. Randy and I decided that there was only one name fitting for this inquiry:

The Martian Chronicles.

If I was right, if these extraordinary objects were, in fact, the artifacts left by someone on this small and arid world, then for all future time those words—"Martian Chronicles"—would refer to this investigation . . . in fond remembrance of a visionary who uncannily described an ancient inhabited Mars many years before . . .

Randy (because of his long relationship with Ren, as well as his offer to assist with the scientific recruitment for such an investigation) was to act as Project Administrator of the Conference. I would be Principal

Investigator—mainly setting out the scientific agenda we would follow. My first entry, dated December 5, 1983—at 11:41 PM!—tells all:

Welcome to a most unusual Activity . . .

### An Inquiry into the Possibility of
### a Former Civilization on
### Mars

For the benefit of those who may be new to this intriguing tale, here is a brief synopsis:

In July, 1976, the VIKING Orbiter snapped a photograph of the surface of Mars . . .

## Notes

1. Hoagland, R.C., "Blivit in the B-Ring," *Analog*, two parts: December, 1982; January, 1983.
2. Davies, M.E. and Katayama, F.Y., "The 1982 Control Network of Mars," *J. Geophys. Res.* 88:7503, 1983.

# THE INDEPENDENT
# MARS INVESTIGATION

"The voice of the intellect is a soft one, but it
does not rest until it has gained a hearing."

Sigmund Freud
*The Future of An Illusion*

The "Independent Mars Investigation" formally began December 5,
1983—almost a decade after NASA secured images of a *most* peculiar
set of objects at Cydonia . . .

And then ignored them.

It ended seven months later, with the "poster session" paper sum-
marizing our results, presented to the Case for Mars II Conference, held
at the University of Colorado, Boulder, in July, 1984.

In between . . .

Seven months of fascinating (and at times, maddening!) dialogue,
discussion and debate—among a team of multi-disciplinary scientists and
scholars representing fields as diverse as geology, anthropology, and art.
And in those seven months, almost every facet of the data assembled by
DiPietro, Molenaar, and myself—and a good deal more—was examined
and vigorously debated—from the evidence present on the original Viking
images to the implications—

If they, indeed, represented a "lost civilization" on another planet.

The Independent Mars Investigation (the name came simply from
the fact that we were truly "independent" of any NASA affliation) was
by no means as thorough as I would have liked; for a variety of reasons
many of my original objectives weren't accomplished—such as the image-
processing of *all* the Viking frames making up the original Cydonia mosaics (both sun angles). But the Investigation did accomplish other things,

not on my original agenda—including the location of a score of *additional* "anomalous objects," scattered almost half-way around the planet!

Corroboration of what I had begun to suspect most: that the evidence at Cydonia was only *part* of an overwhelming larger picture—

Of a planet-wide civilization which, after—somehow—arriving on Mars, subsequently disappeared . . . until we found it.

The most important accomplishment of the Independent Mars Investigation however, in my opinion, was its mere existence: a group of sincere (and courageous) individuals gathered together (if only electronically!) to consider for the first time the evidence for some extraordinary enigmas present on a score of Viking photographs . . .

*      *      *

On the 13th of December, Randy and I met for an afternoon to discuss the goals, methodology . . . and funding . . . of our new-born Investigation. We developed the following categories that afternoon:

1) An "imaging group": consisting of computer experts, such as Vince and Greg, who would furnish the basic data to be analyzed, including any "fancy graphics"—such as three-dimensional "transformations."

2) A "geology group": consisting of experts recruited for their basic geological expertise, who would assess the likelihood of natural forces creating the "anomalous objects," as well as the broader context of Martian history against which any "Intelligence Hypothesis" would have to operate.

3) A "photometrics group": to assess the optical properties of the Cydonia objects at the two sun angles (on the theory that artificial objects would behave *differently* from natural "rocks").

4) A "structural engineering group": who would review the anomalous objects in the context of their solar orientation, loads imposed by Martian surface gravity (to see if their sheer scale was a constraint against their being artificial), and energy properties (if any—derived from the Photometrics Group)—as a means of further constraining any "artifical structure" hypothesis.

5) A "cultural anthropology group": which would take the previous input and see if there was, indeed, a pattern which looked at all familiar—compared to the terrestrial experience of the last several thousand years (well, you have to start somewhere . . . ).

In addition to splitting the Team up into sub-groups, we also worked out a rough "methodology" for approaching the Investigation—which ranged from starting with a review of the pertinent literature in each discipline (including the extensive "Martian" literature available from the Viking mission), to preparing mosaics of the images for distribution to the various teams.

This included running tests of DiPietro and Molenaar's "SPIT" algorithm and preparing three-dimensional stereo "isocontour maps" of the Face and City objects (including the DiPietro and Molenaar pyramid), as well as generating general regional maps—to see if there were other interesting objects we had missed—and for geological and environmental "context."

Each sub-group, using the image data so developed (including the optical reflectance data, the 3-D data, and the maps) would then prepare a report representing each discipline—e.g. geology, imaging, photometrics, structural engineering, etc.—which would be passed to a general review by *all* the Team members (to get the full advantage of the multidisciplinary nature of the problem.)

*      *      *

For example, I envisioned the "structural engineering group" as taking the data from the "imaging" and "photometrics" teams, and investigating the hypothesis that the "pyramids" were, in fact, huge *housing complexes*—

In the manner of the well-known architect, Paolo Soleri!

Soleri, in the late 60s proposed the creation of vast three-dimensional "arcologies," to house urban populations, in place of the two-dimensional "urban sprawl" of modern cities. What made this of more than passing interest to me was the simple fact that, for *structural* reasons, some of Soleri's arcologies were to have—

A pyramidal shape!

Soleri envisioned the construction of these enormous complexes (they were to measure *miles on a side*, about the same as the pyramids on Mars!) in many locations here on Earth, in the Los Angeles Basin, for example—as a way to "protect" the inhabitants from the outside environment (or was it to protect the environment from the activities of the inhabitants . . . ?).

My question: was it possible that "the Martians," because of a deteriorating environment on Mars (rather more severe than living in L.A.!), were forced into an *identical* solution?

It was a fascinating notion . . . and in principle, testable—if one applied a variety of complementary data on the Martian pyramids— reflectance, slope geometry, incident sunlight, orientation, etc.—and evaluated them as *designed structures for the purpose of housing a large, contained population!*

*　　　*　　　*

About a month after our crucial SRI meeting, the production company that produces "Evening (or PM) Magazine"—a general information television program seen around the country in syndication—ran a *three-year old* interview with DiPietro and Molenaar, taped at Goddard. Included were examples of the images processed by these engineers: spectacular closeups of the Face. One viewer of the program was physicist, John Brandenburg—a member of the staff of the Sandia Laboratories, in Albuquerque, New Mexico. Sandia, a well-known national defense contractor, was heavily involved in the nuclear weapons program. Subsequent to the President's 1983 call for a defense against ballistic missiles, it was also at the forefront of the President's Space Defense Initiative—"Star Wars".

Heavily affected by the image of the Face, Brandenburg felt impelled to seek further information. Eventually, by following one lead after another, he found "the Investigation"—and became a member.

A plasma physicist, working with the world's most advanced technology, Brandenburg had access to information of direct relevance to one of the most disturbing observations I had made . . . that the objects at Cydonia exhibited considerably more than "natural degradation"—for any reasonable age.

It all went back to those anomalous "crater counts" (see Chapter VII).

Was it possible (I'd dared to ask myself, after the numbers consistently came out too high), that the evidence of significant erosion on the Face and pyramids—and the abnormally high number of 1-kilometer craters in the area—were a result of some powerful *artificial* agent . . . ?

In other words, had life on Mars—the Roman "God of War"—been exterminated in an all-out *nuclear holocaust*?!

The thought was too fantastic . . . yet, it could explain a lot of things . . . the craters . . . and the other evidence I'd seen; in examining the City under the magnifying glass, I'd viewed direct evidence of *melting* and *flow* on the Main Pyramid within the City. In addition, there was the apparent massive destruction of much of the southeast sides of both that structure, and the equally puzzling "Fort"—*and* of forces which had, somehow, apparently completely vaporized (or blown off) the "roof" which must have originally covered up "the honeycomb!"

Then, there was the strategically placed "impact crater" in the southeastern flank of the D&M pyramid, and the equally suspicious "domed uplift'' distorting its geometry. Was this the work of some kind of ''rocket-borne explosive''—which had penetrated the interior of the pyramid, and then detonated, leaving an exit ''blow-hole'' type crater, and severe internal structural deformation?

Very cautiously, during our second phone call, I explored with John the kinds of craters a nuclear exchange would leave behind, and the ''statistical anomalies'' of so many craters on the objects of most interest—like the Face and the D&M pyramid.

It was his suggestion that, if the craters on this part of Mars were ''artificial,'' there would be a way to tell—because nuclear explosions (as opposed to meteor craters) would be shallower . . . On the other hand, he said (undercutting, in the next breath, my brief hope that here was a *definitive* test!), the craters produced in targets in the lab, as part of the new Space Defense Initiative tests, were deep—not unlike meteor craters themselves—a direct result of the *way* they were produced: by the exotic particle beams ''burying'' their energy deep within the targets.

Now shades of ray guns and Orson Welles' green Martians!

More significant was the fact that John had access to people at the lab who did have information regarding the kinds of exotic *isotopes* a nuclear exchange would likely leave behind within the atmosphere . . .

And, as Viking had landed two spacecraft directly on the planet, and had *sampled* that atmosphere and measured—down to the parts-per-billion-level—its trace constituents, I thought we still might have a way to *test* this hypothesis, regarding a proposed former (now ''advanced''—and very extinct!) civilization on the Red Planet.

What a sobering . . . and perhaps crucial—to literally saving our own world—hypothesis (!) . . . if that had, indeed, happened . . . and was possible to prove.

*　　　　*　　　　*

In the meantime, the prospects of a *funded* Mars Investigation, at least in SRI, were looking considerably dimmer for the near future. Dr. Miller's parting line could be interpreted either as a jesting commentary on our naivité and exuberance, or as an initial willingness to back a ''blue-sky'' project . . . tempered by the reality of committee consultations, ''bottom-line'' analyses, etc. Whatever the reason, SRI's assistance remained confined to photographic reproductions, Dolphin's time, and some technical support.

Other avenues for cash funding, of what was rapidly becoming an expensive ''hobby,'' were also not encouraging.

For, while Russ had finally sold the book (!), the advance was slightly more than cab-fare in New York—a mere $3000—and only half of that up front! (When I was at CBS, fifteen years before, I had been paid more for *articles* than for this book . . .).

On the technical front, however, things were better.

Vince DiPietro, having located a terminal somewhere at Goddard, made his first "official" entry on Jan 9, at 2:21 P.M.: a list of potential additional members for the "Independent Team," composed of several scientists who *had* reacted favorably to his and Greg's work over the years. A few days later, he dispatched via Federal Express, the first of several sets of SPIT-processed negatives to SRI—frames 35A72 and 70A13 (but no SPIT on 70A11, the confirming photo of the City. They'd never bothered with it because it didn't contain an image of the Face, remember?).

In turn, Dolphin produced some stunning new images in the SRI photolab—prints made from *first-generation* negatives—digitally-enlarged—created from DiPietro and Molenaar's excellent computer work, now five years old. Of course, they were all enlargements of the Face (except for one full-frame shot of 70A13, which contained, you might remember, that other "anomalous object" they'd discovered—the "rectangular, buttressed pyramid").

And what a sight: the images, in the $11 \times 14$ blow-ups, were spectacular!

The Face, this mile-wide enigmatic apparition on the Martian surface, where . . . by everything we knew . . . nothing even remotely like it had any right to be, was even more mysterious . . . and captivating. The amount of detail present—the peculiar "adornment" at the corner of the right eye, the so-called "teardrop" below that eye, even the sweeping curve which defined the "cheek"—argued compellingly that—

*This was a work of art.*

All that was lacking was the sculptor's signature!

But something else on the full-frame enlargement of 70A13 caught my eye, something that I had only previously suspected . . .

The "D&M pyramid" was a five-sided—not four-sided—object!

\*　　　　　\*　　　　　\*

Ever since I'd seen it in the DiPietro/Molenaar booklet, something about their "reconstruction" of this remarkable "buttressed pyramid" had bothered me. They maintained that it was "a rectangular, four-sided figure . . . with one long side aimed directly north . . . the other distorted by erosion . . . " (see Chapter I). But to my eye, try as I might, I could not truly see that.

For one thing, the "look angles" from the Viking spacecraft were wrong: the shallow (12 degree) angle of the Viking camera to the surface, coupled with the immense distance (1000 miles above), simply could not account geometrically for the distortion visible in the images they'd reproduced—*if* the object was a "four-sided, rectangular pyramid." On the other hand, erosional distortion would be expected to "soften" the outlines of such an object (by causing "slumping" of the walls, etc.) but not the outright change of geometry required to "explain" a previously rectangular object in the current condition of the D&M . . .

Thus, for a brief moment—as I'd measured everything in sight on the images available to me during the summer—I'd dabbled with the thought that this was in fact a *five-sided* pyramid . . . and then dismissed it, opting instead for the "explosive creation of the impact crater," down on the flank, as the cause of the apparent geometric distortion.

After all, there are no such things as "five-sided pyramids" on Earth, right? (A vivid example of discarding what your mental frame of reference will not allow you to imagine!)

Now, staring at the vastly improved image on the 11 × 14 print from SRI, it was unmistakable: the "D&M pyramid" was definitely five-sided . . . and bisymmetrical!

Visible in the previously shadowed side (on 35A72) of this new high sun-angle version was an unmistakable *fifth* buttress—due east of the first (the one aimed directly at the Face—see Chapter VI). Thus, there were *three* short sides, and *two* long sides (created by the new "buttress"— with a slight adjustment, to take into account the damage on the cratered flank) was the same as that measured by DiPietro and Molenaar: 1:1.6. Furthermore, on closer examination, the center buttress (#1)—appeared to be *another* "bisymmetrical figure"—*three buttresses*—a "trine" (like the three-pronged spear of Neptune!)—with the center member the longest.

The entire bisymmetrical pyramid was precisely "split down the middle" by this central member of the "trine" (#1)—which, in turn, was aligned precisely with The Most Significant Object on The Landscape—

The Face.

But there was something else . . .

For a long time I'd been vaguely aware of a subtle "familiarity" about this structure; the ratios—1:1.6—were, as alluded to before, very close to that celebrated number, the "Golden Section."

But as I stared at its exquisite five-sided bisymmetry, another striking aspect of this "magic" ratio suddenly appeared before me: Leonardo da Vinci's application of these ancient "sacred" proportions . . . to the human form. And suddenly I comprehended an extraordinary possibility: if I superimposed da Vinci's famous figure—"a man in a circle"—over the stark geometric outlines of the D&M—

The two conformed!

The D&M seemed to be *a striking geometric statement of the human-oid proportions*[1]arrayed on an alien landscape almost in the shadow of the central "humanoid" resemblance . . .

The fact that the "D&M" was *another* bisymmetrical "monument" (the first was, of course, the Face itself), located only a few miles away (in fact, within sight of that remarkable humanoid resemblance) was only slightly less extraordinary than the fact that the line of its bisymmetry—its overall orientation on the landscape—was aimed *directly* at this humanoid "monument!" And now, that it seemed to be a geometric "statement" of the humanoid form . . .

Let the critics of the Intelligence Hypothesis try to explain *that* as another "trick of lighting!"

*          *          *

The "honeycomb," by contrast, did not fare so well, and in fact set up a controversy which would substantially deflect the remainder of the entire Investigation.

One of those present at the earliest meetings, some months before, was an image-processing expert named Gene Cordell, who worked in Silicon Valley for a major computer firm which produced advanced imaging and computer graphics systems. At Ren's request, he carefully examined the enlargements of 35A72 I had prepared, as well as the SPIT-processed original (all $5 \times 5$ inches!) from DiPietro. After this examination (which took about half an hour) he confidently asserted that the "honeycomb" was "a mere computer-processing artifact . . . aligned with, and at right angles to, the scan lines of the original Viking image . . . "

Later, I had given him DiPietro's original data tape to examine . . . and heard nothing further.

Now, on a January night shortly after DiPietro had joined the computer conference from the East, Cordell and Dolphin found themselves in the high-tech, computerized surroundings of International Imaging Systems (where Cordell was currently employed)—with screens, readouts, and ways of manipulating images that made the bridge of the *U.S.S. Enterprise* look like the dashboard of a '57 Chevy! It was there that they got their first look at the Viking images newly-arrived on the USGS tapes . . .

The next morning, following this "quick look" at the pictures, Lambert entered the following report in "Chronicles" for those of us who could not attend the "unveiling":

We paid special attention to the "city area" to the west of the Face by looking at 70A11 with (electronic) filters, all reasonable ranges of contrast and magnification, but were unable to recognize any hint of the streets (sic), rectilinear alignments or honeycomb evident in our old 35A72 full frame . . .

What is needed next is further examination of 35A72 to make certain we are not dealing with an artifact (of the imaging process) . . . By the way, the image-processing capabilities at IIS are most impressive. I did not see anything on this tape that improved upon the 70mm negatives Vince sent us last week, especially with regard to 70A11.

By the time I got up later that same morning, DiPietro (from Goddard just outside Washington, D.C.) had already entered a response to Dolphin's entry . . . and some thoughts of his own regarding "the honeycomb"—which he'd examined on a print we had mailed east a few days earlier.

His conclusion: we were looking at "error enhancements" on the Viking data tapes!

In a somewhat technical explanation of the recording and transmission process of the images, from the spacecraft to the computers here on Earth, he pointed out the presence of "massive pixel 'hits' or errors . . ." which showed up on 35A72 as "salt and pepper." 70A11, by comparison, had very few of these "pixel errors." Thus, DiPietro concluded, if we saw "the honeycomb" on 70A11 it was probably real. But, he added, the alignment of the "honeycomb cells" with the X and Y axis of the scan lines of the original Viking image inclined him also to believe it was a computer "artifact" after all; with so many "hits" on 35A72, the likelihood was that "the honeycomb" was merely the result of a "cluster of hits."

DiPietro then followed this with suggestions for preparation of the "raw" data tapes (those newly-received from USGS), which included electronically "cleaning" them (by an averaging technique—where pixel "hits" would be replaced with an average brightness value, determined by neighboring pixels), and subsequent steps, which ended by applying the SPIT algorithm.

The following morning—January 26—Cordell himself made a series of entries into "Chronicles," describing his examination of the images with Dolphin and *his previous examination of DiPietro's other Viking data tape, somehow "lost" at IIS some months before.*

There is much to be said about these Mars images. I am uncertain where to begin. I do have in my possession a digital tape of frames 35A72 and 70A13. These were given to me by Dick Hoagland, who got them from Vince (I believe). My apologies to Lambert. The frame 35A72 may already have been at IIS (the other night), but according to Chris Walker (another imaging scientist at IIS) *they archived it due to its poor quality* (italics added). I did not realize at the time (two nights ago) that it was the missing frame.

Mr. Walker and I spent several hours examining the image (several months ago). The method we used is as follows . . .

Then Gene gave a technical description of the equivalent of Vince's "averaging technique" to eliminate the "noise" present in the frame. This was followed by additional technical details of various contrast enhancements he and Walker had applied to the image, and the various electronic "zooms" used in an effort to "zero in" on the region of "the honeycomb."

Cordell concluded:

We examined the Face and the City in the resulting image. There were *no honeycomb-like structures* (italics added) anywhere in the image . . .

It is the case that the honeycomb structures found in the SPIT-processed image of the "City" are *artifacts of the processing* (italics added).

Then, in a second entry, he continued,

In regards to the Face and the other interesting formations nearby . . .

In all the images I have seen, the Martian landscape is all of a type. That is to say, that we see the same type of geological structure evident in all the frames. There are many plateaus with mounds. In examining these images we found many 'faces,' some small, some large, none as large as the one known herein as the 'Face' . . .

What we are dealing with is another planet about the geology of which we know very little. It is quite evident that it is not Earth-like . . . There is much about the planet that begs for further exploration. There may yet be evidence of other life forms there. Unfortunately, *the 'Face' is not an example of this* (italics added), but a simple permutation of a consistently manifest geological formation.

What was occurring was precisely what I had wanted *not* to happen—the Intelligence Hypothesis "put on trial" on the basis of *one piece of evidence*.

I said as much in my response:

I come back to my basic premise in organizing this multi-disciplinary Investigation: we must observe the interlocking logic of this picture, balancing arguments about 'scan lines' against the extra-ordinary coincidences evidenced by facial features, pyramids, celestial alignments that 'happen' to coincide with the earliest dawnings of the Human experiment on Earth (!), etc.

There is a bigger picture here.

Our purpose should not be to 'prove' there was once life on Mars, merely to make so strong a circumstantial case that 'something very strange happened there' . . .

*That we go back* (italics added)—

With the technology which can answer *once and for all* (italics added) these fascinating questions. And we have less than two years, until the next Martian 'window' opens . . .

My objective is to have built a 'Grand Jury Case' that will permit the sending of . . . the hardware back to an orbit of the planet, for such detail as is required to resolve these 'pyramiding coincidences.'

Dr. Pozos, seeing what was taking place, concurred with my call for a "big picture perspective" on the data. His entry followed:

I must commend you all for the caliber of the discussion to this point. However, before we get further enmeshed in the honeycomb controversy, it will be important to compare and contrast imaging methodologies (and), general features of Martian geology and natural history. In general, I recommend that we spend some time present-ing hypotheses from our different disciplinary perspectives and out-lining their plausibility as a matter of the conference record before we plunge further into the controversies of prioritizing the likelihood of these hypotheses.

*      *      *

It had been clear to me, ever since I had discovered "the honeycomb," that its presence was both a remarkable plus for the Intelligence Hypo-thesis . . . and a potential pitfall—

Which now, if interpreted as Gene Cordell was attempting (with curious assistance from DiPietro), could be the undoing of the entire Inves-tigation, if not the Intelligence Hypothesis itself.

On the one hand, it was an easy "artifact" to grasp: it looked so damn artificial on the prints!!! No natural mechanism could be easily envisioned which could create such a monstrous "parking garage" of regular-sized grids—and place them so specifically in one location on the Martian surface.

If, on the other hand, "the honeycomb" was some sort of "com-puter artifact" (as Cordell and DiPietro now contended), it made it awfully easy to reject the entire Intelligence Hypothesis—if you were someone who needed to "see" a blatantly "artificial object" on Mars, in order to attribute the rest of the evidence to more than "mere chance."

But many Great Problems in the history of science have depended on a more subtle understanding of the Universe. Discovering the truth regarding the Face on Mars seemed, to me, to be the epitome of one of these "subtle problems": especially at this stage in our technology (and existing Viking data)—dependent on a variety of far more sophisticated factors than the blatant discovery of one "unmistakable" building on the planet. Among these factors were the mathematics of the proposed alignments, the orientation of the various "structures" in relationship to one another, and the discovery of significant architectural ratios (such as the 1:1.6) discovered by DiPietro and Molenaar for "their" pyramid which, curiously, was very close to a fundamental ratio used extensively in terrestrial architecture—the so-called "Golden Section."

Or, as Randy put it in an entry:

> Anthropologically, the proper analysis of these Martian land-forms challenges our ability to reach beyond the conceptual limitations of our species. Generally, our criteria for evaluating—let alone the perceptual structuring—of these landforms and other data sets is based on anthropomorphic criteria developed on one planet . . .
>
> The more fundamental question arises. How do we construct and/or interpret data sets from the terrestrial planets with a legitimacy which overcomes our anthropocentric and geocentric experience.

My point here is not to argue for the reality of "the honeycomb"; it is to point out that, considering the pivotal importance of the question—is it "real?"—based on Cordell's cursory examination *and* the highly suspect source of the data he examined—DiPietro's several-year-old tape—*there was no way to definitively know* "in a few hours examination" the answer to that crucial question—

Without a much more thorough and detailed numerical analysis, extending over several weeks, if not months!

For example:

Take the matter of the "averaging technique" that Cordell applied at IIS to the old tape of 35A72. Was it not possible, I asked, that in "cleaning up the tape" the process also "threw away some of the data?" When you "average" anything, you are basically "smoothing out" the peaks and valleys. If the effect you're looking for is subtle, then even a few percent of "smoothing" might, in effect, make it so smooth as to be "effectively invisible!"

*The Monuments of Mars*

But what about DiPietro's point, that the "noise" scattered randomly around frame 35A72 made it likely that this, in the region of "the honeycomb," had fooled the computer into transposing those mysterious "MSBs," creating a "honeycomb" all right—but only in the computer?

I didn't buy it. If you took the trouble to count the "bit hits" in the region of "the honeycomb" (as I did), you discovered something very interesting:

There were only *one or two* in the entire one square mile—significantly *fewer* than in any other randomly selected area of equal size . . .

What, then, made this one square mile *different*? Why was that "difference" readily apparent on even the unprocessed frames—as some kind of "veiling" overlying not only this area, but regions of the adjacent "Fort" as well?

I didn't know . . . but I wanted to find out. Apparently, I was one of the very few who did.

*       *       *

Curiously, many Team members seemed to take Cordell's "pronouncement" that "the honeycomb" was merely a processing "glitch in the computer" as The Final Word; and, in dismissing it, also dismissed the very existence of The City—as a region to be investigated further!

Evidence of this turned up in Lambert's very next entry, dated Jan. 27 at 6:37 A.M.:

> My approach is somewhat different from that suggested by Randy Pozos, in that I feel other examples of interesting landforms on Mars should be gathered and studied prior to formulation of theories about their apparent origin. The more evidence gathered the better the case . . . I gather, for example, that many, many images of Mars have never really been inspected closely, even in a cursory fashion. Since the geology is so non-earthlike, even more geological studies are likely to come up with lots of surprises, even now.

Which all sounded very reasonable and "scientific"—until you understood the magnitude of what Dolphin was suggesting.

There were upwards of *sixty thousand* (60,000) Viking Orbiter frames, taken during the several-year "extended mission" as both spacecraft orbited the planet. If NASA had not had the resources to examine most of these "even in a cursory fashion" over the previous eight years or so, what could we—still an *unfunded* group of neophytes in Martian studies—hope to accomplish in a *mere few months*—in terms of turning up "inexplicable objects" which would further support the Intelligence Hypothesis? Further, what would we look for . . . if we were casually dismissing the objects in the City—for which there existed at least *some kind of mathematical relationships*? And—

Where would we process the tapes of all those other frames, even if we got them?!

But Dolphin, Brandenburg and DiPietro set up a veritable pipeline of new "anomalous object images" from Viking—from as far away as NASA's NSSDC at Goddard, and the USGS photo library in Flagstaff, Arizona. Exactly what we should be looking for, or how we would know it when we found it, no one seemed to really know.

It was during this period—when the Investigation seemed suspended between continuing financial starvation and a creeping "lack of confidence"—that I remembered I was scheduled, months back, to give a talk at the Institute for the Study of Consciousness* in Berkeley. Its title: *The Case of the Face on Mars: A New Investigation into the Origins of Consciousness.*

That night, as I nervously waited for my slides (which were late in arriving from across the Bay, in San Francisco—where Paul Shay's son had done me an enormous favor by preparing them from Dolphin's 11x14 prints), I asked Randy Pozos to make some introductory remarks, from his perspective as an anthropologist. He responded by sketching a most appropriate frame for the images to come . . .

"The Mars project provides several major opportunities to advance research in the central human questions which underlie the humanities and the social sciences.

"Prescinding from the question of whether the landforms are the product of natural or intelligent processes, the question of other intelligent lifeforms in the Universe raises questions about the nature of consciousness and reflexive intelligence . . . "

(Unknown to Dr. Pozos, one of Arthur Young's own books—*The Reflexive Universe*—was an inquiry into precisely such questions. Randy continued . . . )

"Current research in sub-atomic physics, metaphysics, psychology, and anthropology provide new approaches beyond the mind-body Cartesian dualism and indicates that consciousness may be a part of the Universe *extrinsic* (italics added) to individual minds.

"The challenge of the Mars project to the anthropologist is to formulate reasonable hypotheses that can be performed as thought experi-

---

*Founded by Arthur Young, inventor of the Bell Helicopter.

*The Monuments of Mars*

ments and tested through advanced imaging techniques and actual exploration.

"For example, there are three central questions:

"1. Is reflexive intelligence the product of planetary evolution? In other words, is it independently invented?

"2. Is reflexive intelligence something which can be transmitted or diffused?

"3. What is the relationship between the physical development of an intelligent species and the development of the mind?

"Of course, there are many additional questions which arise. Even in the absence of another civilization—or the remains of one—close by in the solar system or the galaxy, the effort required to examine *the plausibility of a society on Mars* (italics added) provides a central crucible for reinvestigating the eighteenth-century infrastructure of the humanities and social sciences which give rise to our (current Western) concept of ourselves as a species, our public policy, and the meaning and purpose of our current predicament which pushes us to the edge of extinction.

"Whether or not the ruins of a civilization are found on Mars, the project provides a significant opportunity to *reformulate* (italics added) the central questions of the humanities and social sciences.

"Of course, if a lost civilization *is* found on Mars . . . "

I couldn't have said it better myself!

But, I tried—and with the actual images from Viking. My talk took two hours.

Afterwards, as I was packing up, two members of the audience stopped by. One of them, Dan Liebermann, was a long-time member of the Institute—and an architect. He was fascinated particularly by the features of the City. And he was *most* impressed by the honeycomb and its apparent "architectural integrity" with the other features on the landscape.

Lieberman's sharp eye picked out other puzzling features in the "complex of pyramids," including some enigmatic "pie-shaped holes" in the surface just north of "the honeycomb." I'd mapped them, and agreed that—whatever they were—they too looked unnatural . . . and deep—

Otherwise, in millions of years, the drifting sand and Martian duststorms should have filled them in. They had not—indicating that these might be entrances to vast underground chambers, or tunnels . . . in keeping with my central hypothesis: that "the Martians" had to go underground and "inside" the pyramids, that they themselves (and thus the honeycomb) were the ruins of enormous housing complexes . . .

Lieberman and I discussed examples of "sculpted architecture" here on Earth (such as in Tibet), where landforms were literally "hollowed out" to provide living space for people. We also discussed his joining the Independent Team—as a member of the "structural engineering group." I left him with an enlargement of Cydonia, and the promise of better ones.

Then I turned my attention to the second individual . . .

Tom Rautenberg, it turned out, was a social scientist from "Cal" (the University of California, at Berkeley), Administrative Director (under C. West Churchman, one of the world's leading "systems scientists"[2]) of something at the University called "The International Program in Applied Systems Design." Rautenberg described it as "an interdisciplinary group . . . within the University's Business School . . . involved world-wide in applying technology and basic systems science to solving economic public policy problems . . . such as Third World development, nuclear proliferation, and similar concerns—requiring a mixture of social science skills, technological expertise, and a systems approach to management." Included in the group, were "a couple of Nobel Laureates, such as Murray Gell-Mann (who'd discovered 'quarks'—a basic building block of matter!), and a number of other well-known scientists like Jonas Salk—the discoverer of polio vaccine." It was an impressive gathering.

Rautenberg finished by handing me his card, saying, "I'm really intrigued by what you're doing. If I can be of any help, just let me know."

The next day I called Randy, to relate what had occurred. He knew of Churchman, had even met him once—a veritable giant on the Berkeley campus. He was no fool. If Rautenberg was associated in some way with him, he must be "real" . . . and pretty good. But in Berkeley, you never knew . . .

We decided to explore the matter further.

Because Randy was Project Administrator of the Investigation, I had him make the call—to arrange a meeting with Rautenberg for a few days hence.

The get-together took place on a Saturday afternoon, a couple of days after I'd delivered my evening presentation; four or five of us sitting around, with 11x14 Viking prints scattered on the carpet, while Rautenberg described his initial reaction to my talk two nights before.

"At first I thought it was some kind of joke, or maybe a complex social experiment being conducted by the CIA—to study psychological reactions to such a hypothetical discovery.

"I mean—SRI involvement. 'Faces' on Mars . . . ? What would *you*

*The Monuments of Mars*

think?!

Randy and I looked at each other. . . .

"But then," Rautenberg continued, "I said, 'Those photographs are *very* eerie . . . maybe they're for real!'

"When I got home that night—I live in a house with a couple other roommates—it was two o'clock in the morning; everyone was asleep—

"THERE WAS NO ONE THAT I COULD TELL!

"I paced around the room much of that night. Finally, I *had* to tell someone—so I called a good friend I know, a psychologist at the University of Pennsylvania, and we talked for two hours—exploring all the possibilities:

"*Was* this an elaborate psychological experiment, sponsored by the defense community? If not, was it just a hoax . . . or, God forbid! . . . could it be *for real*?!!"

## Notes

1. The resulting five-sided figure is strikingly echoed in the proportions of another ancient form, this one found on Earth: the five-pointed "star" of Ancient Egypt! Egypt is unique in representing celestial objects with this specific form—raising haunting questions of possible "connections" between this representation on the Martian surface and what it may have really symbolized in Ancient Egypt, questions we shall examine in much greater depth in Chapter XV.

2. Churchman, C.W., *The Systems Approach*, New York: Delacorte Press, 1968.

# WE PREPARE TO TELL
# THE WORLD AT BOULDER
# ... SOMETHING

"In science, truth is a moving target . . ."

Michael McCollum
*Procyon's Promise*

Stemming directly from my chance meeting with Thomas Rautenberg that night the subsequent months saw the gradual development of *two* overlapping Mars investigations—

The "formal" one led by Randy Pozos and myself, and an "informal one," spearheaded by Tom Rautenberg as his interest in the data grew.*

Even as the on-going "SRI Investigation" (unfunded and increasingly divided by philosophical differences) added several additional participants, I encouraged Rautenberg to make good on his offer to "show my colleagues the data." I felt increasingly that we needed "new blood," and Rautenberg's offer to solicit additional scientific interest in the Martian engimas addressed this perception.

In the course of Tom's extensive travels (to meet with many leading scientists at some of the nation's foremost universities and research institutions—part of his duties under the auspices of Cal's "International Program in Applied Systems Design"), he carried copies of the enigmatic Viking photographs—

And would casually produce them at appropriate moments, with an

---

*For a more thorough treatment of the initial "Independent Mars Investigation," see *The Face on Mars: Evidence of a Lost Civilization?* by Dr. Randolfo Pozos, Chicago Review Press, 1986.

equally casual, "What do you think of this . . . ?"

The results were as consistent as they were encouraging: every scientist (unless they worked for NASA!) in looking at the images would admit to puzzlement and "no easy explanation" for the features—and the need for a more formal, interdisciplinary investigation.

The experts Rautenberg polled in this manner ranged across the board, from geomorphologists at Berkeley itself, to imaging experts both at MIT and the equally well-known Bell Laboratories, to anthropologists at Harvard University's prestigious Peabody Museum. Some reactions were quite memorable.

One well-known space scientist (formally employed by NASA), upon seeing what Rautenberg wanted to discuss, demanded, "Put those back in that envelope! If you're here to talk about that silly 'face,' we have nothing further to discuss! That was dismissed long ago; there is nothing there!"

When Rautenberg attempted to press him on the specifics of this previous "analysis"—including scientific papers in the literature—the geologist refused to (or couldn't) cite specific references. Rautenberg left with his major questions, including some new sociological ones, unanswered.

As part of our continuing discussions, Rautenberg and I examined (as Randy and I had) a possible "terrestrial connection" for the Mars data. If this all was real (the perennial question!) there could well be some kind of "link"—between this tacitly presumed "extinct culture on Mars," and our own cultural antecedents. Was there any evidence on Earth of this "connection"—Von Daniken notwithstanding!?

One night, after one of these "terrestrial connection" conversations, I got an excited phone call from Rautenberg.

"I've just seen the Face!" he exclaimed. "I was sitting here, watching a 'Nova' documentary on Mesoamerican archaeology, and there it was—big as life! Some kind of expedition to British Honduras or something . . . "

A few days later Tom found himself at Harvard University, on University of California business. Taking advantage of the opportunity, he decided to visit Harvard's famed Peabody Museum—known throughout the world for its archaeological collections and research. Carefully, he displayed the Martian images for several archaeologists and anthropologists, asking the key question:

"Is there anything like this, that you know of, on Earth?"

In conclusion, he brought up the possibility of an eerie resemblance in the artifact he'd seen from Central America, documented in the "Nova" program.

Highly skeptical but courteous, the Peabody staff initially had disappointing news: the leader of the Honduras expedition was still in Central America. He wouldn't be returning for some months.

After some further discussion, one or two of the younger archaeologists agreed to instigate in their spare moments a *very* informal search through the Museum's extensive archives, for other possible examples of terrestrial artifacts that in any way resembled "The Face on Mars."

At this point, a colleague just coming into the office was asked jokingly if she could identify the "face," lying on an $11 \times 14$ enlargement on the desk.

"Sure," she replied, "that's Hanuman, the Monkey God, from India . . . " She had eerily picked one of the pantheon of gods known for bringing wisdom to mankind.

Her reaction, when she was told that it was in fact a "mile-wide 'Face' on Mars" is, unfortunately, not recorded.

*  *  *

But was this eerie "simian resemblance" part of a deliberately designed *artistic aspect* of the Face?

In comparing the figure at the two sun-angles—high (30 degrees—70A13) and low (10 degrees—35A72)—I couldn't help but wonder: was the fact that the monument dramatically "evolved" with changing lighting—from the "simian" resemblance early in the afternoon, into the much more "Egyptian" resemblance I had noted in the evening shot, 35A72—part of a deliberate design—

*A visual metaphor for the literal evolution of the* model *for the Face itself . . . somewhere else?!*\*

*  *  *

Early in the Investigation, I entered some thoughts in "Chronicles" regarding the need for an *artistic analysis* of the Face (this was before Rautenberg's visit to Peabody). It read like this:

> All of the discussion regarding the "reality" or "nonreality" of the Face as an artificial figure has neglected one vital element in any such discussion: there are established *mathematical proportions* for human images—ratios of distances between the forehead, eyes, nose, mouth, and chin—that are evident in art around the world (at least this one!).
>
> What we desperately need is THAT kind of mathematical anal-

---

\*See Chapter II, ref. 3.

ysis of this representation of a human face on Mars.

That was one of the key reasons why I (along with Randy) organized this conference: to get people to contribute that analysis to the on-going discussion of this problem. The same mathematics, incidentally, can be tested in the City—as has been briefly mentioned in an Entry submitted by R. Pozos earlier.

In other words, gentlemen, unless we turn this discussion into a NUMERICAL discussion—measuring quantitatively those things we can measure—we'll continue to "just spin our wheels."

Jim Channon was the first artist to answer my request.

Channon is a "concept designer, and illustrator," currently serving as consultant to many of the nation's largest corporations (such as AT&T). Formerly a Lieutenant Colonel in the U.S. Army (and stationed in the Pentagon), Jim had created something called "The First Earth Battalion" some years before we met—a pragmatic proposal to combine the "spiritual warrior" goals of "the New Age" with the pragmatic grounded methodology of the military services, in a search for ways to lessen international tensions through joint exercises and cooperation.

At our first meeting in Los Angeles, Channon intrigued me by sketching—at first sight—many parameters of the Face that I had discerned only through *weeks* of staring at the image and making careful measurements. After a few days, he broke down the elements of the entire Cydonia complex—City, D&M pyramid, and Face—in an equally incisive analysis, and he produced more detailed drawings of specific relationships plus an *artistic analysis* of the central object at Cydonia:

### THE FACE ON MARS
By Jim Channon

Three elements will be discussed to highlight my findings after a two-day review of the photographs provided by Dick Hoagland.

1. Facial Proportions . . .
   Anthropometry

2. The Supporting Structure . . .
   Architectural Symmetry

3. The Expression . . .
   Artistic/Cultural Focus

### FACIAL PROPORTIONS

The artist uses classical proportions and relationships when constructing the human face. The eyes, for example, are only barely above a line separating the upper and lower face. The physical an-

| SYMMETRY | COMPOSITION | ORDER | GEOMETRY | LIKENESS |
|----------|-------------|-------|----------|----------|
| FUNCTIONAL SETTING | ARTISTIC SETTING | DETAIL—EYELID | DETAIL—NOSE | DETAIL—BROW |

EVIDENCE TO SUPPORT CONSCIOUS CREATION

thropologist recognizes a set of classic proportions, that relate facial features in predictable ways. The features on this Face on Mars fall within conventions established by these two disciplines. I find no facial features that seem to violate *classical conventions* (italics added).

## THE SUPPORTING STRUCTURE

The platform supporting The Face has its own set of classical proportions as well. Were the Face not present, we would still see *four sets of parallel lines circumscribing four sloped areas of equal size* (italics added). Having these four equally proportioned sides at right angles to each other creates a symmetrical geometric rectangle. The photo (70A13) with the 30 degree sun-angle reveals that they are clearly formed above the surface of the landscape. These support structures alone suggest a piece of *consciously designed architecture* (italics added).

## THE EXPRESSION

For the artist, there is yet a more precise way to judge the authenticity of this form. The expression expected *from one powerful enough to be so memorialized by a monument of this scale* (italics added) would not be random. The artistic, cultural, mythic and spiritual considerations behind such a work of art would *demand* (italics added) a predictable expression. The expression of The Face on Mars reflects permanence, presence, strength, and similar characteristics in this range of reverence and respect.

The image appears to be a powerful male about the right age to be a ruler. Materials like stone naturally give an expression of this size a slightly lifeless quality. That is usually a function of the engineering requirements needed to translate an expression to the grand scale seen here.

But, it must be emphasized that the artistic attention required to generate an expression like the one studied is NOT trivial. Very slight changes in the eyes could create an *entirely different kind of character* (italics added). The shape of each feature in a case like this is a matter of *precision* (italics added).

### THIS IS NOT JUST ANOTHER FACE . . .

[There] is overwhelming evidence that the structure revealed in the photographs presented to me by Dick Hoagland is a *consciously created monument* (italics added) typical of the archaeology left to us by our predecessors. I would need much more precise evidence at this point to prove the contrary.

Although I could quibble with some aspects of this analysis (Jim's apparent concentration on the low sun-angle frame, 35A72, except for

*The Monuments of Mars*

a measurement of the "platform" with 70A13—thus no discussion of the apparent "simian resemblance"), by and large it was exactly the kind of "fresh approach" I felt the Independent Mars Investigation needed. (And, in a later series of drawings, Channon would indeed capture the "simian" aspects of 70A13—as you can see by examining the later drawings included here.)

Channon's analysis, combined with the extraordinary independent reaction of the Harvard anthropologist eloquently reinforced Greg Molenaar's own differently worded, but essentially identical conclusion:

> If a mountain clump on Mars looks liked a carved humanoid face, the most simple explanation may be . . . that it is!

All that was now missing were the mathematics necessary to *prove* it!

However, with no new tapes to process or specific numerical measurements to discuss in March of 1984, the flow of "Chronicles" turned to matters of a more *theoretical* slant—

Including the former habitability of Mars itself.

As part of this discussion, I wrote a series of entries entitled, "Some Thoughts on Martian Climate": a backgrounder on the current thinking re past Martian geological and climatological history, and how it would constrain the origin and development of any indigenous "Martians" (see Chapter VII).

At the same time, as part of the "hard-copy image sweep" described earlier, John Brandenburg efficiently found *all* available images that Viking had secured of Cydonia in 1976—which included four new low-resolution shots taken from about *ten times greater distance* than 35A72 and 70A13. (DiPietro and Molenaar themselves, years before, had located two of these, but rejected using them because of their very low resolution. One highlight of Brandenburg's success, however, was that two of the new images were *morning shots:* taken when the sun was coming from the east. Thus, they might at least show the symmetry of the base platform of the Face, when compared with the opposite sun angles.)

On March 5 (according to the record of the computer), in the midst of our discussion regarding the discovery of these four additional (though very low resolution) images of the Face, Lambert Dolphin commented that the SRI geologist, Bill Beatty, "having looked at all the controversial images and [having] read the references on Mars cited in the Bibliography, says he would guess there is a 5% probability that some of these artifacts are not natural . . . "

Dolphin then went on to say that he himself "would vote 25-30% probability, on the basis of what I have seen so far." Apparently the discovery of the new photos, and the confirmation of the basic symmetry of the Face in them, had buttressed his convictions back to their "old" levels somewhat . . .

With this, I thought I might as well commit myself too . . . in print:

My perception that objects we are seeing (are) artificial: 80%. (But then I've had about 6 months to look this stuff over, and do a lot of *measuring*.)

Which prompted Brandenburg later that same day to enter the following:

. . . I find myself, despite my efforts not to, reaching a fairly firm, though preliminary, conclusion . . . It is my judgment that this object (the Face) . . . is an artificial object. I consider this highly likely, after seeing our best data.

But even more significant than John's conclusion was the *means* by which he reached it: a point-by-point listing of "considerations." One of these "considerations" caught my eye . . .

. . . the presence of enormous quantities of *ferrous* iron oxide, hematite, on the Martian surface, giving it its red color. Ferrous iron, rather than black *ferric* iron, is found *only where large quantities of free oxygen are present . . . or were present* (italics added). The presence of ferrous iron in precambrian rocks of later epochs, and its absence in earlier rocks, *is considered by geologists to date the change of Earth's atmosphere from primordial to oxygen-containing, being produced by algae* (italics added).[1]

With no more fanfare than this the "geochemical breakthrough" of the Independent Mars Investigation had begun.

It took several days for the significance of John's "consideration" to sink in . . . But when it finally did, it was a true stroke of genius!

Brandenberg had put his finger on the *one*, blatantly obvious question *most relevant* to any proposed past existence of living organisms on the planet:

*Why is Mars red?*

Or, as I recognized belatedly in an entry on March 12th:

It is my perception that we have just passed a critical "watershed" in this Investigation: the identification of an *oxidized Mars* as a critical separate line of evidence (in addition to the artifacts we have been discussing) pointing toward the one-time existence of some kind of ecosystem on the planet—which, as a natural byproduct, released sufficient free oxygen (from the reservoirs of water and carbon dioxide) to have thoroughly oxidized the basaltic lava flows around the planet.

Inasmuch as this concept may become the critical element in putting our "opposition" on the defensive—by requiring them to present an adequate (nonbiological) geochemical explanation for the highly oxidized environment of Mars—

. . . (this) has opened entirely new avenues for this Investigation.

. . . congratulations, John. It is really nice to see each of us making such fundamental contributions toward an eventual "solution" to this puzzle.

Now, guys, I have a new one:

Where did all the *nitrogen* go?

\*                \*                \*

A brief flashback . . .

Pre-dawn July 20, 1976. Viking Lander 1 hurtles through the Martian stratosphere on its final plunge to a rendezvous with the reddish "sand of Mars." A number of us, gathered in the crowded Von Karman Auditorium at JPL, mentally follow a critical experiment Viking is conducting, even as we wait and bite our fingernails . . .

High above the distant Martian surface—over 20 miles—sensors on the falling spacecraft sample the wisp of molecules present even at that altitude, as the spacecraft prepares to discard its still radiating heatshield and "pop" its parachutes, following a fiery entry into the atmosphere from orbit. As all this occurs, a thin radio link sprays a stream of readings towards the "mother craft"—Viking Orbiter 1—for relay by the Orbiter's more powerful transmitters across the millions of miles which separate this "live experiment" from Earth.

Even if the landing fails, this vital data will duly arrive in the appropriate computers back here at JPL—

And fill in one more Martian mystery left over from before the Space Age even had a name . . .

Is there *nitrogen* on Mars?

\*                \*                \*

Nitrogen accounts for something like 75% of Earth's atmosphere (with oxygen about 20%, and the other 5% made up of water vapor and a few "rare" gases). From the first romantic speculations about Mars as "an abode of life," it had been *assumed* that this odorless, colorless gas was at least as abundant there as here.

Then, in 1965, came real data.

Earth-based spectroscopes are unable to detect nitrogen in other planetary atmospheres (because the Earth's atmosphere absorbs the ultraviolet emissions or absorption "lines" that nitrogen creates), so confirmation of the presence of nitrogen on Mars had to await *in situ* measurements—initially conducted as the first spacecraft (Mariner 4) literally flew behind the planet. This technique—beaming Mariner's radio transmissions *through* the thin Martian atmosphere—allowed planetologists to remotely measure (with some assumptions) the average "molecular weight" of the dominant gasses in that atmosphere, by their effect on the radio transmissions—

Allowing these same planetologists to infer (within some limits) a probable composition for the atmosphere of Mars.

It had been previously known from Earth (via spectroscopic data) that *carbon dioxide* made up a lot of Mars' thin air, but the spacecraft readings that July night in 1965 surprised everyone by indicating that carbon dioxide accounted *for over 95%* of the atmosphere of Mars; *nitrogen* was either completely absent, or present in only very small amounts!

The vast predominance of nitrogen on Earth, and with it perhaps life as well, seemed to be a cosmic fluke! (The uniqueness of our nitrogen quotient was highlighted by later, similar measurements at Venus by other spacecraft—including unmanned Russian landers.)

Now, why is any of this important?

Because *nitrogen* plays a key role in the creation of organic molecules—amino acids and resultant proteins, which make up "life" on Earth. With far less (if any!) nitrogen on Mars, would life—and thus the *billion dollar* mission of Viking itself—be possible?

*      *      *

Even as we listen, this night in 1976, to Viking's whispered signals—which left Mars at the speed of light several agonizing minutes ago—the melodic pipings, echoing through the Auditorium, contain the answers to one of the last haunting Martian questions—

Is there nitrogen on Mars?

A few hours later, when the scientists have decoded the results of their first *direct* sampling of the atmosphere of Mars, the answer is a qualified "Yes—but . . . "

The "but" is the amount, only about 2.5%.

Even after correcting for a different geological history, which has led to the overwhelming current percentage of carbon dioxide in the atmosphere (compared to Earth), the amount of nitrogren on Mars still seems to be *less than half* the abundance that would have been expected by comparison (again) to the Earth.

Theories as to "why" instantly come forth, including a very innovative one—that the nitrogen's been "lost to space." But for almost ten years no one will really know . . .

. . . until John Brandenburg points out that Mars is red.

*     *     *

I rushed to the computer terminal and typed:

> . . . in light of our current thinking on the subject, namely that significant geochemical modification of Mars has occurred (via oxidation) from the presence of previously unappreciated biological processes, I would like to propose that the 'missing nitrogen' also has a *biological* explanation:
>
> That the nitrogen has been 'fixed,' via blue-green algae [the same ones responsible for the (former) free oxygen] in the sediments of Mars, and is now 'hiding' far beneath the surface materials available to the sampler on Viking.
>
> If true, this could be a separate geochemical test of the one-time presence of *living organisms on the planet*, since nitrogen forms a major fraction of amino acids and proteins. In fact, if this scenario reflects reality, the current low value of nitrogen in the Martian atmosphere can be used to estimate *how long biology proceeded on the planet*!

Viking, in addition to measuring gross quantities of atmospheric constituents, was also able to differentiate their *isotopes* (elements with the same number of protons in the nucleus but different numbers of neutrons). Isotopes of the same element differ in weight, which allows their detection in a gadget which, in essence, literally weighs individual atoms in an atmosphere—such as that of Mars. [The technological miracle which the Viking engineers performed was building such a device (called a "gas chromatograph-mass spectrometer") to space-rated specifications, including sterilization, packing it into less than a cubic foot of space, then sending four of them to Mars (two in the "entry aeroshells" which entered the high atmosphere protectively wrapped around the Landers, and two in the Landers themselves)—where the latter performed flawlessly for over an Earth year—returning repeated detailed element and isotopic analyses of both the atmosphere, and surface samples of the Martian dust and dirt.]

These isotopic analyses performed by Viking—our only reliable compositional measurements conducted on the surface of the planet—would become critical in our "Chronicles" discussions, regarding the likelihood of life on Mars as well as some of the "wilder" scenarios for how it might have died (see Chapter IX).

Dr. Mike McElroy, of Harvard University (another member of the original Viking Science Team), had published an analysis of the Martian atmosphere in which some key Viking isotope measurements figured heavily—sample runs in which an excess of N-15 (the heavier atom, compared to N-14, the "normal" isotope of nitrogen in our own atmosphere) was noted. Dr. McElroy's conclusion: earlier in Mars' planetary history there had been a much more abundant atmosphere (including nitrogen) "outgassed" by volcanic activity on the planet. As the eons passed, the impact of the sun's solar wind (the highly energetic, very tenuous stream of hydrogen atoms blowing outward through the solar system from its surface) impacted the upper atmosphere of Mars. This led to an acceleration and escape from Mars of more of the lighter isotopes (N-14), compared to the slightly heavier (N-15)—leading, according to Dr. McElroy, "to a selective enrichment of the heavier isotope on Mars, compared to the ratio of these two isotopes on Earth." [2]

In other words, the lighter nitrogen was selectively kicked out of the atmosphere of Mars (because of the lesser gravity—38%) by billions of years of exposure to the ultra-thin (but high temperature!) "extended atmosphere"—the corona—streaming off the sun.

At the time it was first proposed at JPL (a few days after Viking's initial atmospheric measurements came in) it sounded nifty. But, in the wake of John Brandenburg's significant notation of the Martian color, and attribution of this blatant fact to a *past biology* upon the planet, I couldn't help but wonder if the strange nitrogen anomalies on Mars couldn't also have their ultimate answer in *biology*—as opposed to exotic solar physics.

Was there, I speculated, some heretofore overlooked preference in microorganisms *for fixing N-14*, over its heavier cousin, N-15? [Biology on Earth is known to have such a selective preference for certain isotopes—carbon 12 over carbon 13, oxygen 16 over 17 or 18 (depending on the temperature), etc.] But, even if there were no such "isotopic biological preference" in terms of nitrogen, what about the overall reduced abundance of *all* nitrogen on Mars—compared to equivalent ratios on Earth?

Was the decreased abundance of this key element not only a direct by-product of a past, vigorous biology on Mars but, furthermore, a vital clue to a biology which had existed *late in the history of the planet*—at a time when natural vulcanism was dying out (thus not replacing nitrogen

*The Monuments of Mars*

lost by biological processes with new atmospheric nitrogen)?

*　　　　*　　　　*

John's marvelous new insight—why is Mars red?—synergistically triggered Dolphin to ask Bill Beatty the same question, prompting a timely entry in "Chronicles" regarding the role of blue-green algae in liberating the free oxygen we take for granted here on Earth . . . and the need for some similar *past* mechanism to account for "rusted rocks" on Mars.

That, in turn, would nudge memories re the nagging nitrogen anomalies, and *their* possible significance to the critical question of a possible past biology on Mars.

Dramatically, in our search for tell-tale patterns, our attention fell—not on pictures made from orbit as for months previously—but the treasure-trove of hard scientific evidence hitherto buried in the Viking *Lander* data. This evidence, containing a multitude of specific findings on the present (and hopefully, the past) environment of Mars, including vital data on the elemental composition of the surface soils and dust, would be invaluable for constraining theories as to "past biospheres, previously wetter epochs," and the like.

Put together with some equally vital clues from the Viking orbital photography (like a "sea" of "youthful" sand dunes ringing the polar regions of the north), such data could make a convincing case for a remarkable—and perhaps even recent!—transformation in the environmental history of Mars—including a recent epoch of biology.

Brandenburg, following his crucial "Why is Mars red?" entry, produced another winner. Sifting through the voluminous papers published in the wake of the Viking mission, he found the report of Viking's surface chemistry analysis and inserted a brief summary:

> . . . soil full of maghematite, minor mineral on Earth found in red ocean clay, fully oxidized *iron-ferric red* (italics added) . . . soil consists of magnesium, aluminum, iron clays, formation from basalt by weathering indicated. Best match to Earth minerals: nontronite, montmorillonite, saponite and kierserite mixtures of clays . . . *nontronite occurs in deep ocean* (italics added) . . . H. Masursky et al: lots of water flowing in past and intermittent recent epochs. Confused record. 'No beaches so no oceans.' Need to have higher atmospheric pressure in past to allow flows . . .

What struck me about this entry was what John—almost casually—had noted about the iron-rich clays proposed by the Viking team to account for the surface materials sampled by the Viking Landers: that they were also found on Earth "in deep ocean clays."

Suddenly, Mars' mysteries all seemed to be coming together.

*     *     *

If you drive through portions of the Southeast, the western foothills of the Carolinas or Georgia in particular—hundreds of miles from the nearest ocean—you'll find mile after mile of this iron-rich, brick-red clay. It's a very striking, stark contrast to the dark green pine trees that seem to thrive on it.

Hundreds of millions of years ago, before this portion of North America was scrunched up against Africa by the movement of the two massive "plates" that was then carrying these two continents together, this portion of the United States was under water—part of an extensive continental "prism" of deep ocean sediments and muds washed off the continents themselves. In those muds were clays, mixed with iron compounds, with names like "nontronite" and "maghematite," that in the far future some Viking scientists would suspect they had found on Mars . . .

In a couple of hundred million years, as North America and Africa collided and the continental shelves between them were crumpled high above the ocean waters like ripples in a massive rug, the layers of clay and sediments—now transformed into miles-high layers of sedimentary rocks—would become a set of mighty, craggy mountains—only to be attacked by the same rains and elements which had deposited their original muds so many megayears before. Eventually, the assault on these great mountains—the Appalachians—would leave behind nothing but a set of parallel, eroded ridges (the Great Smokies)—mere "ripples of the ripples"—as the sediments that made them, containing the clays and oxides of iron, were continually washed by the rains across the vast, gently sloping red-stained continental edges (the "Piedmont") towards the unchanging sea once more . . .

*     *     *

Brandenburg had apparently uncovered a crucial analogy between this cycle of events on Earth . . . and possibly similar events on Mars; hard evidence that something similar to the unending hydrological cycle I have just described, in chemistry if not geophysics, *had* to have happened on the planet Mars at some time in its past. That, even if Mars does not possess the equivalent of terrestrial plate tectonics ("continental drift"), it at one time—judging by the remarkable similarity of clay-like compounds on both planets—once *must have had something like an ocean*, despite what Viking scientists (like Hal Masursky) claim!

This bit of research on Brandenburg's part, that the "Viking clays" thought most likely to account for the spacecraft's measurements on Mars were apparently "the same ones formed in deep ocean waters here on Earth," was all we needed to put the finishing touches on a host of disparate ideas we in "Chronicles" had kicked around for weeks.

*       *       *

I myself had been pondering some of them for *years*—ever since I'd flown 3000 miles to JPL from New England, many years ago in 1971, to watch firsthand as Mariner 9 shocked the planetary community with its revolutionary views of the "*two* Mars's"—

The one apparently quite ancient, heavily cratered and almost uneroded; the other, only lightly pockmarked, yet with vast level areas of fractured, mottled plains—literally capped by a strange, exotic "sea" of sand encircling an equally exotic layered glacier for a pole . . .

Two hemispheres, with about the only thing in common being that they were on the same small world:

The planet Mars.

Or, as one of Lambert's future entries would phrase it: "Why should half the planet look like the moon and the other half not?"

Now, with Brandenburg's assistance, answers to old questions seemed quite near . . .

*       *       *

Spurred on by the specific identification of the Martian soil with ocean-bottom clays on Earth, I excitedly typed:

> In response to Lambert's latest entry . . . "Today the planet is half cratered and half covered by lava flows with fewer impact craters . . . Why should half the planet look like the moon and the other half not?" . . .
>
> I don't believe in the extensive lava flows postulated to make up the northern plains of Mars (the much less cratered half of the planet). I believe instead that this vast northern plain was once the bottom of Mars' *primordial ocean* (!)—and that sediments eroded from the ancient cratered highlands to the south partially filled this vast basin with material removed from Mars' "other half." Subsequent to these events, the rapid combination of the primordial carbon dioxide with the surface (in the presence of abundant running water) created great quantities of carbonate rocks—which reduced both the pressure (and the "greenhouse properties") of Mars' new atmosphere, leading to a rapid decline in surface temperatures—

which ultimately froze the planet . . .

The absence of "beaches, etc." doesn't bother me unduly, as in several billion years such evidence of an original ocean/land interface would surely have undergone much modification—if only by wind erosion and deposition of the drifts of dust we find covering the planet.

Now I brought in some of the other evidence I'd been quietly accumulating from the orbital photography . . .

. . . of significance to this model, I feel, is the presence of the "great dune sea" surrounding the north polar regions (and ONLY the north pole), which is composed of almost a million square miles of something eroded from somewhere else on the planet. It is my contention that these are the *original sediments* from Mars' vast "northern hemisphere ocean," dried and blown north by the prevailing winds over the immense amounts of time since the ocean *was* an ocean!

. . . thus, in concluding this point, the half-and-half planet is quite likely (in my view) the direct result of the creation of Mars' original primordial ocean. Without plate tectonics to subduct the original sediments eroded off the other half of the planet (even in a million years), this material would simply fill the basin, smothering the ancient craters under a very deep layer of mud!

The beauty of this theory was that all the pieces fit—now that Brandenurg had identified the last missing component: the extraordinary similarity of the dusty plains of Mars to the clays formed in the *known* oceanic conditions here on Earth.

*       *       *

One of the many enigmas posed by the Viking surface analysis (besides too much sulphur and chlorine, compared to samples here on Earth) was the blatant *lack* of something almost everyone had readily expected before the Viking mission: *nitrates* in the soil. Nitrates are compounds formed when nitrogen combines with oxygen (usually during thunderstorms, from lightning, producing nitric acid) here on Earth.

The "missing nitrates" were an embarrassment to both the Viking Team's models of what "should have happened" to the atmosphere of Mars (including McElroy's "escape" model), as well as mine. For, if I was predicting that the missing *atmospheric* nitrogen was really locked up within the soil in dead microscopic organisms, I also had to explain why Viking failed to find them!

*       *       *

Now, neatly, with Brandenburg's linkage of the clays of Earth and Mars, many of these nagging mysteries nicely came together, impelling me to enter in Chronicles:

> . . . one obvious answer (now!) to the question "Why didn't such nitrates show up in the Viking sampler data?" is simply that Viking was measuring the sediments deposited much earlier in "Mars' ocean," long before such a process (biology) took place. The freeze-drying of those sediments, and their eventual exposure to winds which would blow the stuff around the planet, would naturally preclude nitrates mixed in—as the biology (probably, as on Earth) took place on land and in the shallows near the shore!

Viking landed at two places on the planet: Chryse and Utopia, each separated from the other by almost half the planet. But—

Both landing sites were in the *northern* hemisphere, and at least a couple miles *below* the "mean datum" (the arbitrary "zero elevation" line NASA cartographers have drawn around the planet, roughly corre-ponding to "sea level" here on Earth). If you, in fact, filled the portions of the surface below this "mean datum" with an ancient ocean, then both landing craft would have been about two miles *underwater*. [3]

In other words, in the parlance of the ideas discussed here, Viking had landed on and sampled soil from *the bottom of what would have been the ocean*, if Mars had ever had one. To wit:

> . . . the surprising abundance of chlorine in the Chryse samples, not explained by the Viking team, seems evidence to me that Viking was sampling stuff which had formerly been on the bottom of a salty (!) ocean. The presence of so-called "duracrust" around the space-craft, as a chemical "cementing process" from salts seeping upward from below, is also consistent with this concept.

Shades of the immortal Edgar Rice Burroughs' "dead sea-bottoms of Barsoom . . . "

*    *    *

Within days of this discussion in our far-reaching electronic computer Investigation, "Martian Chronicles," NASA's more traditional forum for such discussions—the Fifteenth Annual Lunar and Planetary Science Conference, held each year since the first Lunar Landing in Houston, Texas—hosted researchers in a variety of planetary fields. One of the papers caught my eye: H. P. Jones' "Sedimentary Basins and Mud Flows in the Northern Lowlands of Mars"—

In other words, a paper on *potential former oceans* (or at least lakes!)

on Mars—by one of the "community" itself.

*        *        *

Despite this series of resounding successes for our "Independent Mars Investigation," the relevance of all of this to our original prime focus—the verification of potential "artifacts" on Mars—became, in the rush of new events, increasingly obscured.

Because of a continued lack of funding, "Chronicles" had continued as a "free" conference—supported by InfoMedia itself, to the tune of several *tens* of thousands of dollars of computer and "connect" time. This situation now approached the critical.

One afternoon, toward the end of March, even as I was typing an entry into the computer, the system abruptly disappeared—

Chronicles had "crashed."

## Notes

1. Cloud, P., "How the Air Became Breathable." In *Cosmos, Earth and Man*, Chapter 10, New Haven: Yale University Press, 1980.
2. McElroy, M.B., Yung, Y.L., and Nier, A.O., "Isotopic Composition of Nitrogen: Implications for the Past History of Mars' Atmosphere," *Science* 194:70, 1976.
3. *Controlled Photomosaic of the Mare Acidalium Southeast Quadrangle of Mars*, Miscellaneous Investigation Series, U.S. Geological Survey, U.S. Government Printing Office, 1981; Batson, Bridge and Inge, *Atlas of Mars: The 1:5,000,000 Map Series*, NASA SP-438, U.S. Government Printing Office, 1979.

# XI

# REACTION . . . INCLUDING FROM THE SOVIETS

"& if we do not get up and destroy all the congressmen
turn them into naked men and let the sun shine on them . . .

we will walk forever down the hallways into mirrors and
stagger and look to our left hand for support &
the sun
will have set inside us. . . ."

Robert Kelly,
*The Alchemist to Mercury*

The upcoming July, 1984 "Case for Mars II" Conference was to be held (as the first) in Boulder, at the University of Colorado. Its sponsors were to be the Mars Institute of the Planetary Society; the Boulder Center for Science and Policy; the National Space Institute; and the American Astronautical Society—

In other words: the "who's who" of the traditional planetary science and space-interest communities.

To expect that we—an *independent* research group, and one claiming to have found evidence of a former "inhabited epoch for Mars"— would be welcomed as equals into this tightly-knit assemblage, in the traditional spirit of "scientific neutrality and a willingness to see evidence before passing judgment," would, I felt, be just slightly optimistic—based on the record of this same Conference in the past.

But one can always hope . . .

The first hurdle was the specific form of presentation; scientific conferences are traditionally divided into two main agendas: "conference sessions," where speakers orally deliver their material to a group of colleagues; and "poster sessions," where the papers are displayed on poster boards for "stand-up" presentations to an informal circle of colleagues (usually during quick coffee breaks from the regular conference session presentations!).

Whether a paper becomes a "conference session" or a "poster session" is strictly up to the chairman, in this case Carol Stoker, a Ph.D. candidate in space science from the University of Colorado.

Stoker, however, in her first conversation with John Brandenburg (whom we had decided would actually present the paper) made it very clear that she viewed the topic as "so peripheral to the main topic of the conference—future colonization of Mars," that the best she could do was a poster session for it.

This assignment was, to a certain degree, the result of our own indecision as to what to *call* the paper (part of the even larger problem of what should in fact be in it!)

[John initially sent in a title reflective of a continuation of DiPietro and Molenaar's work—"Unusual Martian Surface Features IV: An Investigation of Some Sites for Manned and Unmanned Exploration." Considering how the initial Case for Mars conference in 1981 had treated DiPietro and Molenaar (by preventing them from presenting even a poster session paper!), I felt this was a mistake; I strongly urged a title reflective of the *new* developments—both from my own work and the Team of new researchers: "Preliminary Findings of the Independent Mars Investigation Team: New Evidence of Prior Habitation?" This title, in my opinion, was both more in keeping with the general theme of the entire "Mars II" Conference, and accurately reflected the results of our own deliberations: namely, colonization of Mars . . . or, more precisely, our evidence for *former* colonization . . . ]

However, rather than cause a fuss regarding our placement in a "mere" poster session, we decided (after some discussion) to let the matter ride. We'd already done far better than DiPietro and Molenaar had in 1981; at least we were *in* the Conference! That kind of "official" recognition had not been accorded the "intelligence hypothesis" in the entire eight years since the Viking mission. We were definitely making progress.

Having passed the first hurdle, we were now confronted with another: the somewhat bothersome detail of what to say about our findings!

Science is supposed to be "objective," "value-free," etc. That's easy when the subject is remote and uninvolving (like measuring the accuracy of atomic radiations in some distant galaxy). When the subject is a little

*The Monuments of Mars*

closer to home, both psychologically and physically—as finding evidence of a former civilization on a nearby planet, and one that looks suspiciously like us!—expect some . . . difficulties.

We had them.

Eventually, however, we reached the next stage of agreement—as reflected in the eventual title of the abstract (condensation) of the paper sent to the Conference in late May, for distribution in the official pre-Conference document:

"The Preliminary Findings of the Independent Mars Investigation Team: New Thoughts on Unusual Surface Features."

Drawn in the best "scientificese" that we could muster around so explosive an investigation, the abstract itself concluded in matter-of-fact tones,

" . . . based on these considerations, it is the consensus of this investigation that, to date, no compelling natural model has been put forth explaining the variety of unusual features discovered on these Viking frames. We go on to suggest: the possibility that these objects are the remains of a past civilization can no longer be discounted."

We were a distinct embarrassment to the organizers of the Conference. Which was very strange; I remembered distinctly that at the first "Case for Mars" Conference the same organizers took perverse pride in being "outside" the Planetary Establishment, in championing an (at the time, in 1981) unpopular idea: that "we should be getting on with getting back to Mars."

And for the (then) most outrageous reason: *colonization*!

They even called themselves "the Mars Underground," a badge of identification for those who shared this crazy dream of the "fringe of the planetary community." But then, it was also these very same self-proclaimed "radicals" who had forbidden official presentation of DiPietro and Molenaar's paper on the Face at the 1981 Conference . . .

Apparently, in 1984 the same "rules of etiquette" still applied.

Reactions to our presentation ranged from the polite ("Oh, that's interesting . . . ") to the absurd ("But . . . where are the roads!?").

Several old friends, whom I had warned by telephone regarding the subject of our paper, acted even stranger; one well-known space expert not only refused to listen to Brandenburg's presentation, he also refused to have lunch with me to discuss the data and its implications . . . yet, months later, in an equally well-known national magazine (*Omni*) he claimed to be conversant enough with the hypothesis to be able to refute it, by claiming " . . . Hoagland's assertions (regarding confirmatory images and sun-angles for the Face) just don't match the facts."

Rautenberg wasn't having much more success with members of the planetary community. He tried to approach Chris McKay, head of the Mars Institute (one of the sponsors of the Conference) on behalf of a follow-on "Mars Investigation Group," to be based at the University of California. But when he asked, "Would you like to join a new, far more formal inquiry into this material?", McKay's reply was essentially, "What could you possibly do with these images that NASA hasn't done already?"

This idea—that if NASA hadn't found anything of interest on these Viking frames, how could we?—was a pervasive reaction of most of the planetary specialists at the Conference.

My response, "Well, first you have to be willing to *look*!" didn't seem to win too many converts.

Even a more substantive "But NASA *hasn't* extracted all the data from these images, as evidenced by what DiPietro and Molenaar were able to accomplish with their limited resources" also failed to climb the hurdle of "the infallibility of NASA."

[The later "imaging teams" of the Mars Investigation Group—the best in the world, but (significantly) *not* employed by NASA—actually discovered new information on these same data tapes, by applying state-of-the-art techniques totally unknown, even to NASA in 1976.]

What seemed behind these dismissals of our efforts was something almost pernicious: not-so-subtle hints from these scientists that we were infringing on "their" territory. The message seemed to be: "only we have the necessary expertise to look at images of Mars . . . and *especially* to conclude whether or not there has been life there!" (Or, as Carl Sagan not-so-subtly told John Brandenburg: It's not whether you're right or wrong, sir. You have not even entered the discussion . . .)

Equally disturbing (to my journalistic sensibilities) was the reaction of the smattering of *reporters* covering the Conference. By and large they seemed content with following NASA's "party line" when it came to our provocative results, as opposed to the kind of investigative reporting (when confronted with NASA's own self-serving insistence there was nothing to our evidence) one might have expected in the wake of Watergate . . . After all, who had the most to gain by denying that we were "on to something," if not the very government agency which had spent a *billion dollars* of the public's money on the "search for Life on Mars" . . . and may have failed to find it? What if we—an "outside" group—had accomplished what all the official effort hadn't? How would NASA (to say nothing of the individual planetary scientists) look—to both the Congress and the public?

But no journalist covering the Conference seemed interested in these potential underlying issues surrounding NASA's continued treatment of the Mars Investigation—which was as an annoyance they hoped would go away.

Instead, the writer assigned to cover the Conference from *Discover* magazine (incidentally, an astronomer himself), in his final write-up of the details of the Conference failed even to *mention* that our paper had been given—despite the fact that the authors represented such reputable institutions as SRI International, the University of California, and Sandia Laboratories. The simple fact was: we weren't "in the club," ergo, we didn't exist. Sound familiar?

[This Orwellian effort to "unwrite" history later extended to the Proceedings of the Conference itself. After requesting a copy of our Team paper, for inclusion in the post-Conference publication, the Conference Papers chairman (Chris McKay) somehow neglected to include it (40 pages, 60 citations) in the final Conference document. It's as if we'd never been there.]

But, I shouldn't leave the impression (before I leave Boulder!) that our experience was negative throughout.

The beauty of conference-going is what occurs *outside* the meeting halls and formal sessions. My real reason for going to Boulder (as opposed to staying home and allowing Brandenburg to present our joint paper) was to get a chance to discuss the data face-to-face with several key people who would also be there—people who could be critical to the next phase of the research—

And to our long-term strategy of getting back to Mars.

One of those was Hal Masursky, senior geologist with the Astrogeology Branch of USGS, in Flagstaff, Arizona.

Hal and I had known each other distantly for almost fifteen years—since the late 1960s, when I joined CBS News and began covering the Apollo and Mariner programs for Walter Cronkite and the Special Events Unit. Masursky had always struck me as a preeminent scientist in the best tradition of the name—a careful, yet intensely curious individual—having both great integrity and courage to explore and defend initially unpopular ideas.

Hal had been at the forefront of the "Martian channel" controversy following Mariner 9 in 1971, arguing forcefully for the one-time existence on Mars of *liquid water* (during a time when even to mention the possibility of running water on the planet was to risk scientific excommunication by most planetary scientists).[1] Masursky even went further, gleefully pointing out at meeting after meeting Mariner (and, later, Viking) images consistent with a former "warm, wet epoch," but even worse, features on those images suggestive of causes identical to those that carve similar river valleys here on Earth—

From rainfall![2]

Since we needed several first class geologists on the new University of California Mars Investigation Group, and at least one familiar with the "standard" Martian geological problems and controversies, my unquestionable first choice was Hal Masursky. The fact that he was going to be at the Boulder Conference, delivering a paper on "Candidate Rover/Return-Sample Landing Return Sites" made it almost perfect; not only would I get a chance to see him and present our case for following up the "preliminary findings" of the Independent Mars Investigation, even if we didn't get Hal to agree to become a "player" on the forming Berkeley Team, we might get a crack at influencing him to recommend a future return mission to Cydonia—if and when the United States ever got around to going back to Mars.

So it was with ill-restrained enthusiasm that I stood beside Hal Masursky in the "poster room," just prior to Brandenburg's 10-minute presentation, and pointed to an 11 × 14 blow-up of the five-sided D&M pyramid, asking quite directly:

"Can you explain *that*?!"

Masursky didn't say anything for several minutes, just stood there—looking at the array of "anomalous objects." After Brandenburg was finished, he turned to Tom Rautenberg and me and suggested a future discussion.

Another interested party was Brian O'Leary, a former scientist-astronaut, selected (ironically) for the astronaut program specifically to go to Mars (when NASA was in the business of considering such adventures, in the heady days of the Apollo program). He also was a planetary scientist of no mean reputation—having served as Deputy Imaging Team Leader of the Mariner 10 unmanned mission to Mercury, in 1974.

O'Leary also had another aspect in his background which prepared him to assess some of the more overwhelming engineering implications of Cydonia: he had once worked as Research Associate to Gerard O'Neill, the Princeton physicist well-known for his studies of artificial "colonies" in space—structures which ultimately were planned to enclose several *cubic miles* of volume.

This varied experience made him (in my opinion) an ideal candidate for a new Study. The more people we attracted with "generalist backgrounds," the faster we would structure an investigation capable of solving this extremely complex puzzle. His expertise on the vidicon television cameras Viking carried (essentially the same ones Mariner 10 had flown

*The Monuments of Mars*

to Mercury) was exactly the kind of "planetary" technical experience the investigation needed. [My hope was that, by comparing the light reflected from the objects at Cydonia at the two sun angles (and by knowing how the cameras would respond to such changing angles), we might actually *see* which objects were reflecting light in an *artificial manner* ("sunglints off metallic surfaces," etc.) as compared to natural rocks.]

O'Leary's reason for being in Boulder was to present an idea almost as radical as ours: the concept that the best way to reach Mars is to forget about going directly to the planet . . . and, instead, plan an expedition to its *moons* . . .

Called the "PH.D Proposal" after *Phobos* and *Deimos* (Mars' two diminutive, asteroid-sized satellites), the innovative concept was the brainchild of another planetary scientist's 30-year interest in the tiny moons of Mars: Dr. Fred Singer. Singer had electrified the previous Conference, by presenting his highly unusual proposal in 1981. O'Leary, in Dr. Singer's absence, was scheduled to deliver a new paper on the subject, as well as lead a workshop for interested participants.[3]

Our "extracurricular discussions" outside the Conference proper extended well into the evenings.

Also present and "raising a few rounds" one night was Duncan Lunan, writer and former President of the Association in Scotland to Research into Astronautics (ASTRA). Lunan has authored several excellent books over the years about our exploration of the solar system, and had himself been briefly bathed in the intense limelight of "extraterrestrial notoriety"—because of his former theory that delayed radio "echoes" observed at intervals throughout the 1920s in fact were "deliberately beamed transmissions from one or more 'Bracewell probes' somewhere in our own solar system" (see Chapter VI).

Lunan ultimately "reconstructed" a message from one of the "probes"—ostensibly a star-map depicting where the craft originated (somewhere in the direction of the constellation Bootes, I believe). Eventually, however, Lunan decided the whole theory was too tenuous and too fantastic—and (most importantly) simply not supported by the data; he officially recanted his "discovery" in the *Journal of the British Interplanetary Society* a few years later. But while it lasted, the fuss and furor around Duncan Lunan's apparent decoding of an extraterrestrial SETI signal caused quite a stir—

(When new images eventually do arrive from Mars some night—spectacular close-ups of the baffling objects at Cydonia—in the event that they prove we've all been excited by a mere "pile of Martian rocks," I hope that I will be as graceful in admitting my own errors as was Duncan.)

Another area of great interest to those of us who had traveled to Boulder was the delicate question: how could we actually get *new* data—pictures!—of the Cydonia region of the planet?

There was one tiny problem. No one—NASA or the planetary science community—seemed interested in better pictures!

There were many scheduled presentations in Boulder on NASA's *sole current plan* to actually return to Mars—an unmanned spacecraft in the early 1990s called Mars Geoscience/Climatology Observer (MGCO), now shortened to simply "Mars Observer." But each of these discussions only pointed up *the one glaring omission* from its proposed complement of scientific instruments:

No television cameras!

Even before I'd gone to Boulder, Mike Carr (former head of the Viking Imaging Team), had expressed his own opinion on the subject to me:

"We have enough pictures of Mars."

So, presentation after presentation in Boulder extolled the virtues of "infrared geochemical scanners" and "radar altimeters and mappers," that would all return tons of new Mars data, some of which could ultimately be assembled into "false color images." But all equally stressed the fact that these false color images would be extremely limited in resolution: the smallest objects they could "see" would be about *a mile* in length.

The entire Face, in other words, would be *one pixel* on these "pictures!"

The plan not to carry a visual (and high resolution) television system on the spacecraft was as much *political* as scientific. Scientists with other instruments (such as atmospheric spectrometers, etc.) over the years had been more and more annoyed with how "imaging" tended to snag not only the major share of public attention . . . but also the communications "downlink" from the spacecraft. With MGCO, they banded together in almost a "palace revolt" (when confronted with the harsh realities of vanishing NASA budgets for *all* unmanned planetary exploration) and deliberately *excluded* a television camera from consideration as "too expensive to be flown aboard this low-cost planetary spacecraft" (MGCO was supposed to cost less than 150 million dollars—compared to Viking's *billion* dollar price tag).

So, if a minority-group of NASA scientists had their way, *no American television cameras were going back to Mars* on the only *funded* mission—a situation that many of us on the Independent Mars Investigation Team felt simply must be changed. The problem: how?

By an intriguing coincidence, just next door to these "official" recitations of the instruments that MCGO would someday carry back to Mar-

*The Monuments of Mars*

tian orbit, I discovered another poster session, this one with the provocative title: "A Mid-infrared Spectrometer and Very High Resolution Camera for MGCO." Authored by Jeffrey Moore, of Arizona State University, the paper went on to describe two proposed instruments that could be flown aboard the Mars Observer spacecraft. One—the so-called "Malin Camera" (named after the Arizona State University geologist who designed it, Michael Malin)—would "have *meter-resolution* (italics added), thus offering an order of magnitude improvement over the best Viking orbiter imagery . . ." We could practically read the license plates with that one!

The paper went on to add, " . . . the relatively low cost of this camera makes it viable for *private interest group support* (italics added) as an alternative to government funding. Moreover, the public interest in Mars exploration would be well served by a camera aboard MGCO."

*Amen!*

A little research (a conversation with Jeffrey Moore) soon revealed that the entire camera—development and production models—would probably cost about $5 million—well within reasonable estimates for "private interest group support."

Now some truly radical ideas, ideas that some of us had mused about before arriving at the Conference, began to seem more real . . . These ranged from the relatively simple (raising public support and funding for inclusion of the Malin camera on MGCO) to the most exciting and extreme:

Raising public funding for a *fully private unmanned mission*!

Could that be done—technically, if not financially? A George Lucas spectacular culminating in a live broadcast of the Face!

It was my impression that it could—provided one found the proper management and technical support.

What is generally *not* appreciated currently is how the "technology of space flight" has matured in the quarter century or so since we've been sending spacecraft out across the solar system. "Private" experiments and spacecraft, from the radio "ham" satellites called OSCAR to experiments designed and built by high school students and flying in the Shuttle, are almost commonplace these days (and for about 3% of the cost of NASA-built spacecraft and space experiments of equivalent complexity!). One enterprising effort based at Rensselaer Polytechnic Institute in Troy, New York, called the "Independent Space Research Group" is even methodically gathering resources and building the world's first *amateur* space telescope—for flight in Earth orbit sometime in the future. This will be the first effort to apply the enormous cost-savings of a "private" space project to strictly space science.

The day after Brandenburg delivered our preliminary findings, one of the managers of Orbital Sciences Corporation, developers of the first privately-financed upper stage rocket-system for the Space Shuttle (for launching spacecraft into extended orbits, including towards the planets) gave a presentation called "TOS: Low-cost Delivery to Mars." Following his presentation, I managed to corner the paper's author just outside the hall: Dr. James Stuart. And I asked him how realistic it would be to undertake a fully funded "private mission" to Mars.

"Well," he said, "I was Project Manager of MGCO—"

I had hit the jackpot!

"—before I joined OSC, last month, so I know something about the difficulties of sending a spacecraft to Mars. Yes, I think you could do it with a 'private mission.' OSC is a private company. It's risking its own resources to develop the first 'private' upper stage for the Shuttle NASA's ever bought.

"In terms of going to Mars, if you had the appropriate spacecraft, you could do it."

I then asked Stuart about another pet idea—first proposed by a NASA Project Manager, John Cassani, for the "Galileo" Program.

"Could you," I asked, "build a *second* 'Galileo' spacecraft out of 'spares' and send it—not to Saturn, as Cassani first proposed—but to Mars?"

Stuart paused, looked at his watch, then said,

"I've got to catch a plane in a few minutes. No, that would be overkill. 'Galileo' was designed to survive the radiation belts of Jupiter, to sample magnetic fields, etc., radiation doses that simply don't exist at Mars—"

"But it *does* have a splendid imaging capability," I interrupted, "which could send back exquisite resolution of some objects on Mars—if it were placed in a low orbit. Furthermore, its on-board fuel and engines could effect several orbit changes, including a lowering of its orbit to take some really excellent close-ups, then have sufficient fuel to raise the orbit back up to the point where it wouldn't decay. Doesn't this capability, *and* the fact that we could basically build it out of existing hardware, argue strongly in favor of doing precisely that—if we wanted to get back to Mars in a real hurry?"

"But it's a very expensive way to do it, it's an expensive spacecraft to run," Stuart countered. "If I were looking for a spacecraft to send to Mars, if all I wanted was some really good imaging, I'd look at a couple of Earth-orbit weather spacecraft currently in bonded storage—like the one RCA just built for NOAA (National Oceanic and Atmosphere Administration) but which the Agency didn't use—which have lots of on-board memory—about 5 gigabits—and which could be outfitted relatively cheaply with the television cameras. You could buy one of those satellites for, say, 25–30 million dollars; a launch would probably run you another 50 million. Operations, maybe another 10. So with such a system you would have a relatively inexpensive mission—compared to the several *hundred million* to operate a Galileo class spacecraft."

Stuart glanced at his watch for about the fourth time and I knew my time was up. We exchanged cards, and I promised to look him up in Washington if the Project ever needed some advice on "private Mars missions."

I was fascinated. A professional, a former NASA manager, had confirmed that with the right combination of talent, management, and luck, even a private research project *might* be able to get a spacecraft back to Mars for those essential images.

*   *   *

Some years ago, the Space Center of the University of Colorado proposed a mission to NASA to explore the upper atmosphere of Earth (called the "mesosphere"). Thus, the "Mesosphere Explorer" was born—a spacecraft built by JPL in Southern California, but with key instruments designed and built by *students* at the University of Colorado. It was to be the prototype of a unique NASA experiment: the first spacecraft to be operated from a University Control Center (as opposed to a regular NASA center, such as Goddard) and by *students*.

The first impression one got on walking into the darkened Control Center was that it was just another NASA Center . . . until you listened to the voices: young and *female*.

The afternoon we visited, one of the student controllers was taking her stint as Flight Director. The shift was about to interrogate the Earth-orbiting spacecraft for a "data dump," somewhere half-way around the world. The voices on the NASCOM (NASA Communications) lines were the familiar Yeager-types we've grown accustomed to from television coverage. The younger voices, answering in equally crisp, matter-of-fact tones from the Colorado Control Center, were equally proficient in "NASA-ese."

After this first, highly subjective impression, other more subtle differences slowly became evident—between this and the traditional way in which NASA operates its spacecraft.

First, the number of people in the Center: there were only two or three, compared to perhaps the dozen you'd find at NASA-Goddard. Second: the kinds of displays glowing softly in the darkened room; there seemed to be more color television monitors than at NASA, with brightly colored data-screens flashing important information in traffic-red or yellow.

Later, when I asked about these differences, the graduate student in charge of the data automation system explained.

"We're much more automated here than at Goddard. We have to be; the students can only work part-time (they have classes, you know!) so if we relied on the traditional methods for getting data from the space-craft, reducing it, and getting it to the experimenters, it would be *months* before they got their information.

"We use a new VAX computer system, with highly automated displays which flag any problem-areas in the spacecraft operation. That allows our small shift complement of flight controllers."

"Could you operate a spacecraft with imaging?" I asked.

"Sure," the student replied, "we're operating at only about a tenth of capacity, and we're about to expand our facilities. In fact, we're about to start contracting our services out to other universities and spacecraft users."

OK, here came the payoff.

"Could you operate a spacecraft in Mars' orbit?"

"Sure," the student said again, as if nothing were beyond the capabi-lities of the facility he ran, "if the communications were through the DSN (the giant antennas of NASA's Deep Space Network). We handle Meso-sphere Explorer through NASA's MSFN (Manned Spaceflight) network; Mars would simply be a longer time-delay and more computer work."

As fantastic as it may have seemed, it was beginning to look as though it might actually be possible to fly a *private unmanned mission back to Mars*—

And have the *students* at the University of Colorado run it!

*          *          *

Boulder was a success.

By any standards, we had accomplished most of what we had intended:

• First and foremost, we had officially alerted the scientific com-munity (specifically, the sub-set of that community known as "planetary scientists") to the existence of a set of "anomalous objects" on the planet Mars, which seemed part of a *consciously-designed geometrical layout*; and had presented complementary evidence in Martian history of a prev-ious "biological epoch"—evidence in the form of geochemical findings from Viking itself.

• We had informed the same gathering of scientists of a *new* inter-disciplinary Investigation into the nature of these objects, one based in an official University environment—the University of California, Berkeley—with a direct invitation to the planetary community to participate.

• And we had gathered considerable information on the various means of "going back to Mars" for the ultimate test of our hypothesis—including the feasibility of an unmanned *private* mission.

That the *reaction* to our findings—by the planetary community—was "less than overwhelming," was a separate issue; we had scrupulously followed accepted procedure, in terms of "where" and "in what forums" scientific discoveries are supposed to first appear. Now it was appropriate for our preliminary findings to have a wider audience—

The general public.

(This need to "go public" at some point was essential for two reasons: to gather the necessary political support to go back to Mars for verification of our evidence; and gradually to prepare people for the implications resonant in our discoveries—if they should be verified.)

Because of these dual objectives, *Discover*'s treatment of our findings was particularly bothersome . . . and curious; the September, 1984 issue in which the editors devoted several pages to a "Special Report: Mars" (essentially the proceedings of the Conference) was also the same issue in which our very presence—to say nothing of the potentially revolutionary discoveries discussed—was assiduously overlooked.

In the same issue of *Discover*, the issue that ignored our Boulder presentation and the Independent Mars Investigation, there was another article on Mars, this one authored by Carl Sagan. Sagan basically supported the contentions of the "Case for Mars II" Conference—that we should be establishing plans for an early return to the planet; but, in a radical departure from his previous position regarding *how*, Sagan for the first time strongly urged a "joint US/Soviet *manned* Mars expedition!" His rationale for this dramatic "switch" was that such a joint manned mission would be a unique and powerful new way to alleviate certain international geopolitical tensions—particularly between the two superpowers on this planet.

But most curious of all—in this special issue which had diligently avoided Cydonia—was one of Sagan's *reasons* for a joint mission: ". . . enigmatic surface markings and regularly arrayed *pyramids* (italics added) on a high plateau—hardly evidence for some ancient civilization on Mars, but nevertheless worth looking into."

The fact that Carl Sagan was now championing a joint US/Soviet manned mission back to Mars was potentially a breakthrough for the Investigation. It was one thing for a small group of "Mars enthusiasts" to gather in Boulder, Colorado, considering ways for "Mankind" to return to the Red Planet; it was definitely another when a "mainstream" magazine ran a series of major articles advocating basically the same thing.

And with the Russians!

What was going on here? Suddenly everyone—including Carl—was talking about "going back to Mars;" just a few months before, when I had started examining the Viking images, you couldn't have *given* Mars away.

One couldn't help but wonder what role, if any, our own potentially revolutionary investigation played in these developments. Not that any of the "players" seemed ready to admit to such a role, but . . .

Tom Rautenberg and I had been discussing involving the Soviets in our research for *months*. It had been our plan to invite official Soviet participation in the analysis and discussion of the Viking "artifacts," as an appropriate prelude to more substantive cooperations later on . . . like a joint manned mission back to Mars to check them out!

Sagan's "coincidentally" making public a replica of our own long-range strategic thinking on the subject, and specifically linking such an expedition—however obliquely—to potential "artifacts" on Mars (even if they were "his" artifacts), and *Discover*'s printing all of that less than a month following our official presentation of similar data to the space community, was pure coincidence, of course . . . or was it?

If our suspicions re the "artifacts on Mars" were right, then anyone could see that, ultimately, public interest in *finding out if we were right* could fuel a thousand missions back (what was that about *a face* that launched . . . ?). Was Carl, as politically savvy a scientist as I have ever known, deliberately hinting at our own provocative evidence—just presented in Colorado—while at the same time staying at a "safe" distance, knowing the inevitable result: a rising curve of vital public interest?

About a month following the meetings, on a Sunday afternoon, I got a call from one of our associates in Monterey.

"Have you seen the August issue of *Soviet Life* magazine?" he asked breathlessly.

"No," I said.

"The Russians," he continued hurriedly, "have published a complete article on *your* discoveries—claiming *they* discovered the pyramids and the alignments first!"

Then, over the phone, he read a few paragraphs of the Russian article, entitled provocatively "Pyramids on Mars?" He was correct; the narrative was eerily close to my own investigation of the Martian "artifacts"—including one very specific reference.

"Artificial structures and their groups differ from natural formations by a higher degree of exactingness in pattern and by definite layout features. But aren't there geometric regularities in the group of Martian figures or analogies with architectural complexes on Earth? As far as I know, nobody has looked for such features . . . "

No one but me, that is! This was the *heart* of my own work: the contention that the geometric and mathematical layout of the Martian objects was a *designed* layout.

At this time, the American press hadn't deigned to print a word about our findings. Now, apparently, the Soviet Union was splashing an eerily similar "discovery" all over its official "window to the West" (which is what *Soviet Life* magazine actually is—an organ of the KGB and the Soviet Embassy in Washington). Another "coincidence?" What the hell was going on!?

Finding an issue of *Soviet Life* on a Sunday afternoon, on the West Coast of the United States (even in Berkeley!), is not the easiest accomplishment. I eventually located a newsstand—Dave's Smokeshop—which at least recognized the *name.*

"Yeah," a voice laconically replied (which I presumed was Dave's), "we had a bunch, but this big Russian wearing those funny baggy pants just bought all ten copies."

Big Russian? Baggy pants . . . ?

No sooner did I hang up, than Tom called.

"I just found ten copies of *Soviet Life* with an article entitled 'Pyramids on Mars?'!"

Tom is well over 6 feet, 4 inches tall, dark, with a beard and mustache, and he had been out jogging, so this was one "conspiracy" that wouldn't pan out.

But the article itself was fascinating—for what it *didn't* say as well as what it did; ostensibly the work of one "Vladimir Avinsky," a Soviet "geologist/engineer," it claimed (following the pattern of "Chekhov," in *Star Trek*) to be the result of Avinsky's own research, stemming from mid-1983.

The article described Avinsky's efforts to understand both the Face (which he calls repeatedly "the Martian Sphinx"), and pyramids. Explaining that the original images were ". . . transmitted to Earth by the U.S. probes Mariner 9 and Viking 1," Avinsky went on to say that ". . . the morphological analysis of the pyramidal figures we did was handicapped by the low quality of the photocopies of the pictures at our disposal."

Avinsky's independent modelling, in clay, of "the City." Note that *all* morphologies are modelled as "pyramids"—including "the Fort." This is another clue to the authenticity of the Soviet claim; better images reveal the Fort is anything *but* "pyramidal!"

[Knowing of the Russian paranoia for copying machines, I couldn't tell if this reference was to actual "photographic copies," or copies on some Russian equivalent of Xerox (although, in fact, Xerox itself sells a lot of copying machines—with special locks—to Moscow!). Later, we determined that, insofar as anyone could tell, no Russian named "Avinsky" ever publicly ordered these Viking images from NASA—although he could have.]

The reason that I mention this has to do with data on the images themselves; Avinsky claimed a certain angle for the placement on the Martian surface of "the Sphinx," as well as for the sun above the horizon. These angles were blatantly in error—as were all subsequent mathematical conclusions drawn from them (including the specific slopes of the "pyramids" themselves). If the Russians had access to the NASA photographs, or copies of JPL originals (which have the key orientation data printed *on the image*), Avinsky wouldn't have made such obvious mistakes.

For this (and other reasons—like the fact that he made no mention of DiPietro, Molenaar, or myself!) I tended to believe the Russian claim, to have "independently discovered" the Sphinx and pyramids on Mars (it wouldn't be the first time). What was fascinating, given this conclusion, was the fact that this investigator—literally half-way around the world, and with a mere set of "photocopies"—had managed to infer a number of key aspects to this "complex" that I myself had not yet divulged—even in our Independent Team's Boulder presentation.

Regardless of the motivations of the Russians for this publication, the fact that the article seemed to verify some of my innermost convictions regarding the Martian "artifacts" was strangely reassuring. We—DiPietro/Molenaar and myself—had wanted "independent confirmation" of our data. Well, here it was—

And from a most surprising source!

The real question was: Why had the Russians published Avinsky's claim to have discovered evidence of an extraterrestrial civilization on Mars—

*One month following the Boulder Conference?!*

Some of the "Soviet watchers" in the Project, one in particular, based at the University of Pennsylvania, came up with an interesting theory.

Was it possible, he said, that they saw a copy of the Abstracts Volume (published about a month before the Conference)? In that Volume, a brief overview of our paper had appeared—including a list of *institutions* where the co-authors "hung their hats." One of those institutions listed (because of Bill Beatty and Lambert Dolphin's employment there) was SRI—a "think tank" well-known in the Soviet Union for several areas of controversial research, funded primarily by the United States Department of Defense.

Suppose, said our consultant, the Russians thought this was an *official* SRI project (suppose, indeed!), and took its impending presentation at the Boulder Conference as a subtle "testing of the waters" by the U.S. Government itself (!) on the provocative subject of the discoveries involved? If that occurred, the Soviets (again, in the tradition of Chekhov on the *Enterprise*) could not allow an *American* claim "to have discovered the first evidence of an extraterrestrial civilization" to pass uncontested—

So they looked frantically around, until they found "Vladimir Avinsky" (who had actually done some work on these same images!) and rushed his conclusions into print—and in the one publication *specifically aimed* at a Western audience where it would not possibly be overlooked: *Soviet Life.*

It was this expert's contention that the Russians were *deliberately* sending us a message, regarding both the presence of these "artifacts"—and what they intended to subsequently do about them!

(And, because of the well-known American habit of publishing magazines a month or so early, the Soviet piece was carefully timed to "coincidentally" appear at the same time as the September *Discover* piece by Sagan . . . )

For the ultimate irony was Avinsky's own conclusion:

"The hypotheses about artificial structures on Mars can only attract the attention of scientists, for all their traditional skepticism . . . To find the answers, the most daring endeavor in human history would be required—an expedition to Mars."

But, as always in dealing with the Russian mind, one finds the most interesting statements not in the article, but bounding it—

Just before Avinsky's enigmatic piece, in which he referred again and again to "the enigma of the Martian Sphinx," another article appeared. Tom Rautenberg first spotted a possible connection . . .

At the beginning was an ancient Egyptian hieroglyphic text:

"When people find out what makes the stars move, the sphinx will break into laughter and life on Earth will come to naught . . . "

The Russians always have had a sense of the mysterious . . . and the absurd. And the absurdity of finding an image of ourselves on Mars was obviously compelling.

Even if we tried to turn our backs on this enigma, *they* would ultimately be going back to Mars to figure out the meaning—

Of the "Martian Sphinx" . . . and what it might ultimately portend for life on Earth itself.

"To find out what makes the stars move."

What is that about "an invitation you cannot refuse? . . ."

## Notes

1. Driscoll, E., "Mariner Views a Dynamic, Volcanic Mars," *Science News* 101:106, 1972.
2. Masursky, H. et. al., "Classification and Time of Formation of Martian Channels based on Viking Data," *J. Geophys. Res.* 82:4016, 1977.
3. Singer, S.F., "The Ph-D Proposal: A Manned Mission to Phobos and Deimos." In *The Case for Mars*, Vol. 57, Science and Technology Series, AAS, 1984; O'Leary, B.T., "Phobos and Deimos as Resource and Exploration Centers," The Case for Mars II Conference.

1. Frame 35A72, low sun angle, NASA batch-processed version. This is what Toby Owen first saw. Note "Face" (a), "city" (b), and D & M pyramid (c).

2. Frame 70A13, high sun angle, NASA batch-processed version.

3. Frame 70A13, processed by SRI International, showing "Face" (a) and
D & M pyramid (b). Note buttressing at corners of the D & M pyramid.
This does not show up in NASA raw version.

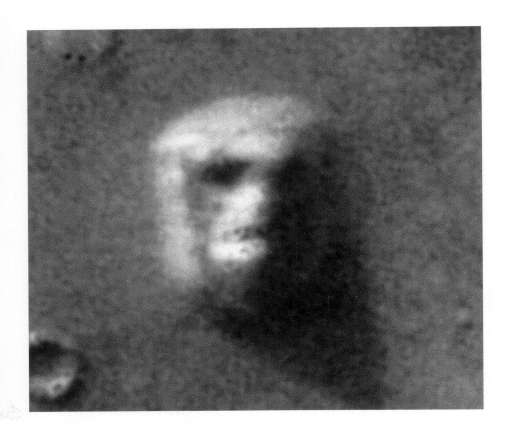

4. Frame 35A72. Computer-enhanced blowup of the "Face" at low (10-degree) sun angle. Note proportion and detail, including "teeth" in "mouth." Photo courtesy of Dr. Mark J. Carlotto, The Analytic Sciences Corporation (TASC).

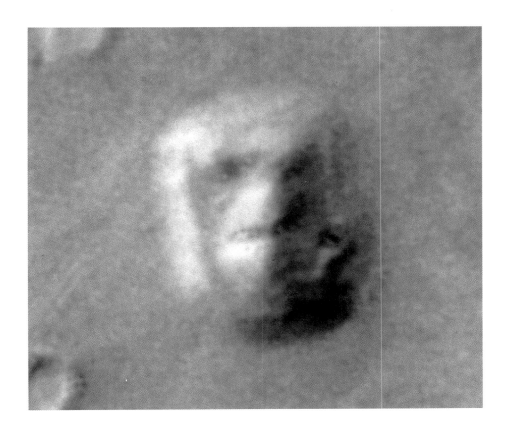

5. Frame 70A13. Computer-enhanced high (30-degree) sun angle photo of the "Face." This second, corroborating image, taken by the Viking spacecraft 35 days after frame 35A72, clearly reveals detail in "missing" shadowed side that was previously obscured. It also confirms a bilaterally-symmetrical "platform" for the underlying structure. Dark feature on lower right side of "Face" is a camera registration mark.

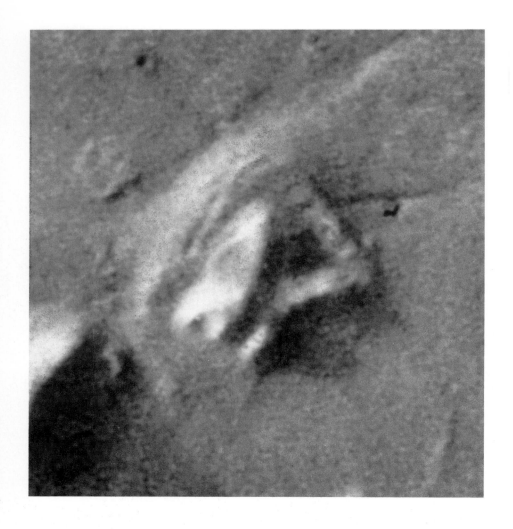

6. Computer-enhanced closeup of the "Fortress," located at northeast corner of the "City. Note three straight "walls" enclosing triangular interior space, and a series of criscrossing geometric striations. What will Mars Observer images, 50 times this resolution, reveal about this remarkably geometric structure?

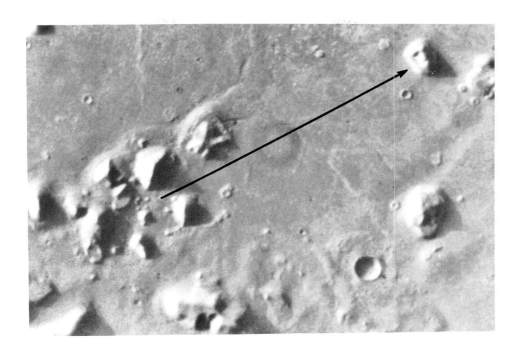

7. View of the "Face" from the "City Square"; a group of features at
the City's exact lateral center which appears somewhat like a "target"
or "cross-hairs." Photo courtesy of Dr. Mark J. Carlotto,
The Analytic Sciences Corporation.

8. Frame 70A11, high sun angle, NASA batch-processed version, with the "city" indicated in brackets.

**MARS-VIKING FRAME 35A72**

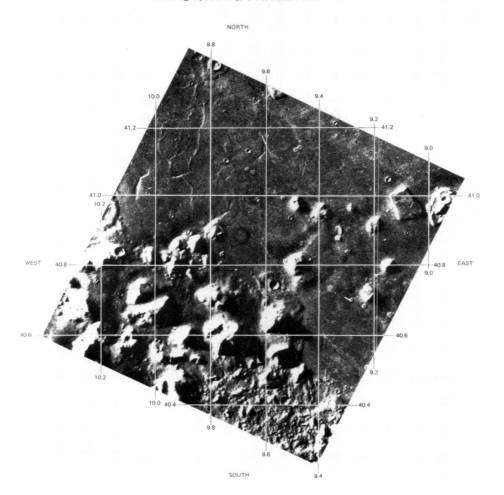

9. Frame 35A72, "raw" version (partial) of original NASA-Viking data. Coordinate grid by Merton Davies, RAND Corporation. Grid has been slightly revised, based on new analysis of original Viking Orbiter navigation information for upcoming Mars Observer mission to be launched September 1992. Photo courtesy NASA-JPL.

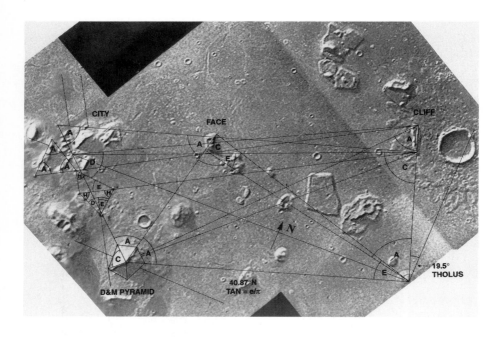

| Angles | | Angle Ratios | Trig.Functions |
|---|---|---|---|
| degrees | radians | | |
| A = 60.0 | = π/3 | C/A = √2 | TAN A = √3 |
| B = 120.0 | = 2π/3 | B/D = √3 | TAN B = -√3 |
| C = 85.3 | | C/F = √3 | SIN A = e/π |
| D = 69.4 | = e/√5 | A/D = e/π | SIN B = e/π |
| E = 34.7 | | C/D = e/√5 | TAN F = π/e |
| F = 49.6 | = e/π | A/F = e/√5 | COS E = √5/e |
| G = 45.1 | | H/G = e/√5 | SIN G = √5/π |
| H = 55.3 | | C/B = √5/π | SIN C = 1 |
| I = 100.4 | | D/F = π/√5 | TAN G = 1 |
| | | | TAN I = -2e |

TAN 40.87°w = e/π

10. The Message of Cydonia. This orthographically-rectified photo-mosaic, composed of a series of computer-enhanced original Viking frames (35A71, 72, 73, 74), illustrates author's latest "Geometric Relationship Model" of Cydonia. Note parallel alignment of structures in the "City," aiming of five-sided "D&M Pyramid" directly at the "Face" and strategic positioning of "Cliff," on ejecta of ancient impact crater, located beyond the "Face" as seen from center of the "City." Note also the "Tholus," enigmatic, mile-wide, 500-foot high mound with exterior "ditch" and interior "spiral groove," located precisely east of the "City Square." Geodetic latitude of the Cydonia Complex (specifically, the apex of the "D&M Pyramid," turns out to be 40.87 degrees — *exactly* equal to the arc-tangent of e/pi, a constant derived repeatedly *throughout* both the "Pyramid" and the "Complex."

Letters on mosaic indicate specific angles (and associated key mathematical relationships), initially discovered by Torun within the "D&M," confirmed by Hoagland throughout the larger "Complex." (See Epilogue and Figure 15). Constants seem to resolve to consistent mathematical "Message of Cydonia" — involving topological properties of a tetrahedron circumscribed within a rotating sphere, and the key predicted latitude (19.5 degrees) of the three vertices opposite the polar vertex. (See also Figures 13 and 15). Viking mosaic enhancements courtesy Dr. Mark J. Carlotto (TASC). Overlay graphics courtesy The Mars Mission.

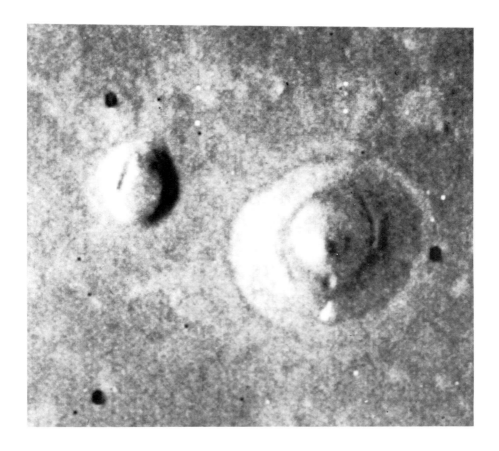

11. The "Mound," or "Tholus," located about 10 miles south of the "Cliff." Its proportions and its surrounding "moat," or peripheral ditch, are characteristic of similar prehistoric structures on Earth. The "Cliff," the D & M Pyramid" and the "Tholus" form a right triangle. Note the similarity of the curious broad grooves on the "Tholus" and its "satellite" mesa. Photo courtesy of National Space Science Data Center.

12. Frame 35A74, enlargement of "Cliff" on ejecta blanket of ancient Martian impact crater. Lack of damage or "blast shadow" around "Cliff," despite major impact event at close range, suggests "Cliff" is of later origin than crater. Equally curious "tetrahedral pyramid" positioned on opposite crater rim further confirms a later, inexplicable, modification of surface features in this area. These facts underscore the importance of tetrahedral geometry discovered at Cydonia, as deciphered in the mathematics now associated with these specific structures (see Epilogue and Figure 13). Unenhanced VIking photo courtesy National Space Science Data Center.

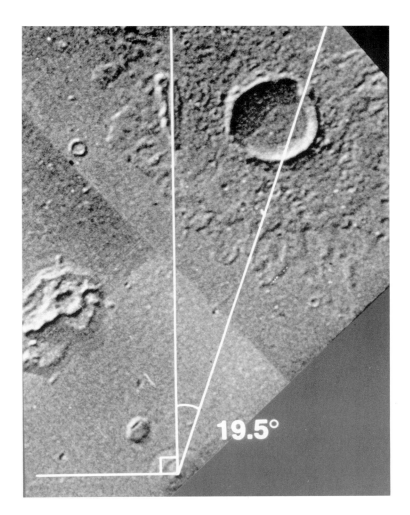

13. Frame 35A74: enlargement of "Cliff," "tetrahedral rim pyramid" and critical linking geometry. The discovery that the "tetrahedral pyramid" lies *due north* of the "Tholus" and is linked to the "Cliff" by a *19.5 degree angle* at the "Tholus," is a major reinforcement of an overall "tetrahedral" Message of Cydonia. Similar 19.5-degree angles are seen at the opposite end of this same "Complex," within the "D&M Pyramid" (see Figure 15). Composite mosaic courtesy The Mars Mission. Tetrahedral graphic and math courtesy Erol Torun.

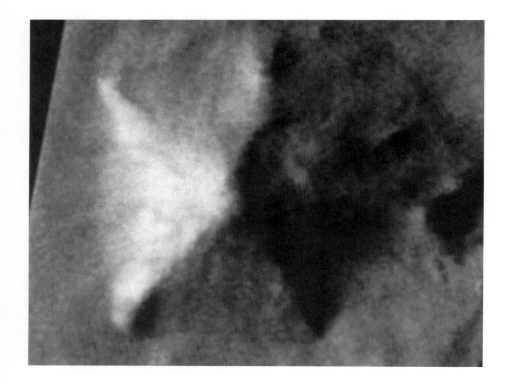

14. The "D&M Pyramid." This unique, five-sided figure, formed of three short sides (each approximately one mile long) and two long sides, is also marked by a peculiar "buttress" at each corner. Note apparent extensive damage to right side and bottom surfaces, and resulting debris flow around the base giving impression of a shortened right "leg." Possible cause of apparent damage is explosive penetration: note "bottomless" (entrance?) hole at right, and apparent domed uplift (result of internal explosion?) just right of pyramid's center.

Recent geomorphological analyses conducted by E. Torun (see Epilogue) argue strongly against any known natural mechanism of origin for this enigmatic Cydonia feature. Geometrical analyses (see Epilogue and Figure 15) reinforce intelligent design as most probable mode of origin, and recognition of this remarkable structure as the "Mathematical Rosetta Stone" of Cydonia. Computer-enhanced photo courtesy of Dr. Mark J. Carlotto (TASC).

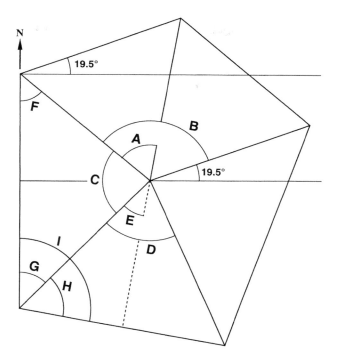

15. The "D&M Pyramid," Mathematical Rosetta Stone of Cydonia. Torun's extensive geometrical analysis of this extraordinary surface feature reveals not only symmetric internal angles but a set of redundant mathematical constants derived from those specific angles. The "message," now overwhelmingly implied by this redundant math, points specifically to the geometric properties of a polar-oriented tetrahedron circumscribed by a rotating sphere. The geometry of a circumscribed tetrahedron is therefore not only underscored by the overall allignment of other structures at Cydonia (see Figures 10 and 13) but begins with the specific siting latitude, specific size, specific geometric shape, planetary orientation, and *specific 19.5-degree internal angles*, of the "D&M Pyramid" itself. D&M measurements, graphics and math courtesy Erol Torun. Graphic composite courtesy The Mars Mission.

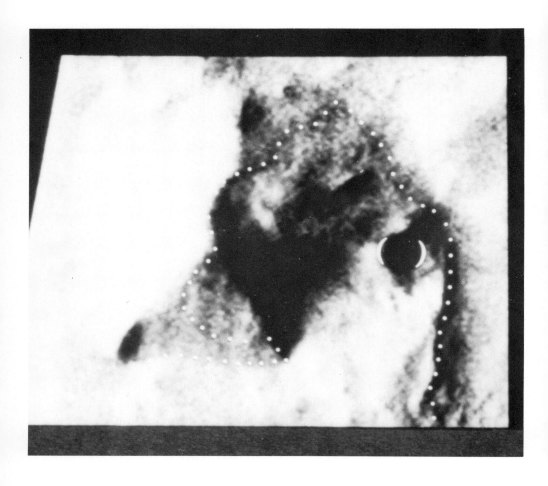

16. The "D & M Pyramid," with outline of apparent structural and surface damage, debris flow around base, and "bottomless" hole about 1000 feet in diameter. This apparent damage has given rise to speculations of "hostile action." Photo courtesy of Dr. Mark J. Carlotto, The Analytic Sciences Corporation. Overlay by Daniel Drasin.

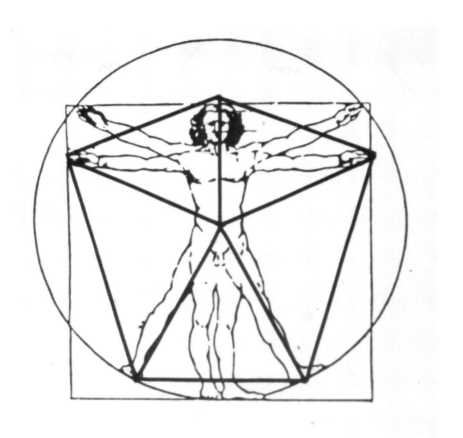

17. Leonardo Da Vinci's "man in a squared circle," which approximates the geometric proportions of the "D&M Pyramid." Since this similarity was initially noticed in 1983, new mathematical analyses of the D&M by E. Torun (Figures 15 and 31) provide a possible key to this remarkable resemblance: the match is the result of the internal geometry of the one geometric figure (the D&M) that can mathematically reconcile the well-known *five-sided* symmetry of living systems (including Man) with the *six-sided* symmetry of non-living crystals, forces and underlying physics (see also Epilogue).

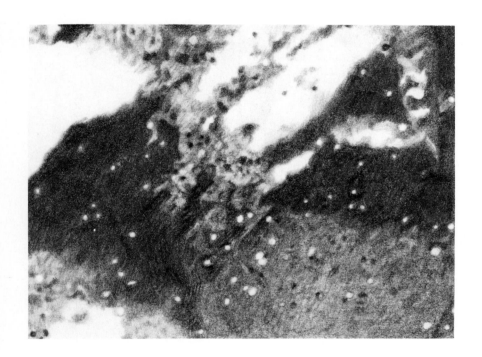

18. The "Honeycomb," located between the Fortress and the City's largest
"pyramid." This is a drawing by artist Kynthia Lynne, after the original
SPIT-processed photograph (copyrighted by Vincent DiPietro and Greg
Molenaar, and unavailable to this publication). In the SPIT version (but
not this drawing), most of the Honeycomb's "cells" parallel the picture's
scan-lines. However, others have an independent orientation suggesting
genuine "architectural" relief.

19. An example of Moiré (pronounced "mwa-ray") patterns, produced when one regular pattern is laid over another. This illustrates the theory that the "Honeycomb" visible in the SPIT-processed photograph may be the product of an actual cellular structure on the landscape, viewed through the "fine mesh" effect produced by the SPIT image-enhancement process.

20. Frame 43A01, low (6-degree) sun angle (see Appendix I). The "Crater Pyramid," a mile-square, wedge-shaped structure perched on the rim of a large (100-megaton equivalent) impact crater. Like the "Cliff" at Cydonia, the "Crater Pyramid" shows no damage or "blast shadow," which suggests that it is of more recent origin than the ancient cratering event. The feature is oriented about 45 degrees to the meridian and is located at 46 degrees north latitude, about 200 miles west of the "Cydonia Complex." The "Crater Pyramid" is also the tallest structure within a 100-mile radius, approximately equal to the half-mile height of the "D&M Pyramid." Computer-enhanced Viking photo courtesy Dr. Mark J. Carlotto (TASC).

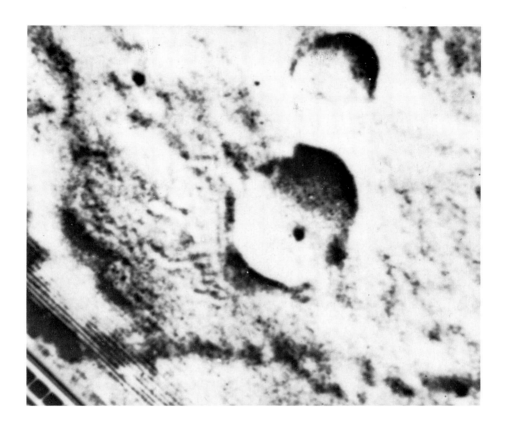

21. Greater enlargement of frame 43A01. On the outer slope of the
crater adjacent to the "Crater Pyramid," an unusual rectilinear
network of cylindrical furrows can be seen. The smaller segments
of this network seem to branch off at right angles to the longest,
unnaturally straight segment, which is oriented due north-south
and runs tangent to the crater's rim. Diagonal lines at bottom-right
are not linear features on the Martian surface but artifacts of
mechanical shutter vibrations. The black, ladder-like markings are
"edge data" printed on this officially-released version of the image.
Photo courtesy National Space Science Data Center.

22. Frame 86A08, high (45-degree) sun angle (see Appendix I). Closeup of the "Runway" complex, an enigmatic linear feature and associated structures located in the Utopia region, at an elevation of about 3 miles on the slopes of a massive volcano, Hecates Tholus, halfway around the planet from Cydonia. Pointer indicates main "Runway" feature, so named originally because it appeared as a tiny straight line on the original NASA Viking photo. It consists of an apparently ruler-straight, 3-mile-long rank of evently spaced "cones" or "pyramids." Both the "Runway" and its nearby "bow-tie"-shaped companion feature (above pointer) appear to be set into unique, shallow basins. Photo courtesy National Space Science Data Center.

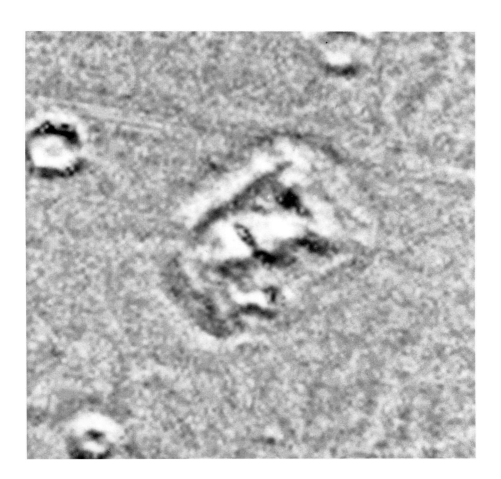

23. The Face as revealed by computer-adjusted "local contrast stretch" of high sun-angle image (70A13). Marked bisymmetry of features — eyes, mouth and "hair" — is confirmed by this technique. Sophisticated placement of shadow-casting pyramidal substructures on underlying mesa is the apparent means of achieving overall facial resemblance at all lighting and viewing angles. Large dark spot on lower right side of "Face" is an ineradicable remnant of a camera registration mark. Photo courtest of Dr. Mark J. Carlotto (TASC).

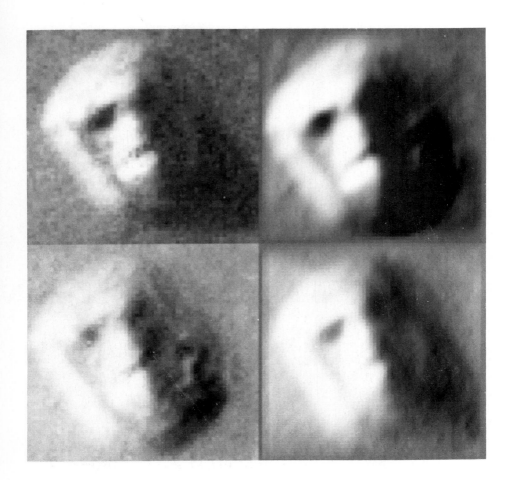

24. Testing the validity of digital reconstructions of the Face. Computer-derived relief produced by Dr. Mark Carlotto, The Analytical Sciences Corporation (right-hand images), ''lit'' at sun-angles matching original frames 35A72 and 70A13 (left-hand images), neatly recreates original Viking data. These results provide confirmation of the basic 3-D mathematical technique. Photo montage courtesy of Dr. Mark Carlotto, The Analytical Sciences Corporation.

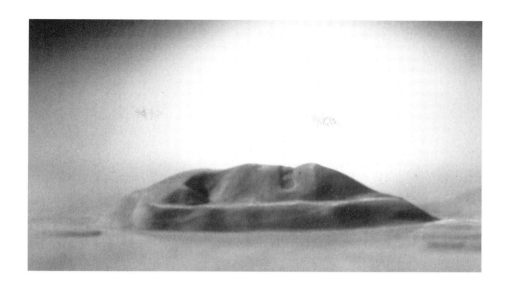

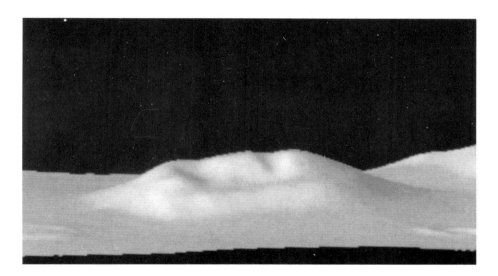

25. Two independent 3-D models of the "Face." Top: Analog clay model painstakingly created by Kynthia in 1984 based on classical sculpturing techniques. Bottom: 3-D "shape-from-shading" computer reconstruction of the "Face" (see Epilogue), created by Dr. Carlotto in 1988. Despite radical differences of reconstruction techniques and investigators (separated by 3000 miles and four years), note remarkable similarity of the results. Model at top courtesy Kynthia; photo courtesy Daniel Drasin. Photo at bottom courtesy Dr. Mark J. Carlotto (TASC).

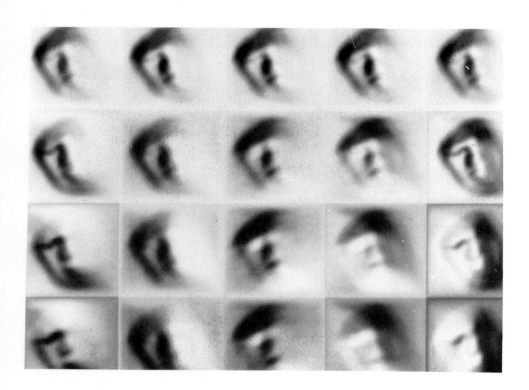

26. Computer-generated mathematical model of the "Face," based on NASA photos of illuminated left side, demonstrating preservation of "face-like" appearance at all sun-angles, effectively eliminating the "trick of light and shadow" hypothesis. Photo courtesy of Dr. Mark J. Carlotto, The Analytic Sciences Corporation.

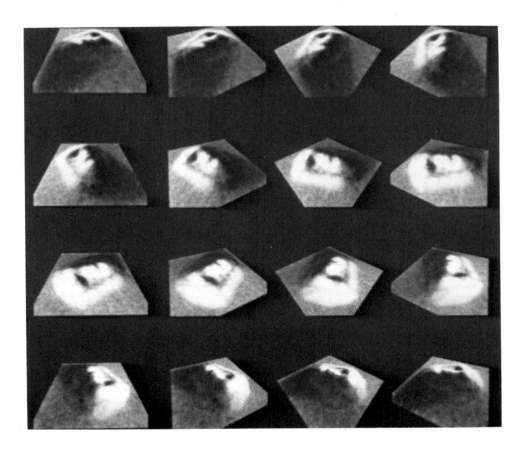

27. Computer-synthesized three-dimensional rotation of the "Face." Photo courtesy of Dr. Mark J. Carlotto, The Analytic Sciences Corporation.

## Table of the Geometric Relationships Exhibited by the D&M Pyramid

| Relationship | Theoretical Value | Measured Value | % Error |
|---|---|---|---|
| (Angle Ratios) | | | |
| C/A | $\sqrt{2} = 1.41$ | 1.42 | 0.71 |
| B/D | $\sqrt{3} = 1.73$ | 1.73 | 0.00 |
| C/B | $\sqrt{5}\,/\,\pi = 0.712$ | 0.711 | 0.14 |
| C/E | $\sqrt{6} = 2.45$ | 2.46 | 0.41 |
| A/D | $e\,/\,\pi = 0.865$ | 0.865 | 0.00 |
| (B+D) /C | $\pi/\sqrt{2} = 2.22$ | 2.22 | 0.00 |
| (Trigonometric Functions) | | | |
| SIN A | $e\,/\,\pi = 0.865$ | 0.866 | 0.12 |
| TAN A | $\sqrt{3} = 1.73$ | 1.73 | 0.00 |
| TAN D | $\sqrt{7} = 2.65$ | 2.66 | 0.38 |
| COS E | $\sqrt{5}\,/\,e = 0.823$ | 0.822 | 0.12 |

Mean error = 0.19%  Standard Deviation of error = 0.23%

## Reexamination of the the D&M Pyramid's Geometric Relationships with 1° Induced Error (Congruent angles C and C') ± 1°

| Relationship | Theoretical Value | Measured Value (For angle C-1°) | % Error | Measured Value (For angle C+1°) | % Error |
|---|---|---|---|---|---|
| (Angle Ratios) | | | | | |
| C/A | $\sqrt{2} = 1.41$ | 1.40 | 0.71 | 1.44 | 2.1 |
| B/D | $\sqrt{3} = 1.73$ | 1.68 | 2.9 | 1.78 | 2.9 |
| C/B | $\sqrt{5}\,/\,\pi = 0.712$ | 0.702 | 1.4 | 0.719 | 0.98 |
| C/E | $\sqrt{6} = 2.45$ | 2.36 | 3.7 | 2.56 | 4.5 |
| A/D | $e\,/\,\pi = 0.865$ | 0.840 | 2.9 | 0.890 | 2.9 |
| (B+D) /C | $\pi/\sqrt{2} = 2.22$ | 2.27 | 2.2 | 2.17 | 2.2 |
| (Trigonometric Functions) | | | | | |
| SIN A | $e\,/\,\pi = 0.865$ | 0.866 | 0.12 | 0.866 | 0.12 |
| TAN A | $\sqrt{3} = 1.73$ | 1.73 | 0.00 | 1.73 | 0.00 |
| TAN D | $\sqrt{7} = 2.65$ | 2.97 | 12 | 2.40 | 9.4 |
| COS E | $\sqrt{5}\,/\,e = 0.823$ | 1 | 1.3 | 0.832 | 1.1 |

For the case of angle C minus 1°
Mean error = 2.7%  Standard Deviation of error = 3.3%

For the case of angle C plus 1°
Mean error = 2.6%  Standard Deviation of error = 2.6%

Conclusion: A change of only one degree from the observed values results in more than a tenfold increase in the standard deviation of error.

28. Geometric Relationships observed in the D&M Pyramid. Data courtesy Erol Torun.

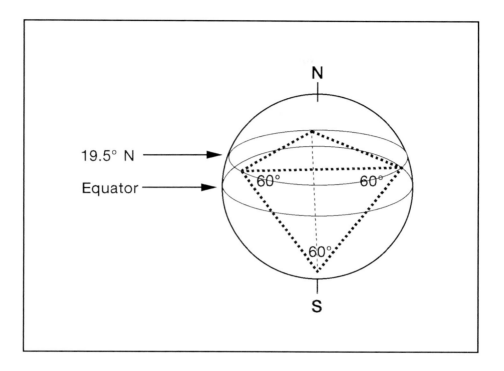

29. A tetrahedron circumscribed by a sphere. The simplest of the five so-called Platonic geometric solids, a tetrahedron is a pyramid with four surfaces and four vertices, composed of planes that intersect at 60-degree angles. When a tetrahedron is placed inside a sphere having a rotational axis, with one vertex placed at either rotational pole, the other three vertices will touch the inside of the sphere at precisely 19.47 degrees of the latitude opposite that pole. In this discussion, this so-called tetrahedral latitude is rounded off to three-significant-figure accuracy as "19.5 degrees." Graphic courtesy The Mars Mission.

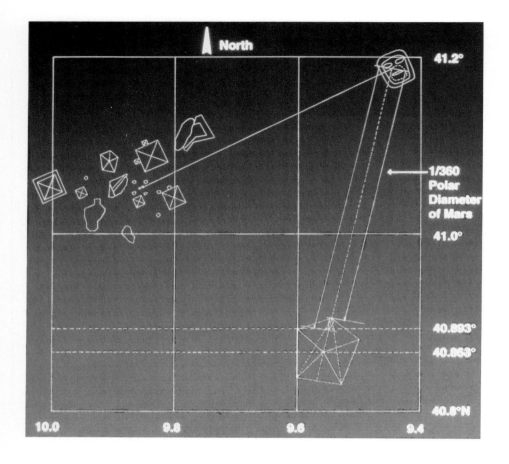

30. This remarkable relationship — a specific spacing between the front "wedge" of the D&M and the "teardrop" on the "Face" marking out precisely 1/360th the diameter of Mars — provides significant evidence now that our terrestrial system of 360 degrees of angular notation of a circle is the system also used (and perhaps even originated) at Cydonia itself. Additional measurements at Cydonia supporting this revolutionary concept had previously been discovered by the author: 2.7, otherwise known as the mathematical constant "e," memorialized repeatedly throughout the complex, in degrees. Later, the possiblity that these measurements could relate to a repeating fraction of the polar diameter of Mars itself (a kind of "planetary-kindergarten" methodology of introducing the specific 360-degree system) was proposed by David Myers. It was subsequently confirmed, through the measurements seen here, by Erol Torun and the author. Original graphic courtesy Erol Torun; modifications by The Mars Mission.

## FIVE SYMMETRY

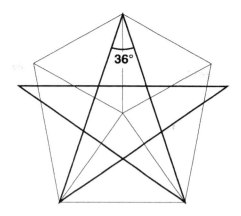

## SIX SYMMETRY

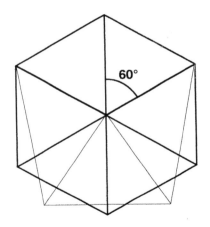

31. The preceding 1/360 polar diameter line connecting the "D&M Pyramid" with the "teardrop" on the "Face," in turn bisects a specific internal angle in the D&M, memorializing precisely *one tenth* of this crucial number, 36 degrees. This redundancy is not only further evidence relating to and specifically supporting a 360-degree system for Cydonia: as shown in the figure, this 36-degree angle elegantly resolves the "five/six symmetry" aspects of tetrahedral geometry as they potentially apply both to biology and to physics. These dual implications of the now successful "decoding" of the Message of Cydonia are also strongly implied through *other* multiple geometric values seen across the Complex. Original measurements courtesy Erol Torun; update by The Mars Mission.

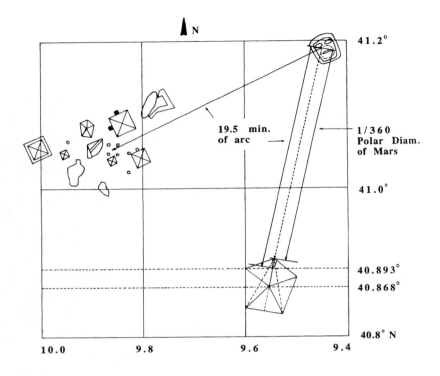

32. The *pièce de resistance* of this "360-degree" Cydonia analysis came with the discovery of a precise *19.5 arc-minutes* of Martian latitude — 19.5 Martian "nautical miles" — measured between the "D&M Pyramid," the "Face" and the "City Square." Since one arc-minute on the Martian surface is (as on Earth) 1/60 of a degree, this can only mean that the Cydonia complex was laid out in *terrestrial* units of our 360-degree system — to redundantly communicate the crucial importance of the circumscribed "tetrahedral" latitude of 19.5 degrees. This, of course, raises profound questions of the true origin of the entire 360-degree system... if not its purpose. Original measurements courtesy Erol Torun and the author; updated by The Mars Mission.

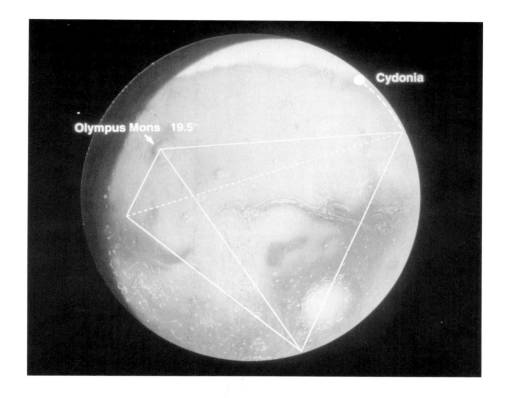

33. Cydonia's own location on Mars is the ultimate reinforcememt of "planetary tetrahedral geometry." Olympus Mons, the largest shield volcano in the solar system, is a surface feature caused by an internal "upwelling event" and is located at 19.3 degrees north Martian latitude. If you place one vertex of a Mars-inscribed tetrahedron on Olympus Mons (at 134 degrees west longitude) the next tetrahedral vertex lies within five degrees of the longitude of the Cydonia Complex at 9.5 degrees west. Thus, the very siting of the Cydonia Complex is elegant reinforcement of the "Message of Cydonia:" that tetrahedral geometry is profoundly important to our understanding of the internal dynamics of real planets — such as Mars itself (see Figure 34). Original longitude measurement courtesy Erol Torun. Mars Artwork courtesy Sally Bensusen, Science Photo Library. Tetrahedral graphic overlay courtesy The Mars Mission.

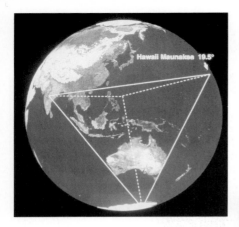

# EARTH

# NEPTUNE

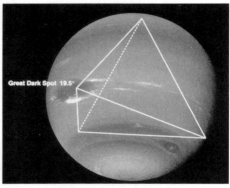

# JUPITER

34. These three planetary images were gathered over the years by a variety of unmanned NASA spacecraft — from earth-orbiting resource satellites to deep-space probes of the outer planets and their moons. They reveal a new, remarkable *solar-system-wide* phenomenon: the largest "upwelling" energy events on a wide variety of moons and planets *seem to occur preferentially at either 19.5 degrees North or South* — or, in slowly-rotating bodies such as the Sun, at *both* key latitudes! This appears to be a heretofore-unrecognized "tetrahedral" pattern, whose underlying physics seems to involve the "hyperdimensional connection" of these rotating "three-space" objects to a "higher-level state-space" (see Epilogue). This illustration is not an exhaustive depiction of all the worlds displaying the effects of this remarkable physical behavior, but merely some of those on which the resulting surface "hyperdimensional signatures" are currently most evident (for listing of individual features on individual planets see detailed description in Epilogue). Earth composite images from Landsat, assembled by Van Sant and the Geosphere Project. Outer planet images courtesy of NASA. Tetrahedral overlays courtesy of The Mars Mission.

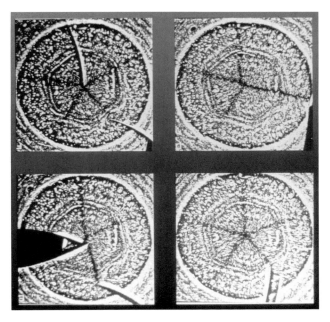

35. The inexplicable hexagonal cloud pattern of Saturn. Published in 1988 by D.A. Godfrey from a series of computer-rectified, oblique Voyager images of the polar regions of Saturn (top), this time-lapse sequence (bottom) reveals a remarkable, highly structured, geometric cloud formation that continues to elude conventional planetary atmospheric analysis. In the opinion of the author this feature, together with the now-classic 19.5 degree upwelling pattern seen on other planets all across the solar system, may be interpreted as the second, "inwelling," hyperdimensional signature discovered in the solar system (see text in Epilogue). Saturn's surprising axially-aligned magnetic field is (in complete contradiction of existing theory) evidence of an almost perfect fluid flow within the planet. This is consistent with conditions that create the polar "hyperdimensional hexagon" in the upper atmospheric clouds. Top photo courtesy of NASA. Bottom photo sequence courtesy D.A. Godfrey, from "A Hexagonal Feature around Saturn's North Pole," *ICARUS*, 76 (1988).

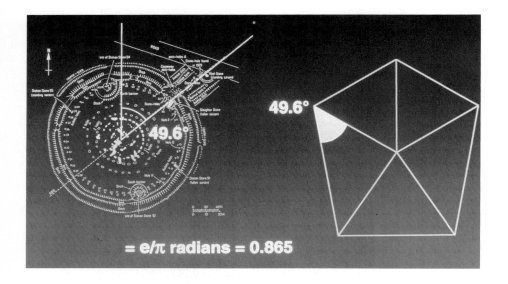

36. The Terrestrial Connection. This diagram of Stonehenge, after calculations by Carl Munck, presents us with an astonishing discovery: the highly controversial angle of its northeast Avenue, as measured from the center of the monument, is none other than *49.6 degrees off True North*. This value is not only identical to that of a key angle at Cydonia (the front buttresses of the D&M), it is equivalent to e/pi radians or 0.865 — the essential "tetrahedral" message of Cydonia. Sir Norman Lockyer, Britain's premier astronomer of the nineteenth century, surveyed Stonehenge and derived a measurement for the Avenue azimuth which deviates by less than 0.2 arc-seconds from this crucial Cydonia value. This is only one of a half-dozen separate measurements demonstrating that Stonhenge — somehow — is profoundly "tetrahedral" (see Epilogue). Diagram courtesy The Mars Mission.

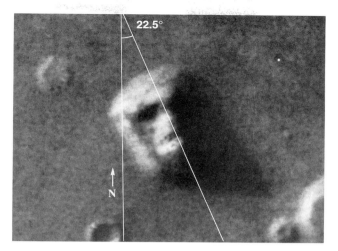

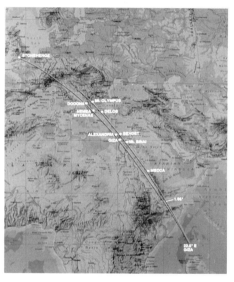

37. Further mathematical "terrestrial connections" with Cydonia. Top: The offset angle of the "Face" from True North, 22.5 degrees, divided by the circumscribed tetrahedral angle, 19.5 degrees, is *also* equal to the central "Message of Cydonia," e/pi! Bottom: This identical relationship is then redundantly communicated back on Earth by the specific siting and geodesic relationships ("great circles") between two pivotal "sacred" archaeological sites: Stonehenge, and the Great Pyramid at Gizeh (see Epilogue). One "great circle," drawn on the Earth's surface so that it passes through the Great Pyramid and crosses the equator *19.5 degrees* east of Gizeh, intersects the Equator at *precisely 60 degrees*, the fundamental angle used in the construction of a tetrahedron. A second "great circle," drawn between Stonehenge and Gizeh, then intersects the Equator at *22.5 degrees* east of the of the Great Pyramid. And 19.5 degrees / 22.5 degrees equals e/pi, none other than the "Message of Cydonia." Cydonia 3-D Viking imaging enhancement (top) courtesy Dr. Mark J. Carlotto (TASC). Overlay courtesy The Mars Mission. Original geodetic measurements (bottom) by David Myers. Graphic representation courtesy The Mars Mission.

38. The Sphinx. John Anthony West (see Chapter XV and Epilogue) is the man responsible for the current geological revolution in the determination of its age. This revolution is opening the doorway to serious consideration of the unthinkable: that "someone" — other than the ancient Egyptians — may have been responsible for Earth's most amazing work of art. In particular, this radical new dating, combined with the discovery of multiple numerical connections between Gizeh and Cydonia (see Epilogue), now must force reconsideration of the fundamental *symbology* behind the Sphinx: the fusion of the "hominid" and "feline." The recent discovery of an *identical* symbology, artfully encoded in the "Face" on Mars — in Avinsky's prophetic terms, a truly Martian "Sphinx" (see Chapter XI) — must raise profound new questions regarding the ultimate "terrestrial connection" for Cydonia... if not for a specific image, now found on Mars *and* Earth. The crucial question can no longer be avoided: why this *particular* symbology... and on *two* worlds? Photo courtesy Caroline Davies.

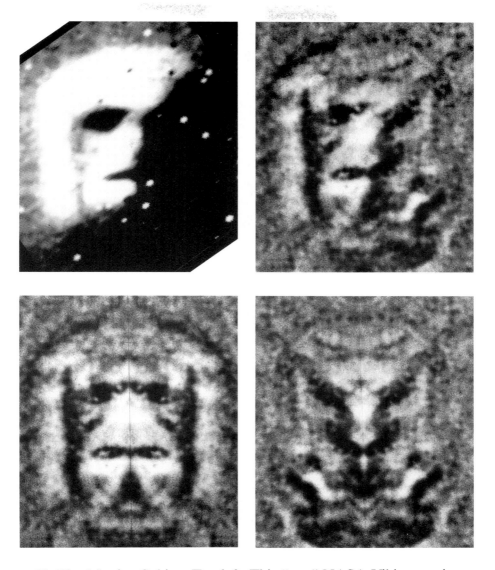

39. The Martian Sphinx. Top left: This "raw" NASA-Viking version of the "Face" reveals almost no hint of the rich imaging detail present in the original data. Top right: the "Face" (high sun-angle view, frame 70A13) after a special form of computer-image processing termed "local contrast stretch." This process brings out detail present on the shadowed right-hand side. Bottom left: the left half of this image "flipped" over to the right and matched. Note the remarkable resembance to a familiar hominid figure present in the terrestrial record of the same time period ca. 500,000 years BC, Homo Erectus. Bottom right: same procedure, with the right-hand side flipped over to the left and matched. Look familiar? (See Epilogue.) Original data courtesy NASA. Computer imaging enhancements courtesy Dr. Mark J. Carlotto (TASC). Original left/right matching by the author. Graphics courtesy The Mars Mission.

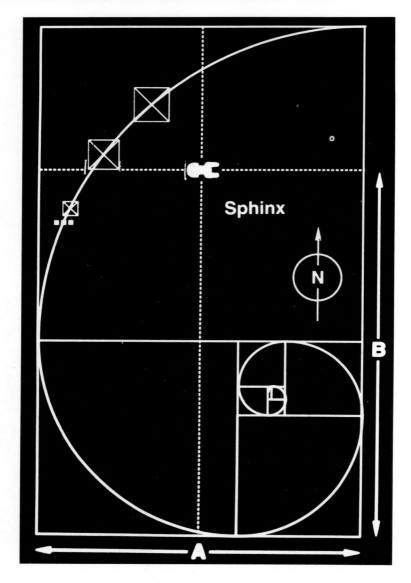

40. The Sphinx positioned on the Gizeh plateau according to the e/pi relationship. This diagram, illustrating the coordinates of the three large pyramids at Gizeh according to a Fibonacci "golden mean" spiral, divides the Gizeh plateau into a series of rectangles. The positioning of the Sphinx in relation to these rectangles and the pyramids turns out to be precisely according to this critical e/pi ratio. $A/B = e/\pi = 0.865$ — again, the "Message of Cydonia." The implications for this precise positioning of Earth's most awesome work of art, in terms of a mathematical code now undeniably linked to another planet, are clear. Original Fibonacci diagram by Rocky McCullum. Discovery of e/pi relationship of the Sphinx in relation to this spiral by Erol Torun. Graphic by The Mars Mission.

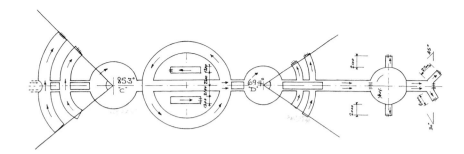

41. The Cheesefoot Head "crop glyph." This striking figure, which appeared August 3, 1990 in central England, typifies "the year of explosive geometric evolution" in the previously simple (if baffling) world-wide phenomenon of "crop circles." The author, upon measuring the central angles of the four "arcs" bracketing the smaller circles straddling the largest circle on the axis, discovered to his amazement that a) not only were identical angles present in the glyph (to two key angles measured at Cydonia — 85.3 degrees and 69.4 degrees), but b) the result of dividing these angles into one another, and examining the radian measure of one angle (69. 4 degrees) revealed e divided by sq rt 5, a "message" that was indisputably redundant, "tetrahedral," "biological" — and identical to that present at Cydonia (see Figures 10 and 15)! Subsequent measurements of scores of additional "circles," appearing both in 1990 and 1991, demonstrate beyond doubt now that "crop glyphs" are clearly (if inexplicably!) attempting to communicate on Earth "The Message of Cydonia" (see Epilogue, and also Figures 42, 43, and 44). [For more details on this crucial "crop circle connection" to Cydonia, see article by Richard C. Hoagland, "The 'Crop Glyphs' and the 'Message of Cydonia,'" *Martian Horizons, Quarterly Journal of The Mars Mission*, Vol. 1, No. 1, Summer (1991). ] Photo courtesy Busty Taylor and The Mars Mission. Angle Measurements by the author.

42. The Barbury Castle "crop glyph." This remarkably "tetra-hedral" geometric figure, found swirled in a field of wheat in central England literally overnight on July 17, 1991, typifies the growing mystery of "crop circles" that has baffled scientists around the world since 1975. "Barbury Castle" measured over 300 feet across between the outside "satellite" circles and "spiral," and in a variety of ways (see Figures 43 and 44) elegantly communicated *all* the essential elements of "the Message of Cydonia" — specifically, through redundant levels of identical geometry and high-level mathematics found coded in the objects grouped around "the Face" (see Epilogue). Photo courtesy Busty Taylor and The Mars Mission.

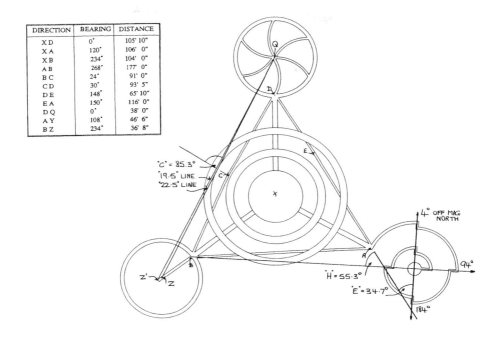

| DIRECTION | BEARING | DISTANCE |
|-----------|---------|----------|
| X D | 0° | 105' 10" |
| X A | 120° | 106' 0" |
| X B | 234° | 104' 0" |
| A B | 268° | 177' 0" |
| B C | 24° | 91' 0" |
| C D | 30° | 93' 5" |
| D E | 148° | 65' 10" |
| E A | 150° | 116' 0" |
| D Q | 0° | 38' 0" |
| A Y | 108° | 46' 6" |
| B Z | 234° | 36' 8" |

'C' = 85.3°
'19.5" LINE
'22.5" LINE

4° OFF MAG NORTH

94°

'H' = 55.3°

'E' = 34.7°

184°

43a. This survey of Barbury Castle denotes not only the scale of the original "glyph," but the appearance throughout the structure of specific angles and dimensions strikingly identical to key measurements taken of Cydonia. The fact that the Barbury Castle "tetrahedral measurements" are in feet and degrees is one of a number of lines of evidence now highly suggestive of a far more ancient origin for these two measurement systems than has been hitherto academically considered — including now a possible origin at Cydonia itself (see also Epilogue)! Original Barbury survey courtesy John Langrish; additional measurements by the author, Colette Dowell, David Myers, David Percy, and Erol Torun.

19.5°/22.5°=Circumscribed angle of a tetrahedron, divided by the "tilt angle" of the Face = e/pi

49.6°  = Front "buttress angles" of the D&M Pyramid and Avenue axis angle of Stonehenge = e/π radians

52°  = The geographic latitude (to two places) of Barbury Castle — exactly the same self-referential "message" as latitude siting of the D&M on Mars

45°  = Precisely one half of geographic distance between equator and pole

45°/52°  = 0.865 = e/π

60°  = The "tetrahedral" angles at each vertex of the "two-D tetrahedron"

60°/69.4°  = 0.865 = e/π

69.4°  = e/√5 radians, the most redundant mathematical relationship discovered at Cydonia — and the most "biologically" significant (see Epilogue)

Relationship analysis by the author.

43b. This table, relating only some of the redundant mathematical relationships that exist among the various angular and linear measurements found in Barbury Castle (see also Figures 43a and 44), illustrates the stunning "tetrahedral" nature of "the message" now communicated by this figure. The overwhelming question raised by this obvious geometric parallel with "the Message of Cydonia" (if not the basic units in which it is expressed) remains: "Who (who also apparently knows what's waiting at Cydonia) is doing this . . . and why?"

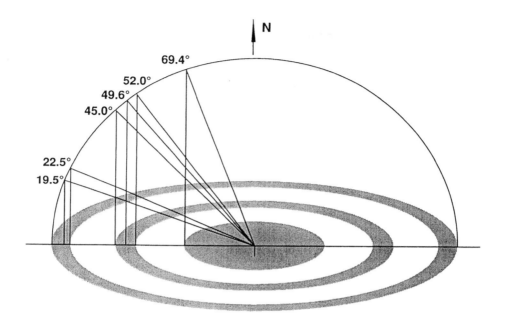

44. The "bulls-eye" of Barbury Castle. The overwhelmingly redundant "tetrahedral" nature of this glyph becomes readily apparent, when one converts the striking "bulls-eye" pattern of the central rings and circle into a cosine polar projection of equivalent latitudes, as in this figure. The resulting "latitudes," including those indicated by the continuation of key "bent" and "structural" lines through the central rings (see survey, Figure 43a, and Epilogue, for fuller explanation), communicate an extraordinarily simple, yet powerful and elegant "tetrahedral" pattern (see Table, Figure 43b). Original computation of polar transformation coordinates by the author; original photographic measurements and graphic representation by David Myers and David Percy; current graphic courtesy The Mars Mission.

**45.** Rephotographed from the South at about 20 degrees off the surface, the "Cliff" appears as a stylized humanoid face, with two eyes, symmetrical cheekbones, nose, mouth and chin. Photo courtesy of Dr. Mark J. Carlotto, The Analytic Sciences Corporation. Observation and rephotography by Daniel Drasin.

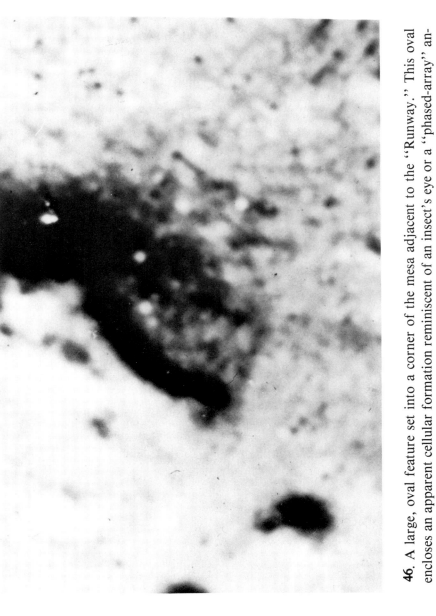

**46**. A large, oval feature set into a corner of the mesa adjacent to the "Runway." This oval encloses an apparent cellular formation reminiscent of an insect's eye or a "phased-array" antenna. Original photo courtesy of National Space Science Data Center. Observation and macro-photography by Daniel Drasin.

**47.** The work of Dr. Vladimir Avinsky, as featured in the August, 1984 issue of *Soviet Life* magazine.

# XII

# LAYING THE GROUNDWORK
# FOR A JOINT MANNED MISSION

"If there really are artificial structures in the photographs of Mars, this will drastically alter our perceptions about the origins of intelligence in the universe."

—Vladimir Avinsky
*Soviet Life*

With the publication of the Russian "interest" in the Monuments of Mars, the whole inquiry had taken a dramatic turn. In a stroke, the field—this fledgling discipline of "Martian archaeology"—had officially been broadened to include a *government*—for, as we've pointed out, nothing could make its way into *Soviet Life* without official Soviet government intentions. In turn, this was a government with probably some very different ideas on what eventually to do with this discovery from ours.

Question: what did this new development portend?

It meant, it seemed to me, that suddenly the stakes had escalated. From a situation where I had started out attempting to interest *this* government in the presence of the extraordinary objects at Cydonia (against influential segments of that same government's determined opposition) we now faced the prospects of another "space race" for the priceless treasures of another planet.

What made the article by Vladimir Avinsky so intriguing was that, despite major errors in the *numbers* which described the artifacts on Mars, Avinsky's conclusions regarding their nature and even possibly their purpose were *identical* to mine—as were his means of determining, at our current level of information, their reality or falsity as artifacts—

By their blatant *geometric layout and design.*

It was obvious to me that, not only were the Soviets serious about their interest in these objects, but that that seriousness was being carefully communicated to the West well in advance of either side's means of going back to Mars. Why?

Once before—in 1956 and '57—the Soviets had deliberately told the world about their intentions—to orbit an artificial satellite in space, including the scientific goals, the approximate time, and the means for achieving this objective.

No one believed them—until the evening of October 4, 1957, when at a carefully timed reception at the Russian Embassy in Washington, a man from the State Department rushed up to the then head of the US National Academy of Sciences, Lloyd Berkner, and stammered,

"They've done it! The Russians have an object orbiting the world!"

We seemed to be witnessing a curious foreshadowing.

Only this time, the artifacts which would electrify the world, which would come under Russian domination if they, indeed, made good their promise for "the most daring endeavor in human history—"

Would be Martian.

And all that that implied.

There had been a premonition of something like this even before we'd gone to Boulder. Rumors had begun to circulate in the technical Western press (*Aviation Week*—that "bible" of aerospace technology) regarding a tentative plan for a new unmanned Russian probe of Mars . . . well, not exactly Mars but Phobos, its inner, lumpy moon, as a sort of "staging base" for an eventual manned expedition to the Red Planet. It seemed to imply a certain Russian "savvy," matching Singer's work, regarding the shortcut Phobos/Deimos offered a prospective Martian expedition.

The day I returned from Boulder, my answering machine contained a message that the Russians had indeed confirmed the rumors—from a "source" whose track for such inside information was unimpeachable: Merton Davies. The message prompted me to place a call to RAND.

"Mert?" I asked, "what's up?"

"I just got back from Washington," was the laconic reply, "where the hot news is what the Russians revealed about their upcoming Mars mission at the COSPAR meeting in Geneva. They've announced they're going to launch in 1988, and some fairly detailed plans to send an unmanned spacecraft to Phobos—to rendezvous and *to shoot it with some kind of laser beam* for information on its composition."

"For the Russians, that's a pretty sophisticated mission," he concluded.

I'd say!

The Russians have never had a lot of luck with Mars. Out of something like fourteen attempts to orbit unmanned spacecraft or land similar robots on the surface, *all* such attempts have failed—save one. And that returned only a few dozen noisy images from orbit before it too sputtered and died—in 1973.

The Russian announcement of their first new unmanned Mars mission in 11 years—and one with such ambitious goals concerning *Phobos*—was as technically audacious as its timing—

Just a couple weeks before Avinsky's article appeared—and asked its haunting question about the Martian "sphinx" and "pyramids." "Could it be that someone has visited Mars and is waiting for us now . . . ?"

The rapidly changing nature of this situation in less than two years—from where no one was even contemplating sending anything back to Mars, to where the two major powers on the planet were suddenly dispatching separate missions, with one of them a sophisticated reconnaissance of the inner Martian moon (a necessary prelude to any early plans to follow soon with *men*)—seemed a bit too fortuitous to be mere chance.

Which once again brought up a question raised repeatedly in "Chronicles": if we were so damn smart, what about the official agencies which had secured these pictures? How could *we* see the evidence of an ancient civilization on the Viking images, yet our own government continue in blissful ignorance?

It was enough to make one a believer in "conspiracy," a game which had blossomed into fashion with a vengeance in the years following the Kennedy assassination. Any number could play—given recent history which included "Watergate" and many other instances of attempted governmental cover-ups. The "rules," when applied to Mars, went something like this: Which was more incredible: that our official space agency had somehow "missed" finding the very life it had sent Viking all the way to Mars to find?; or that we—a handful of amateurs—had succeeded where this vast billion-dollar official effort failed?

My vote tended toward the "dumbness theory": that there are none who are so blind as those who *will not* see—a situation which seemed to apply with almost religious intensity to many of the so-called "scientists" who had been confronted with the image of "the Face on Mars"—and denied the very existence of the data.

Now, from behind the Iron Curtain, came direct evidence of at least one other group which—despite substantial lack of information—had managed to "put it all together," coming eerily close to some of my most private thoughts and theories regarding the "Monuments of Mars."

What did it mean?

For one thing, it meant that I was probably right.

It was a strange feeling to have more apparently in common—in terms of the analysis and conclusions on the Martian data—with an unknown researcher literally half a world away than with some of my own colleagues on the Independent Team. What made it even more ironic was the fact that we were two researchers living in two countries which, for all our lives, had essentially been "enemies"—at least by common definition.

Yet, reading his words again and again, I felt a growing warmth for "Vladimir Avinsky," a feeling of actual commonality which far transcended the petty politics that occupies so much of the attention of the world when "Soviets" or "Americans" are mentioned.

The two of us—Avinsky and myself—were on to something . . .

Something so big, so important, that just possibly it could affect the future of the world—if we did it right. Was it possible, I mused, that instead of launching yet another space race, the discovery of this data—and the fact that the Russians had let us know they *knew!*—might truly forge a *common exploration* of these ancient Martian ruins? Imagine, I thought, the impact of seeing two spacesuited figures—one with the American flag emblazoned on his shoulder, the other with the Hammer and Sickle—opening one of the awesome Martian "monuments" on global television . . .

Could the world ever return to the horrors and the nightmares of paranoid separation that we knew . . . after such a joint experience as finding out *together* why the Face looks so much like us?

It wasn't that one had to enter into such ideas with any intention of relying on the good faith of "the Russians." The simple fact was: we would both be safer—and would undoubtedly *feel* safer—if we shared this momentous exploration jointly in front of television cameras which could watch each other's moves.

But once that much collaboration was a given, what else could happen as we stood—side by side—before the awesome nature of these Martian artifacts waiting in their sandy windblown tombs . . . and the Universe itself . . .

What indeed.

Already, with no direct communication from them, it was evident to me that the Russians had significant contributions to make to this investigation. Avinsky had already glimpsed a lot; what more could happen, what dramatic progress on cracking the biggest mystery of all time, if we *deliberately shared* our data and ideas?

By another "fortuitous circumstance," an occasion now presented itself which allowed me to put some substance behind these fine ideals, to "set the ball in motion" in a way which might result in exactly such a joint manned Martian expedition.

A friend in Santa Barbara—Jane Nobel—had told another friend, one David Woollcombe, a producer, of my findings. Woollcombe, on a visit to San Francisco in late July, wanted to come by my cottage in the East Bay and see the images of Mars.

It just so happened that he was leaving the following day for *Moscow*, where he was involved in negotiations for a "space bridge"—the Soviets' term for a two-way satellite television program between some city in the U.S. and the U.S.S.R.

After seeing the images, we talked for several hours and I showed him the article by Vladimir Avinsky. I then asked David for the following favor:

"Could you," I said, "take a couple of pictures and my paper to the Soviet Union tomorrow? Could you somehow find this 'Vladimir Avinsky' and get them delivered into his hands? I want him to know how amazingly close he's come to getting at the heart of this—considering his lack of data or apparently real Viking pictures."

Woollcombe enthusiastically agreed, and I carefully rolled up one of the last two remaining $16 \times 20$ blow-ups of the City and the Face. If this was what it would take to open a real dialogue with the Soviets regarding these extraordinary objects, it was worth it.

Woollcombe was so enthusiastic over our discussions that he wanted to include us in his own Soviet negotiations—to put us on his "space bridge" (for his production of the "Peace Child" play, which eventually aired on PBS, in 1985). Both Rautenberg and I felt this was premature; any public US/Soviet television programs on this data should *follow* the serious intensive study now forming at the University of California; we both felt that the high visibility of discussing these Martian "monuments" in such a format *before* the essential scientific work had been completed would be counterproductive. We also felt that such premature publicity would be potentially damaging to any such cooperation.

Reluctantly, Woollcombe agreed to keep a low profile insofar as "Mars" and "faces" were concerned and only to attempt delivery of what I'd given him for Avinsky.

Weeks passed, and I heard nothing from my overtures to Avinsky. Rautenberg and I finally decided to open a "second Russian front." According to those experienced in such "second track diplomacy," this was not unusual; not without reason are the Russians sometimes considered a mystery within a mystery. Because of the very informal nature of my "Avinsky channel," Rautenberg and I eventually enacted a more formal mechanism of communications: a copy of our Project Proposal and a letter to the equivalent of the head of the Soviet "NASA," in the Soviet Academy of Sciences. By a fortunate circumstance, another of our widening circle of associates (in what we were now informally calling "the Mars Project")[1] was bound once again for Moscow: Jim Hickman, head of a Soviet Exchange Program at California's famed Esalen Institute. Hickman was a well-known pioneer in such second track diplomatic efforts, and had offered to take our "pouch" along and deliver it to the appropriate Soviet authorities.

We were now speaking, not as "one independent researcher to another," but as "the University of California to the Soviet Academy of Sciences." The Soviets have a very different idea of "independent researchers" than the West. As far as they were concerned, *all* our overtures were ultimately coming from the U.S. government (!), and their hesitancy in responding was probably due to not being able to figure out exactly what they were responding to—some kind of trick or a bona-fide governmental invitation. We hoped the "official" nature of our University communication would make it easier for them to "pigeonhole" the query—and answer.

Meanwhile on the funding front . . .

The weeks lengthened into months and there was no significant response from our prospective donors. It became increasingly clear that convincing individuals of the seriousness of this unique Investigation—even one based at a major university—was going to take time. As Tom and I had agreed that the specifics of the Project Administration be left in his hands until there was funding to initiate additional image processing, there wasn't really very much for me to do.

It was January, 1985 when the next "hit" on the question of potential Soviet participation happened.

I was on the East Coast, a trip which had taken me from Berkeley to New York to North Carolina (to see my parents, brothers, and sisters over the Christmas holidays), then back through Washington, D.C., to New York, and back again to Washington. The purpose of this "twisting itinerary" was simple: the continuing pursuit of funding for the Investigation.

Meanwhile, I had been tipped to an important meeting being held on January 12, at the National Academy of Sciences, sponsored in part by Carl Sagan's Planetary Society. The subject of this meeting was to be "Weapons in Space: Implications for the Civil Uses of Space." Dubbed "the Star Wars meeting" (after the prime debate topic—would the Presi-

*The Monuments of Mars*

dent's "Space Defense Initiative" militarize space beyond the point of permitting peaceful cooperative exploration efforts, including with the Russians?), the gathering was essentially to be a debate between proponents and opponents of the "Star Wars" program—including comments on its expected impact on future US/Soviet space cooperation by Dr. Roald Sagdeev, Director of the Institute for Cosmic Research, of the Soviet Academy of Sciences.

It was a timely place for me to meet face-to-face with members of the Soviet delegation, particularly Sagdeev . . . and invite them to participate in the new Mars Project.

Sitting through the pro and con debates of the morning session, I was struck by a curious similarity between the Star Wars controversies and the essence of The Mars Project: Are the objects at Cydonia evidence of intelligent design?

Opponents to Star Wars had deftly manipulated the numerical probabilities to show that it was impossible to intercept Russian missiles in any effective manner. Further, that it was *always* going to be more cost-effective to shoot up more and more missiles than to shoot them down—which, if true, would eliminate any *economic* basis for such a proposed defense.

The proponents of SDI, just as deftly, manipulated the same numbers and painted a much more rosy picture—showing how "layered defenses" and "boost-phase targeting" could all but make it impossible for any meaningful attack to succeed—thus implying that a potential aggressor would be deterred from even launching one.

Same numbers . . . two *radically* different interpretations.

During the noon lunch break I was standing under the National Academy dome, talking with some friends, when I caught sight of Carl Sagan striding purposefully across the Academy Rotunda. Moments later, he saw me—and made a "midcourse correction" so that he would pass within earshot, calling out during his "flyby,"

"You going to be around for awhile?"

I knew exactly what Carl wanted to discuss . . . and it wasn't SDI.

Some months earlier, after the *Discover* piece, in fact, during fall hearings "on the Hill" called by Senator Spark Matsunaga to consider ways to faciliate Carl's idea of a joint Mars mission with the Soviets, Tom Rautenberg and economist David Webb* had met with Sagan, Hal Masursky, and Louis Friedman (Executive Director of The Planetary

---

*Later, Webb would be appointed to President Reagan's prestigious National Commission on Space.

Society) in Washington, to show Sagan something which might just be relevant to such a mission: the Viking images. During an extensive two-hour private session, Carl minutely examined Tom's large blow-ups (the same ones Tom had "borrowed" from me so many months before), from time to time attempting to call Friedman's attention to features he—Sagan—found particularly enigmatic; Friedman steadfastly refused even to look at the images, calling the whole idea of a former civilization on Mars "utter nonsense."

Carl also studied the "Cal Proposal"—our plan to bring new image-processing techniques and analyses to what was on those images (I know he looked it over carefully; Tom later reported that, out of something like a hundred pages, my comments regarding "his" pyramids of Elysium immediately caught Sagan's eye, in the "Boulder paper" appended to the document).

After this close review, Carl encouraged Rautenberg and Webb to continue laying the foundations for a serious University investigation, but he also expressed extreme reservations regarding the "explosive potential" of the wrong kind of publicity around the Investigation.

Then, curiously, Carl pulled Webb aside and stressed that, if he were ever asked, he "would deny this meeting ever took place . . . that I ever saw those pictures."

Now, under the National Academy dome, Carl wanted to talk to me—the guy behind the fuss. Given the fact that I had known Sagan for almost fifteen years, since before Eric and I had given him the idea for the "Pioneer 10 Plaque," there was little doubt in my own mind regarding what he wanted.

"Are you still interested in this Mars stuff?" was his opening line.

We were both leaning against a massive granite pillar, one of the dozen or so that hold up the National Academy dome. A curious circle of onlookers gazed on this slightly strange tableau from around the Rotunda, attempting to appear as if they weren't.

Sagan was a highly visible public figure and leader of The Planetary Society (which had called this "Star Wars" conference), and his actions were inevitably being watched for clues as to his own thoughts and opinions. That he had initiated a meeting with an "unknown person" in the most visible location one could imagine—in the middle of the National Academy Rotunda—was undoubtedly arousing more than passing curiosity.

"Yes," I replied to his question regarding my continuing interest 'in this Mars stuff,' part of my mind amused at the reaction of our "audience"—if they could eavesdrop on what we *really* were discussing.

"Are you still planning to write a book about it?" was his next question.

Aha! The real reason for this cozy "chance encounter" was surfacing.

"Yes," I said again, realizing now that this was his primary concern—that I would write a book which would bring this "Mars stuff" to national attention.

In a way, it was flattering: Carl had a lot of experience with the tricky field of "science popularization." He had even won a Pulitzer Prize for his success. That he was worried an effort of mine might reach a comparable audience, with comparable—if negative (from his viewpoint) political reaction—was in a strange way reassuring; for that was exactly what I intended:

That the existence of these remarkable objects and the major puzzle they present to modern science reach the widest possible audience.

"You know," Carl continued, "I've offered to review any papers and material on these 'objects' . . . "

"Yes," a one word answer seeming the most appropriate response, for the third time. My mind flashed back to his review of John Brandenberg's version of the Independent Team paper, and caustic summation that John was neither right or wrong, he hadn't even entered the discussion.

"I believe you've already seen my paper on this subject," I reminded him, thinking of his reaction to my comments on the Pyramids of Elysium. "It was contained as an appendix in the 'Cal Proposal—'"

"Yes," he objected, "but that didn't have the illustrations and diagrams."

"True," I said, "that version was bound before we had the illustrations finished. If you want, I can send you a new copy through Shirley (his administrative assistant), as soon as I get back to California."

The visual panorama over Carl's shoulder for this exchange struck me as uniquely appropriate.

Here we were—the acknowledged leader in the field of searching for evidence of extraterrestrial intelligence, and someone who might have finally found that evidence—cautiously probing each other's motives and potential hidden agendas on a potentially explosive issue: was there truly evidence of former life on Mars on those Viking images? Sweeping murals looked down from high above; murals which portrayed the essence of Western thought—Greek scenes of Prometheus stealing the "fire of Heaven," of Apollo driving the sun across those same heavens and other well-known Graecian tales. The mythic cornerstones on which we've constructed an entire civilization, our entire way of looking at the world . . . and what lies beyond . . . were arching overhead, as Carl and I leaned against the metaphorical pillars holding up this entire philosophical perspective . . . and discussed a discovery which may just change our view of who we are and where we came from.

Later, as I delved into the origins of the mythology represented in those paintings . . . which might, if I am right, ultimately trace its origins across space and time to a distant reddish world . . . I couldn't help but recall this strangely appropriate encounter—

Underneath those murals in the National Academy dome: the nearest thing we have to a "cathedral" to our modern myth . . . called "science."

"All I want," I finished, somewhat emphatically, "is to get this data properly investigated. I want to get to the bottom of this; I want to *know* what these objects are and what they're doing on Mars!"

Carl gave me a funny look, as if he didn't believe my stated intentions.

"Well," he finally said, "I guess we'll be exchanging papers on this in the literature."

The "literature" was, of course, a reference to the scientific journals where these types of discussions are traditionally carried on. This too was reassuring; for Carl to say this, was to acknowledge that there *was* something to discuss—a first step in enlisting his unique capabilities in the search for a solution.

In retrospect, on Carl's concerns regarding the public impact of our investigation, I was "right on." On the "reassuring part," I couldn't have been more wrong, as later events would indubitably show.

The high point of the afternoon was my meeting with Dr. Roald Sagdeev, Director of the Institute for Cosmic Research, Soviet Academy of Sciences.

I'd decided to wait until the end of the sessions, before I approached the head of the Soviet delegation with our invitation; this was based on two considerations: one somewhat practical—I was far more likely to get Sagdeev's full attention at that time—and the other born of curiosity: what kind of man was this "Roald Sagdeev?"

By waiting until the end of the afternoon, I would have a chance to observe him on the panel—interacting with other participants in the conference, and responding to questions from the audience. This, I felt, was an important bit of "research," before I took the step of officially inviting his participation in the Mars Investigation.

A couple of years before, I had seen Sagdeev on ABC's "Nightline." Anchor man Ted Koppel's questions, clearly politically-oriented, were inappropriate for Sagdeev's background and position in the Soviet hierarchy. Sagdeev came across more as a true scientist than as a political apologist for the Soviet Union, a man who apparently thought he had been invited on ABC that night strictly to discuss the technical results of the latest U.S./U.S.S.R. Venus cooperation. Instead, Koppel tried repeatedly to pin him down on the fine points of the U.S.S.R.'s *anti-satellite weapons system*, a field which Sagdeev—as a planetary scientist—obviously knew little if anything about, let alone to which he would have any direct policy input.

His agitation at being asked questions about "extraneous matters" was quite evident.

I was impressed this January afternoon with Sagdeev's sincerity and humor—and his evident warm friendship with Carl. Several times Carl would make a point, and Sagdeev would interject a dryly humorous "punctuation." (Sagdeev is on the board of advisors of The Planetary Society, invited to that position by Carl a year or two before this conference, specifically to demonstrate the international nature of scientific exploration of the solar system.)

Sagdeev's comments during this afternoon's Star Wars session were what one might expect: that, from his perspective, one of the most important activities for Mankind was continued peaceful exploration of space; that, as far as Star Wars was concerned, this proposal by the American president placed grave impediments in the path of that continued exploration, particularly future U.S./Soviet *cooperative* efforts.

Although he partially echoed the official Party line, that President Reagan's Strategic Defense Initiative was "militarizing space" (which blithely ignored the reality of earlier Soviet testing and placement in orbit of an anti-satellite weapons system), my perceptions were that his personal objections to "Star Wars" were sincere. Sagdeev obviously felt that major U.S. commitment to SDI, besides being an ultimate technical fiasco, would unbalance the delicate strategic deterrence concept. But equally important, apparently, from his professional perspective, would be the effect of SDI on future planetary exploration: to drain major resources from *both* planetary exploration programs—as the U.S.S.R. was forced to develop its own "Star Wars" systems in response.

Particularly ironic (considering the nature of the Investigation to which I was about to invite Soviet participation) were his repeated references to "the wonders to be found on Mars"—possibly by a "joint manned Mars mission"—should the United States forego deployment of the Star Wars program.

It was my clear observation that Sagdeev's objections to Star Wars were based foremost on personal philosophical and scientific considerations—and on his fear of the recently discovered "nuclear winter" effects of even a limited nuclear war, should deterrence fail.

In other words, he seemed to be a serious scientist moved by exploring the true wonders of the cosmos—and not just a "party hack." I hoped he would welcome the profound mysteries raised by the objects at Cydonia . . . and the positive geopolitical results of a cooperative effort to explore them.

The conference was to culminate in a reception for the participants and press, and I'd decided this was the opportune time to seek out Dr. Sagdeev and introduce myself.

When the crush of those gathered in the main Rotunda once again, drinks and hors d'oeuvres in hand, finally swept us face-to-face, Sagdeev was flanked by several "members" of the Soviet Embassy in Washington (some of whom, I suspected, must be KGB—nonchalantly there to observe any "extraneous" conversations). I plunged ahead anyway.

"Dr. Sagdeev," I began, "my name is Dick Hoagland, and I have here—" holding up the bound Cal Proposal, "a project document from the University of California. It outlines a rather puzzling phenomena on Mars we've found, and a list of scientists who are gathering to undertake further research into the nature of these objects—"

"Yes—" Sagdeev replied, before I even finished, "I have seen this document."

He had . . . ? That had to mean Jim Hickman's "courier service" had succeeded!

"Well, if you have," I continued, trying to appear restrained, "you know what it contains. I am here formally to invite you to participation in this important Investigation. If the objects are what some of us think they might be, such cooperation could be very important for the ultimate resolution of this matter, as well as future relations between our two countries."

By this time I was getting some very strange looks from some of the other "members of the Embassy." I realized immediately, from Sagdeev's own admission, that they *hadn't* known how to "pigeonhole" the original invitation—which was why we'd had no response for all these months. Hopefully, seeing the proposal a second time and having a personal invitation to discuss its contents, from one of the key participants in the Investigation, would trigger the next stage in these "negotiations."

I diplomatically handed the document—with "University of California" emblazoned on its cover, as well as Churchman's name, Rautenberg's and mine—to Sagdeev, who immediately handed it to one of the "members of the Embassy." I then gracefully excused myself, expressing my sincere wish that he would let us know of their response to our invitation "after you've had time to study the Proposal." I then headed for the bar.

On the way, I caught Sagan's eye; Carl had apparently observed the entire exchange from the other side of the Rotunda (hell of a place for "serious meetings!"). I couldn't help but wonder what influence his apprehensions regarding this whole matter would have on his Russian colleagues in the Soviet Academy . . . and on the whole idea of joint Soviet participation in the Mars Study.

Some months later, I would know.

There were some very complex politics at work here.

On the one hand, you had Vladimir Avinsky's eerie perceptiveness—working as he had with little or no data. In the months following Avinsky's article, I had sought answers to this remarkable puzzle—how could Avinksy have come up with some of his "correct" deductions (which had to be "correct," since they coincided so closely with my own!), based on faulty information? Then there was the background on the man himself: who was "Vladimir Avinsky?"

Through a close associate of David Woollcombe's, Anya Kucharev, I found some clues.

According to an extensive file of clippings Anya was kind enough to provide, Vladimir Avinsky was an author as well as an "engineer/geologist," living currently in the Volga region of the Soviet Union. Anya, in fact, remembered briefly meeting him during a book-signing of his latest effort in 1983, at a bookfair in Moscow. Apparently, it was a chapter of this book, *Land and Sea,*[2] which the editors of *Soviet Life* chose to reprint "coincidentally" a month after the Independent Team's Boulder paper first appeared.

According to the clippings from Anya's file, several of which were in Russian from the *Moscow News*, Avinsky had first put forward his hypothesis regarding the Martian "sphinx" and "pyramids" in a newspaper story, in 1983—confirming my suspicions that this was apparently truly a case of "independent discovery."[3]

Unfortunately, the brief biography on Avinsky in the file also confirmed my worst suspicions: Vladimir Avinsky was a confirmed "Von Danikenite"—a long-time believer in Erich Von Daniken's "ancient astronaut" ideas!

Erich Von Daniken was a Swiss journalist who, in the early 1970s, published a highly controversial book, *Chariots of the Gods?* The thesis of this book (and the *hundreds* which came after, as Von Daniken and others cashed in on the feverish interest in these highly sensational ideas) was that Mankind's entire civilization, if not its greatest archaic monuments (like the pyramids at Gizeh!) were not our own—

But had been constructed by visitors from outer space.

The most charitable thing which has been said about Von Daniken and his theories comes, ironically, from Carl Sagan:

"Von Daniken seeks an extraterrestrial explanation for every ancient monument and culture he doesn't understand; since he understands little if anything of what he sees, he ends up ascribing *everything* to extraterrestrials!"[4]

If this wasn't bad enough, in his later books Von Daniken had been caught blatantly making up his "data"—including outright lies regarding his own personal investigation of several archaeological sites supposedly constructed by his "ancient astronauts." This was the final insult to good science; Von Daniken has been considered beneath contempt—let alone serious consideration—by investigators in the field of SETI ever since.

That Avinsky was involved in promulgating Von Daniken's ideas was evident from his having (according to the material in Anya's file) authored something like *500 articles, and several books* on facets of the topic—this and the fact that he was a charter member of Von Daniken's "Ancient Astronaut Society!" All of which was more than enough to put Avinsky "beyond the pale"—at least insofar as the Western scientific community was concerned . . . and quite likely, the Soviet Academy of Sciences.

So, how to explain *Soviet Life*'s deliberate effort to place his "discoveries" on the record, on a par with our own published Boulder abstract?

Answer: the Soviet Union is not a monolithic society, any more than ours is (surprised?). I could envision some KGB agent's frantic phone call back to his superior regarding our Mars paper, and an equally frantic effort in Moscow to find something—anything—which would give weight to a claim of "independent discovery" by Soviet authorities (remember: *Soviet Life* is not a scientific journal; it's a magazine designed to achieve *political* objectives vis a vis the West).

Then, some editor comes across Avinsky's book, from 1983, and has what they've frantically been searching for: at least an equal *Soviet* claim to perhaps the greatest discovery in history.

Speculation? Sure. But plausible.

One reason for Sagdeev's reticence even to discuss possible cooperation on the subject of the Face, could well have been a disdain for Von Daniken's theories (especially, considering the close relationships between Sagdeev and Sagan, and Sagan's emphatically negative views currently on the whole "ancient astronaut" idea).

On the other hand, some years before Von Daniken published his outrageous speculations, a brilliant and well-respected Soviet astrophysicist—I.S. Shklovskii—entertained in print some strangely similar ideas, at least towards the origins of certain very ancient myths (he never proposed, however, that the most eminent monuments of man had been constructed by anyone other than ourselves).

What makes this so ironic, is that Shklovskii and Sagan *co-authored a book* (by long-distance "post") in 1966, titled *Intelligent Life in the Universe*. And in it, Shklovskii not only put forward his "ancient contact" theory . . . but the idea that the tiny moons of Mars were *artificial*—based on several sets of evidence.

Carl cautiously outlined ways to test such "evidence;" he even cited certain very ancient Babylonian creation texts as "models" for possible research—noting, however, that such "contact myths" must be examined in detail before consideration even as *possibilities*. It is highly likely that Von Daniken got new impetus for his ideas from Shklovskii and Sagan's own widely circulated book—something of a classic in the SETI field—which would be doubly ironic, considering Carl's persistent attitude on these ideas.

"Von Danikenism" was now such an embarrassment upon the SETI "landscape," that it was probably next-to-impossible for the Face to have totally escaped its "taint of influence" (witness NASA's presistent attitude for eight long years—that it was beneath "serious" investigation).

Everyone associated with the Independent Mars Investigation was sensitized to this potential problem, and had been extremely careful with all comments beyond the confines of the private computer conference. But one can't keep Von Danikenites out forever.

A Fleet Street "tabloid" started it in the spring of 1985—a feature story so sensational, with inaccuracies and distortions so flagrant that they would ultimately have severe repercussions, both for the Investigation at the University and elsewhere.

A frivolous follow-up appeared a few weeks later, on *this* side of the Atlantic—and in a no less prestigious place than *The Wall Street Journal*. This, despite careful promises from the reporter to do a "serious treatment" of the inquiry, if not the implications.

These, in turn, would trigger world-wide stories on the wire services and a veritable avalanche of "bad press"—which would initiate the following series of events . . .

The University of California, in particular, the Center for Research in Management (during C. West Churchman's year-long absence from the Berkeley campus, to teach at an Eastern business school), came to a new decision: "that a Mars investigation is inappropriate for this department . . ."—forcing the Inquiry to find another institution for administration!

Rautenberg, in a masterpiece of understatement, would later ascribe this to the "University's becoming sensitive to the unusual nature of the Investigation . . ."

## Notes

1. The term "The Mars Project" was taken from Wernher Von Braun's historic plans for the First Martian Expedition. Von Braun eventually published a popular verson of this technical strategy for sending men to Mars, under the title: *The Mars Project,* New York: Viking, 1954
2. Avinsky, V., *Land and Sea*, Moscow: Mysl, 1983.
3. Avinsky, V., "Pyramids on Mars?," *Moscow News*, Weekly No. 2, 1983.
4. Personal communication.

# SECTION III

# ANALYSIS AND IMPLICATIONS

# XIII

## CONFIRMING THE REALITY OF MARTIANS . . .

> "The most creative theories are often imaginative visions imposed on facts . . ."
>
> Stephen J. Gould
> *The Mismeasure of Man*

Thoughtful naysayers (as opposed to those who refuse even to consider the idea) have offered some fairly consistent criticisms of the Intelligence Hypothesis.

"Why just a blurry 'face?'" is typical.

Yes, why not six faces . . . or sixty—set in circles?! If "they" were that smart, why couldn't "they" have done better?!!

We seem ever-inclined to attribute to "aliens" limitless powers . . . if not budgets. These were not (if "they" existed, and built the Face untold ages ago) fantastic superbeings; they were mortal, perhaps (given the gravitational constraints) even frail creatures, operating with some specific goal in mind—but in a *context* we cannot begin to comprehend . . .

One of the fundamental misunderstandings of the complex at Cydonia begins with this subjective impression of a "face." Since we see "faces" everywhere—in stones, in clouds, in moons and even, according to Carl Sagan, the resemblance of Jesus Christ in a *tortilla* chip (!)—this in itself is not much of a reference for the reality of what looks up at us from Mars. The Face may have summoned us to take a second, closer look at this region called "Cydonia," but *by itself* its haunting appearance could prove nothing—certainly not the one-time presence of "intelligence."

It could simply be a "wind-eroded mesa." Why not (Carl again) "one postage-stamp sized mesa—out of the 150 million square kilometers of Mars—that looks a little odd . . . ?"

What was needed was a way past these subjective, and highly unreliable, impressions (which is why, for instance, I was intrigued when I discovered that the Face was merely one aspect of a *complex*, each member reinforcing an accumulating *pattern*). This raises the crucial epistemological question: how would we *know* if this was the product of a deliberate architectural intelligence? What would be the cues—that would *unmistakably* separate the subjective impression of a "face" from structures of doubtless planning and intentional layout . . . from merely weird-shaped hills on Mars?

It was from this need to somehow *systematize* the accumulating impression of intelligence behind this "complex," that I began to make a list—of each of the identified "anomalous features" clustered at Cydonia and, perhaps even more important, their *relationship to one another. Thus:*

• The D&M Pyramid—that unique five-sided, bisymmetrical, "buttressed" kilometer-high object, located but a few kilometers away from the Face (out of those 150 *million* square kilometers of Martian surface)—not only seems to be "aimed" directly at the Face—a humanoid depiction—but seems to possess in the NASA images the *proportions of humanoids* with which we are familiar: 1:1.6 (the famed "Da Vinci relationship" of a man inscribed in a circle and a square). In other processings of the images (Carlotto) the proportions are somewhat "squatter."

• An early recognition that a basic parallelism exists, between most of the major "structures" in the City and the central axis of the Face—which is inclined, you will remember, off the meridian by some 28 degrees (according to a revised estimate of all angles and "control-point" locations at Cydonia, courtesy of Merton Davies).[1]

• Discovery of a resulting axis orthogonal (90 degrees) to this alignment, which allows one to trace a perfect sightline from the precise center of the City (marked out by our now-familiar City Square) across the eyes of the Face, northeast . . . terminating at the northern end of a peculiar cliff situated exactly on the "facing side" of an otherwise intrusive background crater.

• Discovery that a similar sightline, extending from the same central point of reference in the City, passes just below the Face's chin, and intersects the southern termination of this cliff. Perhaps the single most important observation to date, within the "They made it to be seen from the surface" hypothesis, this arrangement makes the "cliff" *the same angular diameter* as the Face, when viewed from one location (and one location only!) on the entire landscape: the previously noted unique array of five objects—the City Square—at the exact center of the City—an even larger complex of remarkably arrayed pyramidal objects.

• The observation that the base of this peculiar "cliff" is also (within errors of measurement) aligned with the central axis of the Face and pyramids (28 degrees off the meridian), yet is many kilometers away and located atop a rubble apron extending in all directions form its associated ancient impact crater. A further observation that an apparent "excavation" exists in this "ejecta apron," on the side between the crater and the cliff (away from the Face), substantiating an overall impression (reinforced by the sharpness of its features) that the cliff is significantly *younger* than the ancient debris on which it is found.

• The observation that the overall morphology (shape) of this peculiar feature (the cliff) seems to be in layers, with a very sharp 3-kilometer long, several hundred meter-wide "defile" defining the uppermost "level" of this object. Furthermore, that the angle of this linear feature to the meridian (20 degrees) is significantly different from the angle of its base (28 degrees).

• Meanwhile, that measurements conducted on the Face increasingly support its *conscious integration* into this interlocking network of apparent deliberate design. The local meridian (the north/south line)—placed across the Face between the eyes (so that it intersects the sightline extending from the City to the north end of the cliff)—also precisely crosses the southwestward termination of the mouth, and defines the southernmost extension of the hairline. For these interlocking relationships to work, the Face must be specifically proportioned *and* specifically aligned (again, by 28 degrees) with respect to the meridian.

• An additional fascinating fact, discovered by Dan Drasin (one of the members of the Mars Project), concerning a mathematical series of relationships *along* the sightline extending from the City to the cliff. Beginning at the southwest "edge of town") defined by a compass placed on the center of the Square), measurements of unique "benchmarks" along this sightline reveal a potentially significant mathematical progression: if the distance from the southwest edge of the pyramid complex to the center of the Square is *one unit*, the distance along this line to the Fort at the northeast edge of the City is exactly *two* such units; the distance to the northeast edge of the Face exactly *four*, and the distance to the cliff exactly *eight*.

• The recognition that the D&M Pyramid, several miles southeast of this extended sightline, nonetheless (as first noted by Tom Rautenberg) shares in this geometry. It lies precisely half-way between the southwestern limit of the City and the Face—so that a line southeast, perpendicular from the Fort as a continuation of its northeastern wall, precisely intersects the D&M itself.

• The observation that the sightline—from the City Square across the Face's eyes, terminating precisely at the northern endpoint of the cliff—also marks a significant *astronomical* alignment: the last Summer Solstice sunrise circa 0.5 million years BP (before the present). Changes in the Martian axial tilt (obliquity) over about a million years cause this alignment to shift dramatically along the northeastern horizon in the space of even a few thousand years. The last Solstice alignment with the Face thus could serve as a relatively narrow "time window" to constrain a period of potential habitation.

• Potentially of equal significance, the observation that orientations of the major "structures" in the City (the main Pyramid, the Fort, etc.) reveal an equal geometric "recognition" of the *Winter Solstice* sunrise point, as well as Summer Solstice *sunset*—all for *the same obliquity epoch* of the Martian cycle.

• Finally, and perhaps most curious and potentially significant, the recognition that the 20 degree deviation of the uppermost "defile" atop the cliff is—coincidentally?—*the same* as the obliquity (tilt) of Mars upon its axis—when the Summer Solstice sightline coincides with the line running from the City to the cliff across the Face!

*      *      *

Almost a year after the close-out of the Independent Mars Investigation, as we were continuing the painstaking process of assembling the new Mars Investigation Group, one of the physicists we approached—a Nobel Prize winner—dismissed any basis for the Investigation with the offhand comment,

"The chances for life on Mars are a trillion-to-one—against!"

What impressed me about this statement was not the eminence of its originator, but his certitude despite *never having examined the preceding evidence.* And I began to think: what are the *real* probabilities that this maze of interlocking relationships, between the Face and other unique objects on this one tiny spot of Mars, are due strictly to chance?

And as I reached for some kind of mathematical "model" to explore the significance of these relationships, I realized that this Investigation shared a lot in common with the other "search" for evidence of extra-terrestrial intelligence: SETI.

The basic paradigm of SETI (the Search for Extraterrestrial Intelligence) is that "E.T. will call us on the phone."

For over 25 years, since the classic papers of Morrison and Cocconi in 1959,[2] and the subsequent historic "bootlegged" radio-listening experiment at Greenbank, West Virginia, in 1960, conducted by Frank Drake,[3] the basic methodology of all searches for evidence that "we are not alone" has been a search for intelligently-generated *signals*. Later, U.C. Berkeley physicist C. H. Townes proposed expanding the spectrum for those signals to include a search for *lasers*—light, infrared and ultraviolet. And, in the early 1970s one such search was tried briefly by a University of Michigan investigator, using a NASA orbiting telescope called *Copernicus*—with no results.

But despite exotic hardware and new wavelength regions, the paradigm still remained the same[4]: that sending electromagnetic information is easier than sending spaceships—if you want to contact someone else across the interstellar "deeps." (Notice that planets in our own solar system were no longer even under consideration as potential targets of The Search—having long since been revealed as hostile to our kind of "advanced" life.)

In the early 1980s, after almost twenty-five years of bootlegged listening around the world (a few hours here and there on radio telescopes, crowded in between more "scientific" research programs), the small band of original SETI enthusiasts from Greenbank finally reached the "big time": NASA officially launched a SETI program of its own—including the development of an array of sophisticated new signal processors, computers and radio receivers which would eventually be capable of monitoring *millions* of radio frequencies simultaneously—

Searching for that one faint spark of artificially-generated energy . . . which would prove that "we are not alone."

For almost a quarter of a century, nothing in the basic SETI paradigm had changed, except the hardware . . . until the Mars Project.

In our discoveries (starting with the work of DiPietro and Molenaar, and including the intuitions of Avinsky), the best evidence to date that we are truly "not alone" has now shown up—not as a mysterious interstellar filament whispering across the night—but as a set of *ruins* on images secured by an unmanned probe of the surface of a nearby world . . . a development in *complete contradiction* of assumptions made by SETI—

Starting with the gross improbability of finding "someone just next door," and ending with the complete absurdity of that "someone" looking anything like us!

To see just how subversive "our" Mars and the probabilities for its former "inhabitants" might be—in the light of "the SETI paradigm"—let us examine for a moment the basis of the paradigm itself:

The now-classic "Drake Equation."

At that historic meeting of SETI pioneers at Greenbank in the Spring of 1960, including such future proponents of The Search as Carl Sagan, Phillip Morrison and others, one young enthusiast, Frank Drake—then a member of the staff of the National Radio Astronomy Observatory—attempted to devise a scientific method of expressing the likelihood that any searches of the heavens with a radio telescope would turn up something . . . His logic, as expressed in an equation which would later come to bear his name, was straightforward—if in actuality a long way from being possible to quantify.

Drake started with a question: "How many inhabited planets, capable of transmitting a signal powerful enough to be picked up by a 1960s radio astronomy receiver, currently exist within our Galaxy?" [5]

The answer turned out to depend on a string of variables—quantities which could change the answer depending on determination of their actual numerical values—most of which in 1960 (and even now!) were essentially unknown. [6]

Take, for instance, the first variable—the number of stars in the Milky Way Galaxy which currently have "Earth-like" planets (the assumption being that any "communicative civilization," like us, will likely be based upon a planet—suspiciously, one very much like Earth!).

Well, in attempting to pin down this one variable, astrophysicists were really being called upon to estimate a string of "hidden variables"—such as the *rate* of star formation in the Galaxy (because you probably needed stars to get planets!) and, of course, the rate of stellar *death* (because . . . well, that should be obvious). Then you needed somehow to determine the number of stars that actually form planets (because planetary formation is probably not efficient—some stars might form a lot, others none at all); then, the number which have a planet *at the correct distance* from the star (for the simple reason that too close, and water boils; too far away, and it forms ice; just the right distance is required to permit the origin and subsequent evolution of any kind of life dependent, as ours is, on *liquid* water).

OK, *then* he needed the number of those stars below a certain mass—because the bigger (thus brighter) stars simply "burn" too fast; the biggest use up their nuclear fuels before life even could originate (in less than a million years!). Life needs a smaller, dimmer star—which by burning slower and thus living longer—will permit evolution to continue to "intelligence" (if such an endpoint is even an inevitable result of evolution, which, of course, is *another* total unknown in the equation!)

So far, all Drake's variables were simple (!), depending on basic quantities from astrophysics. If these weren't known with any precision (or simply not known at all), future technological developments—like space telescopes—would someday provide answers to some of these parameters—such as, for instance, an accurate survey of the number of stars in our own galactic neighborhood that have actually formed planetary systems.

Drake was really groping in the unknown, however, when it came to the *tough* questions—the biological (and, ultimately, anthropological) ones, like: if intelligence eventually appears, what is the probability that it will then move in the direction of technology as we know it (without which, you simply don't get radio transmitters and receivers) . . . ? Or, the likelihood that such a technological society, once formed, will not also invent *nuclear* technology—which (according to some current pessimistic thinking) in short order puts an end to the entire string of probabilities!

In other words, Drake's simple question crucially depended on *all* these probabilities—a multiplication of the likelihood of each individual physical event by all the others—to estimate the overall probability that someone else is out there . . .

It was the fact that the probability for each of these events was estimated in most cases as a tiny *fraction*—of a star's having planets, of one of those planets being at the right distance from their sun, of that planet's actually evolving simple lifeforms, etc.—that made the final product of these probabilities so very low—something like "one star in a million" in the entire Galaxy, which *might* have a technical civilization with a longing to communicate . . . And with numbers that low, even the nearest of our "neighbors" was in all likelihood *several hundred light years distant*—certainly not as near to us as Mars!

Which all looks suspiciously like our problem—with one enormous difference:

The SETI estimate of potential numbers of inhabited planets was strictly *theoretical*; they had *no data*. Our problem was determining if an observed set of *complex data* was real.

It was in the course of pursuing some way—any way!—to determine the reality of these *relationships* that I hit upon the idea for something like the Drake Equation—

Modified for Cydonia.

In place of quantities expressing the current (estimated) fraction of stars within the Milky Way which have an Earth-like planet, multiplied by another (estimated) fraction of those that orbit at the correct distance from their stars, etc., what if I attempted to estimate the probability of *each relationship I had discovered at Cydonia*—against it being merely random—

*Multiplied* by the probability of each additional relationship?

The product of the total number of these identified relationships would thus give an "overall probability" for assessing if what we are seeing at Cydonia favors a design—or merely chance.

Potential pitfalls for this were neatly summarized by Churchman: "What does 'probable' mean? Since there are an infinite number of ways an assumption can be false, what then? . . . The really crucial problem [is] never addressed: this is the problem of selecting the right hypothesis to test. Compared to this problem, the problem of the right theory for testing a *given* hypothesis seems trivial."[7]

Since no one could solve *this* problem, the intelligence hypothesis for Cydonia cannot rest on statistical analysis alone, i.e., as one positive case against an infinite number of possible negative ones. But we can work with what positive data we have. After all, even if a given theory (probability) for testing a hypothesis is wrong, it is much more important to have chosen a meaningful hypothesis to test (that the objects at Cydonia are artificial).

Examples:

The existence of multiple unique objects at Cydonia—in terms of overall morphology—is striking; within a very tiny area we find a remarkable bisymmetrical resemblance to ourselves—the Face; a five-sided bisymmetrical "buttressed" pyramid with humanoid proportions—the D&M; a multi-leveled, sharply defined feature uniquely associated with an ancient impact crater—the cliff; and a distinctive triangular object with straight "walls" and an interior "containment"—the Fort. All of these are located within a few kilometers of each other—out of the now-familiar "150 *million* square kilometers on Mars."

What is the probability for this being merely a random situation?

To be strictly accurate, since we haven't examined each of those 150 million square kilometers, we cannot say these objects are unique to this one region [although we have "spot-checked" several hundred images randomly around the planet—courtesy of John Brandenburg's efforts during the Independent Mars Investigation—and nothing like, for instance, the "cliff" (on its crater), the Fort or the D&M showed up—certainly nothing with the careful proportions of the Face!].*

So, we can *estimate* (like SETI!) the degree to which this region is remarkable—for its strange association of morphologies.

Why *are they* remarkable? (After all, Nature can erode Martian mesas into anything—which is what the critics of the Intelligence Hypothesis inevitably claim.) They're most remarkable because these shapes have special *meaning* for us—particularly the Face and the D&M. Furthermore, that meaning is strikingly enhanced by their *association*.

So, another element we should somehow enter into this probability analysis is the odds of several *meaningful* morphologies—all clustering together on the Martian landscape.

But we have to introduce a new idea—relationships—to get into the real problem. For, while it is unlikely that Nature could have randomly created a set of *unique* morphologies in one tiny location on the planet, if we can demonstrate that in addition they are *mathematically and geometrically related*, then we have a problem of a whole different order.

Let us begin with the Face and D&M.

They are unique; as far as we have been able to determine, no other features with these specific morphologies or measurements lie elsewhere, anywhere on Mars! Yet, even more remarkable, they lie *within a few kilometers of each other* . . . and share a unique symbolic "bond": they are both expressions of something we call "being human."

It was my belief that if real, this "uniqueness"—of form and association—should be expressable in numbers.

*       *       *

Imagine for a moment that these two "morphs" are all alone on Mars— that the planet is a perfectly smooth sphere and we have somehow "plunked down" these objects randomly somewhere on its surface. What are the odds that we would find them located *together* on the planet?

To express this mathematically, we simply find the area represented by a circle drawn around either the Face or the D&M—with the radius of the distance in between. The area of this circle (about 100 square kilometers) represents the "region" they inhabit.

Now, consider the entire surface area of Mars. The odds against randomly finding these two objects within the smaller area (the 100 square kilometers) must stand as the proportion between this area and the total surface area of the planet—that 150 million square kilometers mentioned

---

*For a survey of the intriguing "anomalous objects" we did find, see Appendix I.

before.

Or, one chance in *one and a half million!*

Some critic will immediately leap up and shout, "But, that's assuming a strictly random process. Geology is *not* a random process; it's a series of coherent forces in a region, which produce surface 'anomalies' all the time!"

Which is perfectly true . . . and totally irrelevant to the basic argument presented here.

It is the *uniqueness* of this association which those numbers demonstrate, not any one model to explain it. Apart from the proposal that Intelligence has set this up, any strictly *geological model* will also have to address the enormous odds against this remarkable association—and also discover some form of causal relationship for why they should be together—before we can consider this association adequately "explained." (So far, all glib attempts to do so—as the result of faulting or wind erosion—break down on examination of the *specifics* of each object and their geometrical relationship to each other—which we'll consider further in a moment.)

The point of a statistical approach to this problem is an attempt to express the *degree* of remarkability contained within this "complex"— regardless of the cause!

So, we have these two unique objects located side-by-side on Mars, with about "a million-to-one" chance of this occurring randomly; what this is really telling us is that this is the product of some complexly related formative process. Whether that "process" is geology or intelligence is a question at another level of this analysis, which probability theory can also help us with (as we'll see later).

The next step is to examine the "geometric relationship" of this association—for these two objects are also *not* randomly oriented with regard to each other on the surface; the D&M is "aimed" squarely at the Face!

What are the odds against that alignment also being due to chance?

This one's also easy. Depending on the "error bars" one assigns to this alignment (in other words, "where" on the Face the precise alignment actually occurs) one can simply divide the angular diameter of this point into a full circle—to derive the odds of its occurring.

In the most conservative case, one can simply take the apparent angle subtended by the entire Face as seen from the D&M; if the axis line of the D&M's bisymmetry aligns with any aspect of the Face, we can consider them "aligned." The odds of this are simply the total angular diameter of the Face (about 7 degrees, when viewed from the D&M) divided

*The Monuments of Mars*

into a full circle of 360 degrees—or about one chance in 50 (a bookmaker makes a lot of money with far less odds!).

To be aligned with one another is obviously another level of "involvement" for these two features (as against their being merely physically nearby). Alignment further implies some kind of "connection"—which again, can either be geology (and I'm including meteorology—wind erosion—in this category) or—

Design.

The decision between these two alternatives must come from a blend of "Occam's Razor," known geological processes . . . and a dose of common sense (no copouts like, "But it's a completely alien planet; the geology *should be weird*!"). Alignments usually occur because of fault-lines. Alignment of completely *different* morphological entities—like the Face and D&M—is remarkable; how remarkable we can see in what follows.

To determine the overall probability of these two objects randomly being beside each other on the planet *and* being geometrically related, we simply multiply the two preceding probabilities together—

Which gives less than one chance in *a hundred million* that this unique relationship—between the Face and D&M—is random!

When we include the qualitative factors,* that these two objects separately and together have another unique "connection"—the essence of that indefinable called "being human"—the true remarkability of this association must hit home . . . and raise profound questions regarding how far we are willing to entertain extraordinary geological models for this implausible occurrence. If we are looking at multiple levels of connection and association, Occam's Razor would tell us to choose the simplest model for it—which here appears to be that we are looking at Design!

*     *     *

But that's not all. Any ultimately geological explanation for this singular phenomenon would have to address the problem of *the entire "complex"*: the spatial and geometric improbabilities of finding the Face, the cliff, the D&M, the Fort, and all the other features we have described congregated within the same 100 square kilometers—and all *specifically related* through measured spacings and alignments!

For instance, take the Fort.

This (again) unique morphology is an integral part of the previous

---

*Even "quantitative" science is always measuring degrees of some *quality* or other; in the real world nothing is ever purely quantitative or qualitative.

relationship—as one wall is aimed directly at the D&M (specifically, the "head," that central buttress with the three-pronged shape), while the *other* wall is aimed directly at the Face.

What are the odds against that *randomly* occurring?

Geological models to explain these alignments will confront severe problems—as the objects are not aligned along linear "faults" in the surface crust, and are morphologically completely different. They are also not oriented predictively on the landscape, making extremely difficult models which would use prevailing winds to erode facing slopes, etc.

Or, take another feature: the City Square.

Here we have five objects, four elongated ones spaced around a circular fifth one in the center. They are all located within several hundred meters of one another and form another unique unit—in that four of them are *parallel* as well as arrayed at *right angles* to each other. In addition, they are located at the center of the larger City.

What are the odds against this series of relationships being accidental?

The first relationship—that four parallel objects, forming a unique "feature," would be found *randomly* together on the landscape—suggests another "area probability" solution, similar to the one we performed for the Face and D&M. To be fair, since we can't rule out that other such clusters may exist on parts of Mars we haven't looked at, we'll restrict our current calculation to the area covered by the mosaic photographed by Viking—something like 13,000 square kilometers around the Face. This will provide a *very* conservative lower limit for the unique occurrence of such a feature.

The area covered by the City Square is roughly a third of a square kilometer (about 1000 square meters.) So, the probability of finding these four objects (related by alignment) *together* in such a tiny area is—again, very conservatively—the area of the Viking photographs (13 billion square *meters*) divided by the area of the City Square itself!

Or, *one chance in 13 million* (if you doubt it, you can check the math.)

The second set of relationships—the alignment of the four exterior objects with each other—can be treated in terms of the odds of each member being "parallel" or "anti-parallel" with the other three. The odds against *four* objects being randomly aligned—while being associated as well—turn out to be pretty incredible: one chance in 180 (with a measurement error of plus or minus half a degree) for any two objects, *multiplied by this same probability* for each of the remaining objects (180 times 180 times 180)—

Resulting in less than one chance in almost *6 million* for the random alignment of all four!

The product of the two preceding probabilities—the alignment calculation and the apparent physical association of the objects—leads to an overall probability of less than one chance in 70 *trillion* that this grouping is the result of merely random forces!

Again, there will be those who are highly skeptical of these rough calculations, if not incredulous at the results. These critics will point out that, since these objects lie almost connected, within a tiny area that's *circular*, we are more than likely dealing with *a simple impact crater*—whose broken and eroded walls account for the closeness of the "objects" as well as their alignment.

This reasoning breaks down, however, under any real scrutiny; nothing like this is observed anywhere else on the estimated 13,000 square kilometers we are considering (which have dozens of comparable-sized impact craters strewn across their area). Furthermore, erosion of a crater's walls produces *radial slumping* into the floor of such a crater (as is readily observed on other parts of Mars, as well as on the Moon and Mercury), not four equally sized, discrete, faceted, and *parallel* "structures"—with two of them tangential to the edges of the "crater." And definitely not four features *arrayed at a precise right angle to each other* (the odds of which we'll let you, the reader, calculate!).

But perhaps the strongest single argument against this "explanation" is the simple fact that this "impact crater" lies in the *precise* lateral center of the complex.

Again, the probability of this being merely another "accident of nature" is calculable: simply the area of the association (1000 meters square) divided into the total City area (about 3 million meters square)—or, about one chance in 30 thousand against its lateral central position being something random. (This is for an "error" of about 1000 meters; if we use the actual, measured precision of the location of this object—central along that line to within a few *tens of meters*—the odds against its being just some cosmic pot shot dramatically go up—to one in several *hundred thousand*! This illustrates, hopefully, that we are using rather conservative numbers for these estimated probabilities.)

Again, if we multiply these improbabilities together, we discover that so far our City Square has only one chance in a *million trillion* of existing—in the form and location we observe!

But we are not yet finished: for this apparently unique "pinpoint" on the Martian surface (as I hope we have demonstrated with the preceding calculations) is also the *one point on the entire Martian landscape* where an observer can stand and see the Face in profile—

Projected against the backdrop of that "cliff"—*another* one-of-a-kind feature on the surface!

What are the odds on this too being "mere coincidence"?

The calculation turns out to be the product of the angular diameter of the Face, as seen from the City Square (about 5 degrees), multiplied by the angular diameter of the cliff (which is the same as the Face itself, when viewed from our now famous City Square)—both then multiplied by the added probability that this cliff will even appear on the same side of the crater as the Face. The result of this calculation is revealing: one chance in 70 times once chance in 70 (for the Face/cliff alignment), times one chance in 70 (for the cliff happening to overlap the azimuth of the Face, viewed from the center of the crater)— *

For a combined probability of over 300,000 to 1 in favor of my "backdrop theory"—and against this lineup being anything like chance (which, in this case, is direct support for the Intelligence Hypothesis—as no meaningful geologic model can account for the presence of the cliff *precisely where it is*).

Oh yes, we must now multiply this probability by all the rest—to derive the true "uniqueness" of the City Square. The actual calculation produces the following incredible result—

---

*This latter expression is actually a shortened form of *two* additional probabilities: the chances of the cliff even being on the side of the crater "facing" the Face (derived by taking its observed angular diameter, as viewed from the center of the crater—30 degrees—and dividing it into a full circle of 360 degrees), or about one chance in 10, *times* the probability that this would overlap the much smaller angular diameter of the Face, when also viewed from the center of the same crater, which is about one chance in 14. The resulting combined probability—or one chance in 140 for the "cliff" overlapping any aspect of the azimuth of the Face—is the complete form of this part of the previous probability calculation of the total alignment: City Square, Face, and "cliff."

For simplicity's sake, however, I have used an somewhat arbitrary combined "one chance in 70," which is about a factor of 2 *lower*. This was initially a much simpler calculation, (for probabilities which are only order of magnitudes anyway!), simply 70 *cubed* and resulted in a conservative estimate of this crucial total probability. But for those purists who might object to an apparent discrepancy in deriving these important quantities, I have been urged to explain my "shortcut." In fact, these are only estimates. Their overwhelming magnitude is what's so convincing, regardless of a factor of 2 or 3, here or there.

*The Monuments of Mars*

Less than one chance in a *trillion trillion* that the objects clustered on this infinitesimal spot of Mars—from where you just "happen" to be able to see the Face in profile *and* against its backdrop of the cliff—are there by accident!*

By now you must begin to get the Big Picture: this collection of objects on the Martian surface not only *appears* unique, it turns out to be *expressibly* (if incomprehensibly) unique—in terms of more than ample data (stark contrast to some of SETI's basic problems). Even if you quibble with a factor of a thousand here or there, the overall probability is overwhelming—

That what we are observing—in the combination of the City and the Face—is *not* a fluke of nature . . . but designed.

To believe otherwise, when confronted by the staggering improbabilities of even the handful of interrelationships so far described, is to believe that we have stumbled across a "geological anomaly" which cannot be expected to exist *more than once on all the planets in a million gal-*

---

*Some critics have objected that the "cliff" and its associated crater cannot even be seen "from the 'City Square,'" because of the sharper curvature of the Martian horizon. It is true that, for someone "standing on the ground" (visual observing height: approximately 2 meters), the distance to the Martian horizon (because Mars is a much smaller planet than Earth) would be only about 4 kilometers.

The equation which allows us to predict the distance to the horizon (on Mars) with any altitude, looks like this:

$$D = 2600 \times \sqrt{h}$$

Where D is the distance (in meters) to the horizon
2600 is a constant (in meters)
And h is the height (in meters) of the observer.

From any substantial height (the upper stories of a surface structure, for example) the distance to the "observer horizon" dramatically increases. Example: from the upper floors of a 30-meter (~100 ft)-height structure, the horizon would be 48 kilometers (28 miles) distant—easily creating the "problem of the crater," along the Face's line of sight. Since there *are* several structures, exceeding 100 ft. in height, making up the City Square, the critics' objections are made moot on this point. In fact, it seems logical to infer that the "cliff" was probably constructed *after* the objects making up the City Square—when it was discovered that the view of the Face no longer took place over an unobstructed horizon! For some further recent discoveries, see Appendix II.

*axies—*

Let alone next door!

Yet that is precisely where we find this strange collection.

At this point, we must step back and consider some deeper philosophical implications of the preceding logic and simple calculations.

Faced with such overwhelming odds, the question which arises in my mind is simply this: how many relationships must we discover before we know with *certainty* that what we are seeing is truly artificial?

For in the end, if and when we get new spacecraft images of Mars (hopefully, from at least the Soviet Orbiters in 1989), and they are good enough to show the proverbial "hieroglyphs," what will we *really* be seeing?—

Which will allow everyone to gasp with certainty,

"My God! It really *is* an alien civilization!"

My answer: simply more relationships—a *lot* of them.

Gregory Bateson, the brilliant anthropologist, once presented a class with a similar *epistemological* problem (epistemology is roughly, "the science of how we know what we know."). His example seems curiously relevant . . .

Bateson showed the contents of a bag, a freshly cooked crab, which he placed on the lab table with the following challenge:

"I want you to produce arguments which will convince me that this object is the remains of a living thing. You may imagine, if you will, that you are Martians [!] and that on Mars you are familiar with living things, being indeed yourselves alive. But, of course, you have never seen crabs or lobsters. A number of objects like this, many of them fragmentary, have arrived, perhaps by meteor. You are to inspect them and arrive at the conclusion that they are the remains of living things. How would you arrive at the conclusion?"[8]

Bateson later explained,

"I was asking: What is the difference between the physical world of *pleroma* (the nonliving), where forces and impacts provide sufficient basis of explanation, and the *creatura* (the living), where nothing can be understood until *differences* and *distinctions* are invoked?"

Continuing, Bateson said,

"By putting them on an imaginary planet, Mars, I stripped them (the students) of all thought of lobsters, amoebas, cabbages, and so on and forced the identification of life with living self: '*You* carry the benchmarks, the criteria, with which you could look at the crab to find that it, too, carries the same marks.'"

Bateson coined a phrase which succinctly generalized the process he was seeking to illuminate. He called it,

"The pattern which connects."

I believe that that same pattern is resonant in the unique array of objects at Cydonia—the pattern of "the living."

Repeated five-sided shapes in nature—unless they're *biological*—are nonexistent. Bilateral symmetry presents almost the same situation. Yet at Cydonia, we find *both* of these together on the landscape, not once but several times in the forms of several pyramids within the City, to say nothing of the D&M—and specifically *related* through unmistakable alignments. Bateson's definition must remain: what is the "pattern that connects" these objects, if not a higher-order *context* than the relatively simple forces which carve rocks? A context we call "living?"

For that was Bateson's other message: all meaning is *in context.*

At Cydonia, there is a "pattern which connects" the Face and D&M. The Face should not exist on Mars, but we can prove that it is "real"—a three-dimensional model, accurate to remarkable degrees, to our own image. Furthermore, it is associated with another bisymmetrical object, which echoes our own *dynamic morphology*—the form of living things. Geology—capable of explaining isolated aspects of this array of anomalous morphologies—cannot explain their underlying *context*—

Which is "human," as inexplicable as that may seem.

Thus, my final "vote" on what we are seeing at Cydonia—based on both the probabilities and the underlying *pattern:*

We are seeing "the products of Design" . . . and all that that implies.*

## Notes

1. Davies, M.E. and Katayama, F.Y., "The 1982 Control Network of Mars," *J. Geophys. Res.* 88:7503, 1983. *ibid.*
2. Cocconi, G. and Morrison, P., "Searching for Interstellar Communications," *Nature* 184:844, 1959.
3. Drake, F.D., "Project Ozma," *Physics Today* 14:40, 1961. Copyright © 1961 American Institute of Physics.

*This conclusion comes, of course, under the heading: what we think we know. If we are seeing something in nature that defies current understanding, then of course it would represent an aspect of the universe we cannot presently even imagine. Any way the problem is approached, if this turns out to be a "pile of rocks," it is (to paraphrase Loren Eiseley) a wondrous pile of rocks indeed!

4. Goldsmith, D. and Owen, T., *The Search for Life in the Universe,* Menlo Park: Benjamin/Cummings, 1980.

5. Drake, F.D., "How Can We Detect Radio Transmissions from Distant Planetary Systems?" *Sky and Telescope* 19:140, 1960.

6. Sagan, C., ed. *Communication with Extraterrestrial Intelligence (CETI)*, Cambridge: MIT Press, 1973. *op. cit.*

7. Churchman, C.W., *The Systems Approach and Its Enemies*, New York: Basic Books, Inc., 1979.

8. Bateson, G., *Mind and Nature*, New York: E.P. Dutton, 1979.

# XIV

# SEARCHING FOR
# THEIR ORIGINS AS WELL

"Interstellar flight is not a problem in physics . . ."
but a matter of biology."

—Eric Jones

Well, dear reader, having traced the twisted trail of this discovery through interwoven layers of evidence and logic, we are now ready to confront what I believe is the central enigma of this tale:

From whence—if they really did exist—came "the Martians?"

For of one thing (and one thing only) regarding this unfolding odyssey I am almost certain:

Whoever they were . . . they did not come from Mars.

As I have tried to show in previous sections of this book (Chapters III and VII), given the "knowns" of early Martian history—a brief epoch of "warm and wet" atmospheric evolution, followed by eons of glaciated cold—one has a hard time imagining Martian scenarios which permit the origin of life itself, let alone the appearance and evolution of something as complex (and, on Earth, as late-appearing) as *intelligence*!

On Mars there simply wasn't time.

Granting for a moment that we are perhaps wrong in this pessimistic assessment—the difficulty of getting even simple lifeforms to put in an appearance on the planet (after all, our very sketchy knowledge concerning the origins of life comes strictly from a single "data point"—the Earth!)—this still leaves as inexplicable the totally improbable development of "conscious life": Intelligence.

Even if we throw out *all* current conceptions of the possible—and grant that, by some almost miracle, the earlier appearance of intelligence on Mars might be a result of accelerated evolution (from environmental forces currently unknown—for instance, the apparent lack of a Martian magnetic field and thus a greater background of mutating cosmic radiation)—we are still left with an apparent development impossible to reconcile with anything we know—

The presence of the Face.

By any stretch of the imagination, an artifact constructed in the likeness of whomever once inhabited the planet—"the Martians"—should emphatically *not* resemble us! Yet that is precisely what we find . . . a true paradox if we insist on an indigenous origin for "Martians," and the Face as "a work of art" intended to represent themselves—as I mused in my initial "model," back in Chapter II.

An alternative suggestion (Chapter VII), is that the Face was constructed in the likeness not of those who might have made it ("Martians" or otherwise), but of *ourselves*—as a signal to be found when we arrived. like the famed Pioneer 10 Plaque—now heading for the stars.

The crucial difference between the terrestrial technological example (Pioneer 10) and the situation of the Face should be apparent: the Plaque carries an image of its makers (us) to an audience which it will most likely never "find," and who (to its creators) will quite likely forever be unknown; in striking contrast, the image on the Martian surface, in this model, was specifically *intended* for its future audience, and thus constructed in "their image."

The necessity of this scenario I am unfolding (i.e., not only a non-Martian origin for the architects of the Face, but a terrestrial explanation for its humanoid character) is inexorably prescribed by our current model of basic evolution.

By everything we think we know, the *independent* appearance of a humanoid image *"anywhere else in the observable Universe"* (according to orthodox interpretations of evolutionary theory) is utterly impossible.

That such an image would have the temerity to appear, of all places in the vastness of the universe and its planetary environments, "next door"—on Mars—is certainly more than can be accommodated by any existing evolutionary theory . . .

Which is why my own early attempts to find an explanation for the Face always returned to the "message" model—as a means of preserving the basic paradigm regarding evolution (which has, after all, some redeeming value!), while at the same time not discarding increasing evidence for the reality of this wondrous Martian feature—as a deliberate . . . "monument" . . . to someone.

But, however satisfying this "explanation" first appears, one invariably develops, on further surmise, a nagging suspicion that it still

*The Monuments of Mars*

doesn't answer the key questions raised above—

Because the central problem with the "message theory" is those pyramids—and their intricate connection with the Face. The sheer scale of engineering behind the construction of this massive "complex" argues compellingly *against* the message model. It's one thing to inscribe a set of geometric lines on a 20-centimeter panel of aluminum; it's quite another to array quite precisely and geometrically a set of discrete objects, some measuring *kilometers* across, tens of kilometers around an alien landscape—all ostensibly for the simple purpose of communicating with a remote audience . . . who might never come!

No, the Face *by itself* might be a message, but the rest of the associated features—and their inextricable geometric linkage with the Face—argued strongly for an *indigenous* reason for their presence—and all that that implies.

Which brings us back to where, for me, all this had truly started: the pyramids themselves . . . and a Terran genius named Soleri.

*     *     *

Paolo Soleri (see Chapter VIII), one of the best known of the "environmental architects," was among the first to propose a set of *monumental architectural constructions* as the ultimate solution to "the urban problem" (another was the legendary designer of the "geodesic dome," R. Buckminster Fuller).

In 1969, a compendium of Soleri's designs appeared in a large-format book. A lavish series of drawings illustrated his multi-leveled concepts of "macroengineering": three-dimensional structures—termed "arcologies" (*arci*tectural *ecologies*)—capable of housing *entire populations of some terrestrial cities inside a single structure*!

Soleri based his constructions on the ultimate necessity of an "organic" view of habitation; in his opinion, "the sprawling, essentially flat cities and suburbs that are eating up the surface of the earth are 'utopian' in the negative sense that they are absurd and unworkable."[1]

In dramatic rebuttal to the horizontal spread of current terrestrial urban complexes, Soleri proposed (according to the blurb on the cover of the book), "a population *implosion*—the flat stretches compacted in many folds into a true solid, a *city building*—a work of total architecture, a fact of neonatural ecology: an arcology . . . (italics added)."

The description continued: "Complex, *insulated from entropy, self-sustaining, miniaturized*, the city and its people become as one, an involuted/evoluted superorganism. Nature at large, at the doorstep and immediately accessible, returns to its 'natural state,' undefiled and in harmony with its own ecology (italics added)."[2]

Soleri envisioned arcologies consisting of many hundreds of levels, their heights commensurate with their horizontal dimensions—some of which were to be measured in *kilometers*—and designed to be large enough to provide life for *millions* of inhabitants. Yet their urban densities were calculated at only about 300 individuals per acre or 50,000 per square kilometer (it's probably more appropriate to speak in terms of units per *cubic* kilometer, as Soleri envisioned his inhabitants fully utilizing and enjoying the manifold benefits of a *three-dimensional* existence).

Or, as he himself concluded:

"In the three-dimensional city, man defines a human ecology. In it he is a country dweller and metropolitan man in one. By it the inner and outer are at 'skin' distance. He has made the city in his own image. Arcology . . .

"The city in the image of man."[3]

*      *      *

The thing which struck me at once about Soleri's millennial vision was its exquisite answer to the central riddle: why *pyramids* . . . on Mars?

What better solution for maintaining several hundred thousand inhabitants against the currently inhospitable Martian environment, I realized, than to house them in a series of vast, artificially constructed *enclosed environments* . . . arcologies.

"The city in the image of man" . . . perhaps *literally*—considering the proportions of the D&M!

(Contrast this vision—of a truly sophisticated civilization, "complex, *insulated from entropy, self-sustaining, miniaturized* . . . "[4]—with Carl Sagan's cursory examination of Mariner and Viking photographs, in search of "Los Angeles" . . . )

My fundamental reevaluation of the scale and complexity of the Martian pyramids—from seeing them as analogs of their Egyptian "cousins" (Chapter II), to seeing them as potential "arcologies on Mars"—did not come overnight, but was nurtured by a combination of new discoveries on the images themselves and a review of what we know about the planet.

One key reason for the evaluation was my discovery of the "honeycomb" that memorable night (Chapter V), which instantaneously cast grave suspicion over any simple "neolithic" interpretation of the pyramids' construction—primarily because of engineering problems which would inevitably be encountered (even under one-third gravity!) in erecting "transparent" spans over three *kilometers* long. (We'll return to the famous controversy regarding the reality of "the honeycomb," and some

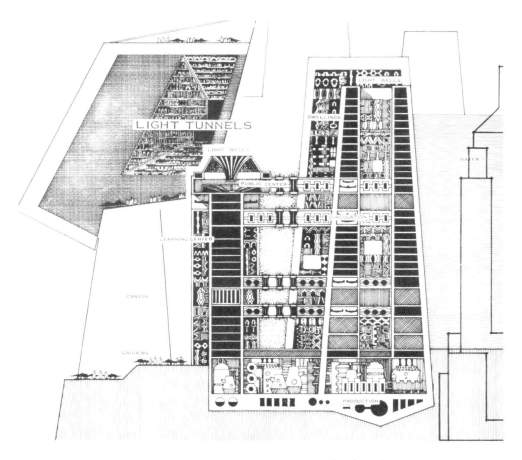

Paolo Soleri, THEODIGA

new insights into what its image actually might represent, in a moment.)

Complicating the "honeycomb discovery" was my detailed reconstruction (Chapters III and VII) of the environment of Mars—which, as I stated at the beginning of this section, effectively seemed to rule out any chance for either the indigenous origin of life on Mars or, in the unlikely case of primeval Martian "biology," its subsequent evolution to Intelligence.

Which meant, if we were seeing intelligently constructed "pyramids on Mars," they *couldn't* be in the prototype of the "Egyptian model"— enormous limestone tombs; the environment simply couldn't have supported the biological developments which on Earth preceded such "primitive" stratified civilizations and their religious architecture. That left only one reasonable alternative:

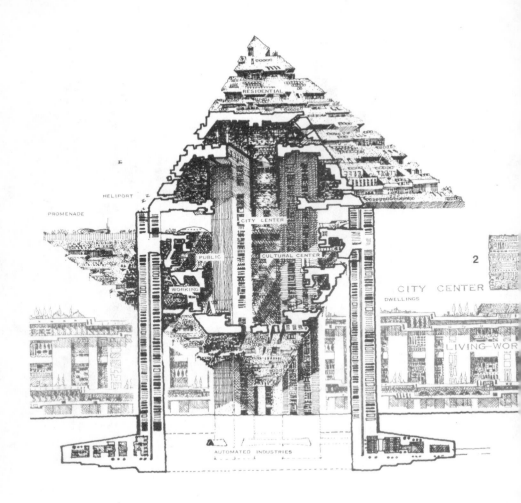

Paolo Soleri, HEXAHEDRON

Migration. By beings obviously not only more advanced than Egyptians but more advanced than us.

If the Martian pyramids were, indeed, a series of intricate, contained environments in the middle of a "hostile" landscape, then the obvious function (for structures that vast!) would be in sustaining an *exogenous* population which required a life support unavailable on Mars. In other words, if you came from someplace else and—for whatever reasons—wanted to inhabit Mars, you would need to prepare a self-contained "colony" or "base"—

Precisely like what we apparently were seeing!

One of the critical criteria of such a potentially tall tale was that, for it to be accepted even as a tentative solution, the "arcology model"

*The Monuments of Mars*

should be *testable*—in keeping with the other elements of the Intelligence Hypothesis.

A key measure seemed to present itself immediately: the "arcologies" themselves (in Soleri's engineering) were enclosures dependent on *solar heating* for their homeostatic thermal regulation. Mars, being farther from the sun, and having a much thinner atmosphere, meted critical limits for the *orientation* of such "deep greenhouses" on its surface; they should *face the rising sun*—particularly in the deadly days and nights approaching the morning of the winter solstice (which, even by Martian standards, is somewhat chilly).

Imagine my astonishment when I discovered that the major structures were, indeed, oriented *with the summer and winter solstice sunrise*

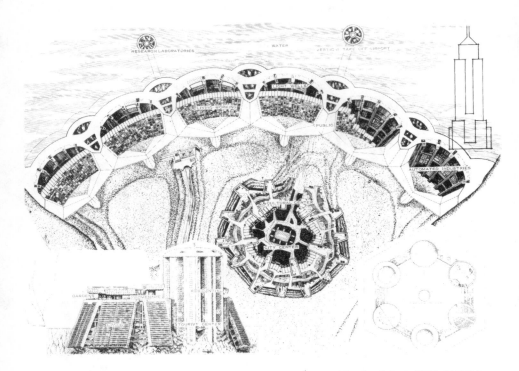

Paolo Soleri, VELADIGA

*sightlines* first determined for the Face. This was precisely the kind of evidence needed to support the arcology theory! (Not only that; it was highly unlikely that a mere "pile of rocks" could "know" where the sun should rise on those specific mornings, in *the same epoch* as the Face. This was another gross "improbability" in favor of the general Intelligence Hypothesis!)

Fine measurements of these alignments, from pyramid to pyramid within the City, revealed another set of interesting data—which could be interpreted as favoring both the "arcology hypothesis" *and* supporting a *long* period of habitation—

Slight but definite, systematic differences, visible on large-scale prints (up to a few degrees), in the alignment of several different structures.

Did this indicate an attempt to "track," despite the obliquity shifts, the constantly shifting sunrise on the horizon? [The Egyptians incorporated periodic re-alignments in their architecture, first in successive pyramids, and later, by alterations to specific temples—in order to track both the solstice sunrise and important stellar risings, to respond to motions caused by the significantly smaller shifts that occur in Earth's celestial axis.]

*The Monuments of Mars*

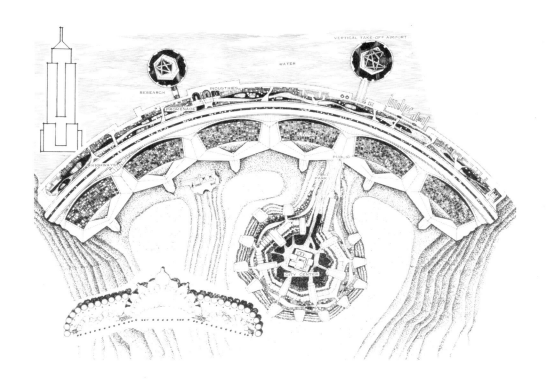

These subtle alignment differences of structures on Mars were consistent, then, with the theory that the Martian pyramids had been erected at slightly different epochs ("slight" being a few thousand years apart!), as the majestic "nodding" of the entire planet inexorably moved the solstice sunrise along the horizon. They further indicated that whoever had constructed these massive works of engineering had not done so "in an afternoon . . . " but in the course of spending considerable time inhabiting Cydonia!

Now, some of Dan Liebermann's keen observations suddenly made sense—like those mysterious deep "holes" he had noted in the ground just north of what I called "the honeycomb" (Chapter IX).

If the vast surface structures were, indeed, environmentally contained arcologies, then it stood to reason that much of their structure was probably *underground* (interconnected below the various apparently separate pyramidal surface features). Soleri's own arcologies had "roots," in many instances as deep as they were high, along which everything from factory complexes to basic air-conditioning operated. (The residences were, of course, located higher—where windowed sunlight would be available from specially-designed "lightwells" incorporated deep within the several hundred levels.)

This design was particularly attractive for the Fort—the one structure in the City that is decidedly non-natural *and* non-pyramidal. The Fort seemed to be a very complex structure, assembled around some kind of central "courtyard" . . . (reminiscent of a "lightwell" in Soleri's drawings). In addition to pyramidal arcologies, Soleri conceived a host of others, designed in a variety of complex topologies. The Fort seemed more like some of these exotic environs than the "standard" Martian pyramid (perhaps flagging some abrupt environmental reason for the change).

However, like the pyramidal structures, the Fort also seemed to "know" where the sun would rise and set; the northwest "flat," side of this peculiar object faced the *summer* solstice sunset; the pointed southeast section was aimed almost directly at the *winter* solstice dawn . . .

And all for the same epoch as the Face!

These alignments seemed too accurate to be mere chance. Rather, they also suggested an attempt deliberately to orient this structure for solar insulation. The adjacent "honeycomb," extending southwest, likewise.

The "honeycomb."

No single feature that I'd found on Mars had caused such excitement . . . or such controversy (Chapters V and IX). If it was what it *appeared* to be—a series of cubical "cells" arrayed in a deliberate architectural configuration against the Fort (judging from independent analysis of the shadows cast by its three-dimensional relief)—then it was almost certainly an *artificial structure*—

And thus the single most impressive confirmation of the whole Intelligence Hypothesis!

If, on the other hand (as Gene Cordell and Vince DiPietro maintained), it was a mere "computer glitch," then its presence was not only a distraction, it had done more harm than good (Chapter IX).

*     *     *

It wasn't until the very evening before John Brandenburg's presentation of our "Boulder paper," that I finally figured out the "honeycomb" . . .

It was a moiré.

Let me quote from Gregory Bateson,

"Interesting phenomena occur when two or more *rhythmic patterns* (italics added) are combined . . . the combination of two such patterns will generate a *third* (italics added). Therefore, it becomes possible to investigate an unfamiliar pattern by combining it with a known second pattern and inspecting the third pattern which they together generate."[5]

The best known examples of *audio moiré patterns* occur in music—where two frequencies interact to generate a "beat frequency." It's because of this phenomenon that two very high notes (which would otherwise be beyond the upper range of human hearing), while beating together, may generate a third tone *lower* on the scale—which *can* be heard.

On television (which is a little closer to the Mars situation), the optical equivalent of the above phenomenon is generated by the regular *scan lines* of the video camera: by the imaging process itself.

When a TV camera is pointed at a second regular pattern—for example, the slats of a Venetian blind—what you see on your home screen is a series of wavy, broadened bands (usually running diagonally across the screen), the result of the *moiré* between the two regular line patterns—the one in the camera and the one the camera is shooting. The same phenomenon occurs with window screens, picket fences, modern skyscraper architecture—anything which presents a *regular pattern* to the television camera's *regular* scanning system.

Thus, we can say as a general rule that when any two regular patterns interact, they will produce a third—which on occasion can be much *larger* than either of the two originals.

How does this apply to the "honeycomb" on Mars?

In a way, Gene and Vince were right . . . The "honeycomb" *is* a computer-generated "artifact"—but one also produced by an equally real "artifact on Mars." It's a moiré, generated (I now believe) by the interaction of the regular scan-line pattern in the Viking camera, looking down on a *regular pattern* on the surface of Mars, in between the Main Pyramid and Fort.

What apparently was required to truly generate the "honeycomb moiré pattern" was the special kind of computer algorithm devised by Vince and Greg, called SPIT (Chapter I). By dividing each original Viking pixel into nine new "subpixels," SPIT applied *another* rhythmic pattern to the data—which apparently was just enough to cause a moiré in this *one* region of the photograph. The essential point is this: *creation* of such a moiré via SPIT would have been impossible—

*Unless* there was a smaller (higher frequency), regular pattern on surface to begin with—which could most likely *only* come from artificial causes!

The fact that this was the only place on the entire Viking image which generated such a pattern, and the fact that it architecturally made sense (coupled with the peculiar "light scattering" properties along its width on even non-SPIT images), all made me believe this was the scientific explanation for "the honeycomb." It fit all the data on the problem . . . and, it was a *testable* idea.

The second image of "the honeycomb" region (70A11) apparently did not reveal the same pattern. This fits perfectly into this "moiré model"; 70A11 was taken at a *different* distance from the planet and at a different azimuth, thus the projection of the camera scan lines onto the regular pattern far below the spacecraft would be completely different— resulting in a "non-beat situation" inconducive to the generation of moiré!

In a sense then, the absence of "the honeycomb" on 70A11 was a *negative* test of the "moiré theory." But what we really require is something more positive, like—can the theory reveal anything about the nature of the "honeycomb" itself? In principle it can, by applying Bateson's logic above:

" . . . it becomes possible to investigate an unfamiliar pattern by combining it with a known second pattern and inspecting the third pattern which they together generate."[6]

Since we know the mathematics of both the Viking camera and the pattern of the "honeycomb" we observe on the SPIT images, theoretically we can invert the problem—to derive the actual size of the "regular pattern" on the surface *causing* the moiré!

If this mathematical analysis is carried out (as it will be in the course of the Mars Investigation), we can, in turn, subject that analysis to an *experimental* test: by preparing a scale model of the Fort and "honeycomb" (with screening of a variety of different "meshes") and shooting it with a version of the camera used in Viking!

Finally, even before this new "conceptual breakthrough" on the problem of "the honeycomb," both Lambert Dolphin and I had independently noted (in "Chronicles"—Chapter IX) what seemed to be "extensions of the 'honeycomb'" *across* the Fort itself: hints of many different levels, seen as sharp boundaries which, on crossing other features, suddenly lightened underlying shadows—as if an overhanging layer was "veiling" or "scattering" light from deeper levels of the structure. What we *didn't* see was any actual "'honeycomb'-like detail," indicating (if the moiré theory is correct) that the screening pattern of this "decking" is of a different size from our familiar "honeycomb" to the southwest. (Shades of Soleri's "several hundred levels" . . . )

Thus, by a combination of a new idea, optical analysis, and experimental model-making, the "problem of the honeycomb" might ultimately be resolved—hopefully, in favor of its being our clearest single piece of evidence for the "arcology nature" of the City . . . and thus the entire

Intelligence Hypothesis.

*   *   *

All of which argued compellingly *against* the "message" model.

For, if these increasingly sophisticated and complex structures were evidence of some kind of permanent Martian habitation, and the Face was inextricably embedded in that Plan (as demonstrated in Chapter XIII)—then the Face *had* to share somehow in this *indigenous* purpose for the existence of the Complex.

(Incidentally, its own placement and orientation on the surface, such that it was perfectly "framed" by that peculiar background "cliff," was compelling argument alone for an indigenous purpose for its presence.)

Which returns us to the single overriding question: who were "the Martians?"—

Who had not—*could not*—have originated on this "wan and wasted world," but who had left a series of megalithic architectural mementos to their former presence—including one that looked suspiciously like us.

A clue came, intriguingly enough, from the possible presence of the "arcologies" themselves.

The blurb on Soleri's book stated that "the nearest ancestor of the arcologies is . . . *the ocean liner* (italics added)—another highly structured solid that self-contains a total environment for human life."

Some years before, by a fascinating coincidence, I'd found a different—haunting—reference to this "ocean liner model" . . .

*   *   *

The scene was from an Arthur Clarke short story . . . a wondrous vision, seen by a Polynesian boy paddling a small outrigger canoe out beyond the reefs at twilight . . .

"Every two weeks, the liner would pass by—a vast, glittering city in the darkness. This night, he paddled farther than he ever had before, waiting for the magic moment when it would come again . . .

"Then, there it was: a magnificent array of color, all aglow, the throb of mighty engines driving it from over the horizon.

"He paddled faster, straining to catch up with this celestial vision, this fabulous place he longed to be a part of. He imagined he could hear laughter, the strains of music, people shouting, 'Hurry!'

"Yet, even as he strained with every fibre on the paddle, he realized that this wondrous vision was moving too fast for him to catch it; it was leaving him behind!

"Soon the night was quiet once again; the mighty engines were gone, the lights of the floating city disappearing over the horizon. And he was left, one small boy, alone . . .

"Alone on the vast ocean, underneath the dark and starstrewn sky . . .

"For all of man's brief history on Earth, when the word 'ship' was used, it applied to such ocean-going vessels. But for all future time—for a history which would spread across a million worlds and perhaps as many years— the word would apply to that unique class of vessel which had spread humanity across the stars . . .

"The *star*ship."

\*　　　　\*　　　　\*

With apologies to Arthur, I've had to reconstruct this scene from memory; somehow my copy of the original was lost in the intervening years, since I first was captured by his memorable image—of "ships" flying on forever through the stars . . .

It was an image so powerful that it prompted a fundamental question now regarding the artifacts I'd found: did a combination of Clarke *and* Soleri's visions generate the "how" and "why" of what we were witnessing upon the Martian landscape?

Isaac Asimov (another writer who's shed an occasional illuminated thought on some of these same subjects . . . ) once had an interesting observation regarding those who would ultimately go to the reaches of deep space. He said,

"It won't be men and women from the Earth itself who explore the outer planets and, someday, the stars . . . It will be those who've grown up in space colonies or in bases on the Moon, who've never known the 'wide openness' of Earth. Those voyages will require people who are *comfortable* within close quarters, within the severe limits of a closed-ecology—which is what life aboard a deep space exploration vessel will impose . . . "[7]

Asimov's point, made years ago during the 1970's flap over Gerry O'Neill's space colony ideas, stuck with me . . . to re-emerge in concert with Arthur's haunting South Pacific vision—as I pondered the *reason* for vast, contained ecologies on Mars.

Suppose, I thought—

(My inner watchdog warning, "Watch it! You're really speculating here . . . ")

—that the originators of the Martian "ruins" were, indeed, migrants from another planet. The logical next question was: "Which one?"

If you look around the solar system, it quickly is apparent that, other

*The Monuments of Mars*

than the Earth, there's not a lot of habitable real estate to come *from*. [8] Consider the other "terrestrial" planets of the inner solar system . . .

Mercury is hot enough to melt lead (and is airless). Venus is currently as close to "Hell" as it is possible to come, in a planetary context: unbreathable carbon dioxide at 100 times Earth's current atmospheric pressure; a liberal sprinkling of pure sulphuric acid thrown in; simmering at about 900 degrees F. (and most likely, it has been this way for a long, long time—too long to have created its own intelligent planetary "migrants!").

The outer planets are just as inhospitable; buried under *thousands* of miles of equally unbreathable and crushing atmospheres, with one other "goody" thrown in: there's literally no place to stand (for someone who would one day be interested in walking on the "sands of Mars!"). These massive outer worlds are essentially massive spinning *oceans*—only of such exotic stuff as *super-heated liquid hydrogen* and *helium* (and, as if this weren't enough, these elements, even with traces of "impurities," don't have an inclination to form organic molecules of the variety required to form "life").

Which means then, if the "colonists" of Mars didn't come from Earth (for reasons we previously discussed in Chapters II and V), we are left with only one reasonable alternative—

The stars themselves.

\*     \*     \*

To understand the significance of this conclusion, you must understand the true dimensions of this thing we casually term "space." Space is vast—vaster than anything within our comprehension. The only way even to approach an appropriate appreciation for its vastness is by analogy: using miniature models (usually composed of fruits, vegetables—and even household "notions!").

Example:

If the Earth is represented by the head of a common sewing pin, orbiting at its normal distance from the Sun, the Sun in this miniaturization can be represented by a grapefruit (see!)—placed about 50 *feet* away.

In this model, the "outer marker" of the solar system—the orbit of the very distant Pluto—is something like 40 times as far away from us as the center of the solar system, the Sun. Thus, it is another pin-head (though only about a quarter the size of Earth's!)—slightly less than *half-a-mile* (about 2000 feet) away.

Let's consider time.

In the "real universe," light travels at the familiar 186,000 miles per second (300,000 kilometers per second, for you metric fans). Even at this awesome speed, it takes light 8 full *minutes* to flash across the gap between the Earth and Sun! It takes slightly more than five *hours* for that same light to cross the orbit of Pluto—over 4 billion miles (6 billion km) further out. (I used to get a kick out of telling Walter Cronkite, back when he was doing the CBS Evening News at 7:00 P.M. and Harry Reasoner was doing the 11:30 version, that when Reasoner came on he, Walter, was "just crossing the outer boundary of the solar system—at the speed of light!")

The distance of the nearest star is almost a *million* times the distance of the Earth from the Sun (800,000 to be precise) . . . which in our "pin and grapefruit model" translates to a whopping 8000 *miles* separation—the diameter of Earth itself!; if we had two grapefruits floating 8000 miles apart, that would accurately represent the comparative separation between the Sun and Alpha Centauri, the star nearest to the Sun. (And, of course, as almost everyone has heard, it would take over four *years*— traveling at the speed of light—to cross that separation.)

Distances to a representative sample of the other stars closest to the Sun are even more daunting: the scattering of suns within about 20 light years (which astronomers sometimes quaintly call "our Galactic neighborhood") would, in our model, be an array of "grapefruits" floating in an emptiness represented by a volume equivalent to over 60 "Earths"— each grapefruit about one Earth diameter from those around it!

To talk realistically about travel between stars within this "neighborhood" (no "hyperspace" or "warp drives," please!) is to talk in nothing but superlatives—both in technological advances required even to begin to solve the problems such journeys represent; and in terms of the sheer *time* they would require. For, we are not presently talking about speeds which approach even a modest fraction of the mythical "velocity of light;" rather, even considering incredibly sophisticated (though engineeringly predictable) propulsion advances—such as "fusion" or "anti-matter" systems—we are usually talking velocities which are only about a *tenth* the speed of light—which immediately translates to at least a *half-century* travel-time between Alpha Centauri (slightly over 4 light years away, remember?) and our Sun!

Half a century . . .

It's no wonder, then, that someone once called the distances between the stars "God's quarantine regulations . . . "

Which is why one does not lightly propose that someone, sometime in the solar system's yawning past, crossed those cavernous *trillions* of very lonely miles, to light upon a speck of sand orbiting our very average

*The Monuments of Mars*

sun . . . unless inexorably *forced* to that conclusion.

*     *     *

Now that you know something of the awesome barriers to interstellar flight (which caused one physicist some years ago to exclaim, "All this talk about travel to the stars belongs back where it came from—on the backs of cereal boxes!"), we can get down to considering the real problem of the Martian artifacts:

If those who made them didn't originate on Mars, and they didn't come from any other planet in *this* solar system, then how *did* they cross the apparently insuperable barrier between the stars? Here we come face-to-face with the stark reality of "interstellar flight . . . is a matter of biology . . . "

With virtually limitless distances to traverse, and the very finite capabilities of the best of foreseeable propulsion systems, we are driven to conclude that flight between the stars is basically limited by the *biology* of the intrepid individuals who set out to make the crossing . . .

Which (in terms of human biological constraints) presents us with basically two options: the "slow-boat, generation ships" (made popular initially by science-fiction writer, Robert Heinlein,[9] and recently expanded on in some detail by such investigators as Dr. Eric Jones, at the Los Alamos National Laboratory[10]); or a method whereby the potential "colonists" are placed in *suspended animation*, to be "awakened" centuries later . . . when the ship ultimately arrives.

Both of these alternatives are the result of the *brief lifespan* of human beings. If our proposed "visitors to Mars" had similar limits on *technology* (relatively slow, sublight interstellar speeds) but very different *biological* constraints, the real possibility is opened up that the reason we have found an apparent "colony on Mars" is because the individuals who journeyed to this solar system lived *a much longer span of years—perhaps even millennia!*

[Recently, NASA (and the U.S. Air Force) have conducted systematic studies of these various technical alternatives for achieving "sub-light interstellar flight," some of the best carried out by Dr. Robert Forward, of the Hughes Research Laboratories and others.[11] These studies have confirmed what decades of science-fiction writers presumed: there are no "insuperable barriers" to flight between the stars—there is just a lot of hard engineering (and perhaps genetic science!)]

What we can conclude from this is that interstellar flight *is* technically feasible . . . provided you have a LOT of energy and time—and a firm handle on constructing truly closed life-support ecologies. And, should

*Searching for Their Origins As Well* 251

the lifespan of those who make the journey be significantly longer than our own, then the *sociological* aspects of the problem change dramatically, with far-reaching implications for rapid colonization *of the entire Galaxy*—a conclusion reached by another physicist (and science-fiction writer) who's done some first-class work in this relatively untouched area, Dr. David Brin. [12]

Which is where Arthur Clarke's grand metaphor and Paolo Soleri's equally impressive vision come breathtakingly together on the reddish sands of Mars . . .

*   *   *

For, to me, what we apparently were seeing at Cydonia was a fascinating, almost inescapable *extension* of the overwhelming social and technical necessities of interstellar flight: an entire culture which, after flying through interstellar space for—who knows how long?—had created an "insulated . . . self-sustaining, miniaturized . . . superorganism" on the planet they eventually chose to live upon—

A *duplicate*—socially and environmentally—of the starship that had made their journey possible . . . Which brings us to the next emphatic question: why—with Earth so close and so available—

Why did they choose *Mars*?!

*   *   *

It is difficult for "Earthlings" (like myself) to imagine someone going to the trouble of crossing interstellar space, finding a lush "paradise" like Earth (uninhabited and so invitingly at hand) . . . only to wind up choosing Mars as a place for habitation! The current SETI paradigm regarding potential interstellar denizens (Chapter XIII) envisions a host of "Earth-like worlds" within the Galaxy, where (presumably) life originates and climbs—slowly, painfully—up the "evolutionary ladder," until it attains the capability for communication with the stars (nominally, by radio, you will remember). It is almost incomprehensible, in terms of such a model, why such beings, originating in such a similar environment and deliberately undergoing the extraordinary expense and hardship of an interstellar *expedition* to find another "home," would willingly forego an "Earth"—in favor of a "Mars!"—

Unless . . . you remember those intriguing thoughts of Isaac Asimov: the *type* of individual (and where he or she might have lived before) who would in all likelihood volunteer for such an interstellar mission. Then, at least part of this strange pattern starts to come together.

Our analyses, our projections of the possibilities for life on other planets, how that life might one day travel to the stars, are subtly, inevitably, conditioned by our own experiences here, on Earth (just examine our references here, to "generations," "he/she" etc.!). Earth is a lush and verdant world with a plenitude of things we take for granted—water, oxygen, resources . . . and, above all, *room*. It is almost impossible realistically to project the biological, technical, let alone the *sociological* parameters of those who might one day choose to leave such "luxuries"— for a long and dangerous journey to another solar system.

In some such (very *terrestrial*) models, the individuals who start also know they will never live to finish—but will hand down to generations of descendants the purpose for the flight . . . and what to do one day when their great, great, great etc., grandchildren finally arrive (the "generation ship"). Under these conditions, with dozens of potential generations growing up *inside* the closed environment of such a comparatively tiny vessel, the possibility arises (if it is not deliberately *designed in!*) that the extreme *predictability* of such a totally controlled existence comes to represent "normality" . . . and an "external" planetary existence (if that was, indeed, the environment the colonists even left behind) comes to represent an . . . undesirable . . . situation.

Imagine, then, in the "generation-ship" model, with the preceding psychological conditioning (be it unintentional or deliberately designed), what might happen when these *extraordinarily remote* descendants of the original pioneers finally arrive within the target solar system. Imagine the potentially *terrifying* nature of a "normal" planet (!), such as Earth—

With storms and earthquakes, a million forms of alien life, and above all, *seasons*—all elements that these potential "colonists" are totally unable to control . . .

Now, envision the possibilities for extreme pressure to duplicate the "safe" environment experienced for all their lives aboard the starship . . .

And you begin to get an inkling of why such colonists might ultimately—against all "common sense"—choose Mars instead of Earth to live on: as the *one terrestrial planet in the entire solar system* which could provide essential raw resources to construct a series of "perfect" closed environments, yet at the same time, *not threaten* with external environmental factors (storms, oxidation, etc.*) the permanence of these vast "ar-

---

*While it is true that Mars is capable of producing planet-spanning duststorms, because of the much thinner atmosphere ($\frac{1}{100}$ Earth's), the "windloading" is insignificant. According to Viking's years-long surface measurements, the actual environmental effects of Martian duststorms are very small—insignificant, compared to Earthly weather.

cologies"—as a duplicate of "home" . . . the starship!

*　　　　　*　　　　　*

Viking itself beamed back vital information supportive of this model: Mars, of all the places we have visited across the solar system (the Moon, Mercury, Venus—to say nothing of the outer planets and their moons) is the *one* place, other than the Earth itself, *with all the raw materials essential for the maintenance of life and high technology!*

If Earth was excluded from "colonization" for the reasons cited earlier (basically, as too "unpredictable"), Mars would make the perfect substitute for such an enclosed "base" . . . which could ultimately expand into a "pyramidal City" . . . and beyond . . .

[Of course, one of the curious questions this entire model raises is: why didn't the "colonists" *remain* in space (which, technically, is both predictable and quite benign—in terms of engineering), setting up huge orbital colonies, pulling energy and resources from the entire solar system— as in scenarios envisioned by O'Leary and O'Neill? Perhaps the fact that "they" did not implies something equally profound about the difficulties or desirability of current terrestrial scenarios for developing the resources of the solar system . . . ]

*　　　　　*　　　　　*

Speculation? Certainly. But it is careful speculation, starting with the evidence before us in the Viking data—both in pictures and environmental surface measurements—and following (hopefully) some equally compelling logic—

To these (very!) *tentative* conclusions.

I do not say this *is* what happened; I merely say that it is one logical extension of the things we know. It's a scenario that is (as physicists are fond of saying) *consistent* with the data. Unfortunately (for those who want their answers simple!), at our present understanding of the evidence it is not the *only* scenario that fits . . . as you will see.

*　　　　　*　　　　　*

Let's return to the fundamentals of the so-called "SETI-paradigm": that the Galaxy is *filled* with other Earth-like planets, many of which have not only given birth to life but *technically-inclined life*—with a propensity for curiosity about their neighbor stars.

This assumption—that the Galaxy is veritably teeming with what I have termed "Miami-planets" [those which, at first glance (or first breath!), it would be very hard to tell apart from Earth]—stems directly from the careful efforts of one man: Dr. Stephen Dole, formerly also of the RAND Corporation. In 1964, Dole published a painstakingly-researched report—"Habitable Planets for Man"—in which he sought to determine the number of such planets in the Galaxy. His conclusion, based on a variety of separate factors (which were described in some detail in Chapter XIII, in terms of the famous "Drake Equation") was that there are about 600 *million* "Earth-like planets" in the Galaxy at large.

[One interesting fallout from Dole's study: Alpha Centauri, the star system closest to the solar system, had *the highest probability* of such an Earth-like world out of all the stars within about 20 light years of the Sun. This was the direct result of the fact that Alpha Centauri is actually *three* stars orbiting around each other—technically called a "multiple star system." Two of the components are yellowish stars comparable to our own Sun, separated by enough distance to allow independent planetary formation and evolution, without disruptive tidal influences by the other member of the system. The third component is a tiny, reddish, dim, and very distant (from the other two), cool dwarf star, whose planets—if any—must be very frigid worlds . . . ]

In 1984, an Australian astrophysicist named Thomas Donaldson published a radical new study,[13] based on much better data regarding star formation in the intervening 20 years—particularly, how stellar composition varies with both *age* and *location* in the Galaxy. The impact of this new composition information on theories of overall planetary formation—sizes, masses, and percentages of particular classes of stars which will form planets—was Donaldson's special area of interest.

Donaldson's basic assumption was this: that planet formation is a function of *the percentage of heavy elements incorporated in a forming star.* ("Heavy elements" are what astrophysicists call elements with more protons and neutrons in the atomic nucleus than helium contains—a list which includes nitrogen, oxygen, silicon, aluminum, iron, gold, uranium, etc.) Since "terrestrial planets" are composed primarily of five of these (oxygen, silicon, magnesium, and aluminum—wrapped around an iron core), it stands to reason that, all other things being equal, the amount of these elements in the original "solar nebula" probably was the prime determiner of the *masses* of the eventual "terrestrial planets"—

Mercury, Venus, Earth—Earth's Moon—and Mars.

Donaldson's "common sense assumption"—that the masses of these rocky planets are directly dependent on the observed percentage of heavy elements in the existing star (as opposed to the massive "spinning oceans" in the outer reaches of our solar system, composed primarily of hydrogen and helium)—becomes a powerful tool for *inferring* the number and types of "terrestrial" planets orbiting a variety of stars [which, even in the world's most powerful existing telescopes (including the someday-to-be-launched Space Telescope), remain mere "points of light"—with their planets (if any!) currently invisible to direct detection methods].

If his assumption is correct (a big "if!") Donaldson's theoretical technique allows us the first real mechanism to estimate (before we have real data!) roughly where in the Galaxy "Earth-like" planets will be found, and how young or old the star will be that they are orbiting. This, in turn, is of prime importance in attempting to estimate the stage of biological evolution on those worlds . . . and the possible presence of intelligence.

*      *      *

Donaldson's numbers, when he goes through the calculations, are *almost a direct refutation of the existing "SETI paradigm."*

"Earth-like" worlds turn out to be almost *vanishingly* scarce. But—and this is the real shock—MARS-TYPE PLANETS SHOULD BE LIKE GRAINS OF SAND STREWN ACROSS THE INTERSTELLAR NIGHT . . .

To quote Donaldson directly:

"Even though Mars-type planets are overwhelmingly common, planets like the Earth, old enough for complex life and an oxygen atmosphere, are very rare indeed . . . our estimates for likelihood of 'Earthlike' planets have decreased dramatically (since Dole's original RAND study), *to less than 0.002% of all stars* (italics added). In particular, the nearest 'Earth-like' planet, immediately habitable, is probably more than 200 LY from the Earth; at a velocity of .1c ("c" being the speed of light) we would need a *minimum* of 2000 years to reach it. Very likely humanity will have founded many colonies, and actively traveled the stars for several thousand years, before human crews land on a planet with its own native complex life . . . "[14]

Which makes all the more perplexing why someone would travel all the way to this "oasis" in the dark . . . and wind up choosing *Mars*!

For, if "Earths" are rare and "Mars's" are plentiful across the almost limitless Galactic night, then to find both of them *together* in the same system must be an incredibly rare event.* Why would anyone in

---

*Assuming *every* system with an "Earth" automatically also has a "Mars," the percentage of *total systems* with both is limited by the percentage of expected "Earths"—0.002%. Since only 15% of all systems (in Donaldson's model) will have "Mars," the actual percentage containing both is about 0.0002%!

*The Monuments of Mars*

their right mind, when presented with such an apparently rare *choice*, choose Mars . . . ?

But stop and think: the statistics of planetary systems, if Donaldson is correct, must overwhelmingly favor *Mars-type worlds* as those which usually orbit within the "life zones" of their respective stars. This means, in turn, that *these*—not other "Earths"—might be the preferred sites for the origin of life across the Galaxy—

If not the *evolution of intelligence* itself!

*We* might be the true "anomaly" . . .

This radical conclusion (which is mine, not Donaldson's) is drawn from the overwhelming numbers; Donaldson estimates fully *15 percent* of all planets orbiting the stars are "Mars-types!" With so many "Mars's," in comparison to so few "Earths," a certain undeniable percentage of the former will inevitably be formed much closer to their respective parent suns—with profound implications for subsequent environmental *and* biological evolution.

(Which, incidentally, includes the probability that the planets of the nearest other suns—those of Alpha Centauri—are more like Mars than they are like the Earth . . .)

The thought, forming in my mind, was, "Suppose, whoever came this way *didn't* originate upon an 'Earth' . . . but came from the equivalent of Mars . . . ?" Several "standard" scenarios, some looking suspiciously like plots for grade-B movies, immediately flitted past my eyes—including the "dying world" scenarios favored in the 1950's, where a race must emmigrate to find another planet when theirs becomes uninhabitable, for any of a number of depressing reasons . . . (Shades of Percival Lowell!)

Remarkably though, this tentative idea—that "they" may have come here from another version of "our" Mars—was supported from two *completely separate* lines: Donaldson's new theoretical statistics on planetary formation (based, in turn, on *observed* composition differences between thousands of Galactic stars) *and* my own observations of the "artifacts" on Mars itself.

Which still did not explain (in this scenario) why, at the end of such a long and dangerous journey, "they" apparently chose Mars instead of Earth—which, in *this* solar system, *is* the favored planet orbiting within the stellar "life-zone"—

Unless . . . there was an overwhelming reason not to.

Slowly, incredibly, the single overriding necessity for that bizarre selection—of Mars over an "inviting Earth"—began to dawn . . . A reason so simple and at the same time so profound, that I felt it *had* to be close to the reality of what we were observing. And it could be expressed in just one word:

*Gravity*!

Imagine: of all the things which could be conquered by a sufficiently advanced technology, able to travel literally to the *stars*—the vast distances, navigation across trillions of empty miles, life-support in the literal middle of nothing, the stupendous energy requirements, etc.—the *one* parameter which could not be altered (within any physics that we know!) would be—

The *gravitational fields* of the planets they would find!

And—uncontrolled—gravity would completely dominate the course of one's attempt to colonize another world: from affecting the very minerals and geology (through compression and the ease of plate tectonics, which concentrates ores and minerals for mining); to methods of construction (if not the scale!); to surface transportation networks—

To the most critical parameter of all: *biology.*

*       *       *

We have discovered (in the last twenty or so years) that gravity, more than any other single environmental aspect, subtly affects all organisms. Studies under reduced gravity are difficult to carry out on Earth (for obvious reasons!—which is why space station data will be so critically important in the future); but *higher-gravity* studies on various continuously running centrifugal systems in both NASA laboratories and others, seems to indicate one overwhelming effect, on everything from growing seeds to animals subjected to the runs:

It kills them . . . sooner than they otherwise would die, from "natural" causes. [15]

It's not too difficult to understand some of the basic mechanisms, starting from the simple physics of the situation. In any higher organism, higher gravity means that the heart must pump blood "uphill" against a greater force, putting more strain upon its muscle, resulting in literally more wear-and-tear with time . . . until the muscle fails. Other, far more subtle effects—on everything from blood chemistry to ion balances within the cells, to fallen arches (!)—have been discovered in these, admittedly preliminary studies—leading to the overriding conclusion: higher gravity is not good for living organisms.

*       *       *

*After millennia of dangerous travel as a veritable self-contained miniature world, the starship ultimately arrives in a small and distant solar system.*

*The voyagers soon discover to their delight that, unlike the fuzzy data relayed back centuries before by an unmanned robotic probe, the system actually contains* two *potentially habitable planets! One—the planet inward toward the sun—is a vast and sprawling paradise of chlorophyll, of open rolling oceans, and highly oxygenated air; the other—farther out—is a small glaciated world, a dying desert turning slowly into rust beneath the unfiltered ultraviolet of a distant searing sun—breaking down the very air . . . a wisp of carbon dioxide wrapped around the reddish sands and mesas of this barren, cratered world . . .*

*Then, with shocking suddenness comes the stunning bitter news: the gravity of the lush, green planet is a staggering* three times *as high as that on the surface of the battered, rusted one. The prospective colonists stare unbelievingly at the images of this lush world, forever taken from them by events which occurred* eons *before they ever entered this distant stellar system . . . by a chance accretion in a spinning solar disc of dust and gas, and their own accident of birth upon a smaller, lighter world, now light years distant.*

*Sadly, fighting back the agony of disappointment—of having come so far only to have the reason for their coming snatched away by one of nature's most fundamental laws—the pioneers resolutely set about becoming "Martians." They carefully dismantle and transplant to the surface of the frozen, dusty sphere, the closed-ecology aboard the ancient starship. And they follow this by mining crucial minerals from the blowing crimson soil, for the creation of the first of several self-contained environments—duplicates of the ordered but all-too-sterile vessel they've endured for all too-many-years. Yet some of them, even as the first of these "arcologies" are going up against a rust-pink sky, begin to plan . . . for a very different world . . .*

*While a shining blue-green star which will one day be called "Earth" beckons ironically above the Martian dawn—a Promised Land these castaways can never hope to enter . . . on their own.*

*   *   *

Which explains everything . . . except the presence of the Face.


## Notes

1. Soleri, P., *The City in the Image of Man*, Cambridge, Massachusetts: MIT Press, 1969.
2. *ibid.*
3. *ibid.*

4. *ibid.*

5. Bateson, G., *Mind and Nature*, New York: E.P. Dutton, 1979. *op. cit.*

6. *ibid.*

7. Asimov, I., *Extraterrestrial Civilizations*, New York: Fawcett, 1979.

8. Bova, B., ed. *Close-up: New Worlds*, New York: St. Martin's Press, 1977 *ibid.*

9. Heinlein, R., "Universe," *Astounding Science Fiction*, June, 1940.

10. Jones, E.M., "Discrete Calculations of Interstellar Migration and Settlement," *Icarus*, Vol. 46, 1981; Jones, E.M. and Finney, B.R., "Interstellar Nomads," in press.

11. Mallove, E., Forward, R.L., Paprotny, Z. and Lehmann, J., "Interstellar Travel and Communication: A Bibliography," *Journal of the British Interplanetary Society,* Vol. 33, No. 6, 1980.

12. Brin, G.D., *Quarterly Journal of the Royal Astronomical Society* 24:283, 1983; Brin, G.D., "Xenology: The New Science of Asking 'Who's Out There,'" *Analog*, May, 1983.

13. Donaldson, T.M., "The Galaxy Before Man," *Analog*, September, 1984.

14. *ibid.*

15. This conclusion is tentatively confirmed by Kenneth Souza, a biologist at the Ames Research Center and Principal Investigator on the Russian "Biosputnik' spacecraft series in 1976. According to Souza (personal communication) generational research in "hypergravity" (centrifugally-simulated higher gravitational fields), with mice and chickens over the past twenty or so years, have produced indications of shorter lifespans at higher gravities. Purely gravitational effects, however, are far from clear—in research carried out to date; mitigating and confusing factors seem to include everything from diet to the ability of some species to adapt (in succeeding generations) to moderate gravitational increases. More research will be necessary before any firm *generational* conclusions can be reached; however, experimental results do indicate a sudden and dramatic shortening of lifespans of the *first generation,* if gravitational fields exceed certain threshold levels. This carries intriguing implications for the thesis expressed here: that "the Martians" were forced to choose Mars because of gravitational constraints on the initial generation of "colonists" . . . and on succeeding generations.

*The Monuments of Mars*

# XV

# A TERRESTRIAL CONNECTION?

> "The rash assertion that 'God created man in
> his own image' is ticking like a timebomb at the
> foundation of many faiths . . ."
>
> —Arthur C. Clarke

The indigenous presence of a human-looking countenance on any other planet violates the cornerstone of current evolutionary theory, as espoused by Simpson, Stephen Jay Gould, and others. *If* we are the product of "a million random factors, operating within the Earth's environment alone," then indeed we have every right to think we are unique; no other time or place (or planet!) could have perfectly generated all these factors— and in the identical order—to produce the sentient outcome of those factors: us.

Yet, the Face is there . . . and, if it is artificial, an actual humanoid artifact rather than a damnable mirage of faraway rocks undergoing an inexplicable geological and meteorological process, how do we explain its presence, within the framework of a "migration hypothesis," *and* within the framework of contemporary science?

My simplest hypothesis, consistent with everything we know, is also the one hypothesis that everyone seems most eager to ignore:

There was—somehow—a direct "terrestrial connection."

Admitting even the possibility of this, however, instantaneously transports us across a sort of "magic line"—a vast gulf in current science separating "real" research from a scientific "no-man's-land" of dismal "data," wild speculations, and even wilder accusations! It is simply not "polite" in current scientific discourse to propose that "extraterrestrials" had any connection with our planet, let alone ourselves.

But, if we systematically eliminate all other reasonable explanations for this majestic Martian "monument," what are we to think?; are we not *driven*—inevitably and inexorably—by nothing less than Occam's Razor, to the one hypothesis which (seemingly) eliminates a major component of the paradox of finding an image of ourselves on Mars—

By proposing that the Face is, indeed, a monument to *us*—or to some extraordinary link between our two respective worlds . . ?

There was a time, and not so long ago, when such a proposal would have been quietly—nay, eagerly!—explored by "mainstream" science—as recently as 1959. Shklovskii, in *Intelligent Life in the Universe*, cites the work of a Soviet ethnologist, M.M. Agrest who, in the 1950's, explored the possibility "that representatives from an extraterrestrial civilization have indeed visited our planet . . . [it is conceivable] that perhaps a number of events described *in the Bible* were in reality based on a visit of extraterrestrial astronauts to Earth (italics added)."

Shklovskii's own collaborator, Sagan, had some equally unfettered thoughts on this potentially explosive topic as recently as 1966. Following Shklovskii's preceding quotes from Agrest, Sagan wrote:

> Some years ago, I came upon a legend which more clearly [than Biblical material] fulfils some of our criteria for a genuine contact myth. It is of special interest because it relates to the origin of Sumerian civilization . . . [1]

He then quoted substantial portions of three different, but cross-referenced, historical accounts of "a legend [that] suggests that contact occurred between human beings and a non-human civilization of immense powers on the shores of the Persian Gulf . . . in the fourth millennium B.C. or earlier."

The legend Sagan related came from three Greek writers[2] in classical times—Alexander Polyhistor, Abydenus, and Apollodorus—and referred to a series of encounters, by peoples living on the shores of the Persian Gulf, with "beings" simply identified by Sagan as "the Apkallu" (although in one account an individual is also named: "Musarus Oannes"). Oannes and his fellow beings, according to Sagan's telling of the legend, came *to teach Mankind the basis of civilization*—from law to architecture!

The original, which survives in translation from the Greek, can be traced to an individual named "Berossus," a priest of Bel-Marduk, living in Babylon at the time of Alexander the Great. Berossus was attempting to write a history of his own culture—including the origins of human beings and civilization—based on access to ancient cuneiform tablets and pictographic materials (cylinder seals) passed on through a succession of much more ancient cultures: the Babylonians, the Akkadians . . . and the Sumerians. Each had inhabited the Persian Gulf region for a time, thousands of years before Berossus' writings. Civilization itself had started (according to Polyhistor, quoting Berossus)—abruptly—with Sumer itself; this was the only civilization then known which "suddenly appeared."

Sumer had no comparable precedent. The first public schools appeared there, the first representative government, the first written legal codes, the first massive public works, the first gridded cities, and much more. This has always been something of an embarrassment and a mystery to archaeologists. One of the world's foremost Sumerian experts, Dr. Samuel Noah Kramer, probably phrased it best in an expression that was later to become the title of his classic book:

"History *begins* at Sumer."[3]

What did Sagan make of all this, in 1966?

> These cylinder seals [from which Berossus drew his history, in part] may be nothing more than experiments of the ancient unconscious mind to understand and portray a sometimes incomprehensible, sometimes hostile environment. The stories of the Apkallu may have been made out of whole cloth, perhaps as late as Babylonian times, perhaps by Berossus himself. Sumerian society may have developed gradually over many thousands of years. In any event, a completely convincing demonstration of past contact with an extraterrestrial civilization will always be difficult to prove *on textual grounds alone.* But stories like the Oannes legend, and representations especially of the earliest civilizations on Earth, *deserve much more critical study* than have been performed heretofore, with the possibility of direct contact with an extraterrestrial civilization as one of many alternative explanations (italics added).

So, what happened to poison this appropriately skeptical, yet open, scientific atmosphere on the subject of "extraterrestrial visitations" . . . between then and now?

The answer is as simple as it is tragic, for true scientific curiosity: Erich Von Daniken.

The damage caused by this one individual, to the very *concept* of serious investigation of claims of ancient visitors to Earth from "outer space," has been incalculable. Suffice to say that, whereas, in the "pre-Von Daniken Era" a scientist like Sagan—imaginative but careful—would openly entertain ideas of visitors from other cultures and from other worlds (one of his earliest memorable works on this subject was a scientific paper investigating the expected frequency of such "manned interstellar visits"—within the bounds of current astrophysics), in the "post-

Von Daniken Era'' even Sagan cringes at suggestions that there may have actually been any form of contact with "past visitors."

Or, as anthropologist Richard Grossinger observed, in a discussion with me on this very subject:

> . . . progressive science doesn't want to see the Face, because the humanoid-populated universe has been specifically appropriated by the Chariots-of-the-Gods/Lost-Continent-of-Mu contingent in what is implicitly a fascist and white-supremacist scenario, at least as portrayed by them. The invaders from outer space become the explanation, after the fact, for why non-Western cultures could not have built their own temples and statues or even created their own languages and number systems. The humanoid scenario, with its post-fascist associations, is simply bad company; no one wants to hang out there.
>
> So the face on Mars is not free to be like the Easter Island statues or the Sphinx, or any other artificial megalith; instead, it has to be like the Von Daniken legend behind those structures—that is, a mirage, a fabrication. [4]

This is not the first time in the history of science that an area of inquiry, or a body of data, has become "too highly charged" for rational discussion.

According to Thomas Kuhn, the main obstacle to general acceptance of Charles Darwin's ideas on evolution "was neither the notion of species change nor the possible descent of man from apes. The evidence pointing to evolution, including the evolution of man, had been accumulating for decades, and the idea of evolution had been suggested and widely disseminated before . . .

"[No], the belief that natural selection, resulting from mere competition between organisms for survival, could have produced men together with the higher animals was the most difficult and disturbing aspect of Darwin's theory. What could 'evolution,' 'development,' and 'progress' mean in the absence of a specified goal [presumably specified by God]? To many people, such terms suddenly seemed self-contradictory." [5]

If a straightforward "terrestrial connection" as the ultimate explanation for the Face continues to be assiduously ignored (when and if its reality is established beyond question), scientists will have to reach for other explanations . . .

Including—quasi-scientific/quasi-mystical scenarios on "parallel evolution in radically differing environments," such as those of Rupert Sheldrake.

British biologist Sheldrake has in recent years popularized the concept of "morphogenetic fields"[6]—that there exist pervasive, non-physical "templates" somehow embedded in the very structure of the Universe, capable of guiding evolution through non-material means. These include feedback of morphology from the manifest to the unmanifest, suggesting an intrinsic tendency in the universe to reinforce successful forms—though without, necessarily, a physical connection among their various manifestations.

Sheldrake's initial experiments were designed to test one corollary of this theory—that organisms in widely separated environments can somehow transmit behavior *outside of known communications systems*. For instance, animals trained to run specific courses in mazes have apparently been observed somehow to pass on this behavior to a control group . . . which has been kept in total isolation!

If such transmission were possible, it would obviously transcend our Western conceptions of space and time. Thus, the "dislocation" of our image to Mars might be conceivable without any material agent. The Martians (or Martian immigrants) might have fabricated (in the sense of "constructed") our replica *without even knowing who we were* . . . Farfetched, to say the least!

There is, however, another level of Sheldrake's theory which bears more profoundly and universally on the Face. Departing from the limitations of standard theories of "genetic codes," and "DNA-controlled heredity processes," Sheldrake seeks to understand *embryology* as an additional effect of morphogenetic field theory: that even the *form* of living things—in widely separated environments and under widely differing conditions—could share the same "habit pattern" to create *the same basic form* . . .

Perhaps including two *humanoid* species, evolving independently on two completely different planets!

Let me be very clear on this: I raise Sheldrake here *only* to underscore a point. If a direct "terrestrial connection," as explanation of the Face, is consistently derided, its replacement will inevitably revolutionize current concepts of terrestrial biology . . . if not Martian evolution.

In other words, there will be no such thing as a "trivial explanation" of this mystery—the presence of a "human" Face on Mars; some answers—such as a "mere" physical connection between planets—may in the end be a lot easier for us to understand and (Von Daniken notwithstanding) ultimately to accept . . . than others.

Grossinger summarized these problems later in our discussion:

> If humanoids also evolved elsewhere, then there must be some basic humanoid structure in nature. The force need not suggest only

the psychologically archetypal, as with Carl Jung; in fact, it is far closer to something biologically or astrophysically archetypal, going right into the heart of the creative process of the universe itself. It suggests a possible translation from basic atomic morphology to bio-morphology to psychomorphology, a route that is the earmark of a whole vitalistic science the abandonment of which is one of the axioms of this century.

So to have a humanoid, and a humanoid on the very next planet over from Earth, suggests one of two things, both of them very disturbing to progressive liberal scientific mentalities. You are suggesting either that there is an intrinsic "humanoidizing" force in the universe (which goes against the atheistic basis of science itself), or that humanoids have come from elsewhere and been associated with this planet, this solar system. If it's the former, then you are very close to spiritualizing the universe. If it's the latter, then you are giving aid and comfort to the whole ancient-astronaut fringe which is so distasteful to modern astronomers.

If we put Sheldrake aside for the moment, and also tentatively reject the extreme notion that ancient members of our own species visited Mars before a cataclysm, destroying Earth's civilization, we are left with only one possibility—that of a cosmic connection, initiated by beings from another solar system, consummated at the dawn of our own epoch, and somehow involving connected settlements on worlds.

This is, admittedly, an outrageous proposal, and, although it doesn't bring an archetypal "force" to the heart of biology, it does challenge everything we know about our history—not just the history of the West or the history of the East (or the mysterious histories of South America and the Pacific), but the history of the whole Planet Earth and the Solar System . . . hence, our historic genesis—if not our very destiny!

If you think you hear footsteps of Von Daniken here, know that, yes, we are treading on very dangerous ground. On the one hand, I would like nothing more than to put Von Daniken to rest—as a sloppy scientist, a charlatan, and perhaps even a fabricator of data; on the other hand, in the process of burying him, I fear that I am resurrecting him in an even more deadly form. There is no way I can point to the obvious parallels between "pyramids in Egypt and on Mars . . ." without giving birth to a whole new tapestry of Von Danikenesque images. So powerful is his conceit that any form I create will be immediately sucked into his infamous dioramas; any Egypt-Mars scenario I propose will instantly become more his than mine, for his glitzy notion has captured the imagination of our entire generation . . .

There is, in short, no way around Von Daniken.

*The Monuments of Mars*

In order to look for the shadow of Cydonia on Earth (however faint the shadow), I must turn to the Monuments of Egypt. In order to suggest transport between worlds, I must invoke images of "humanoid" astronauts and giant starships. As soon as I do, they will find their way into blond, white space explorers and (possibly) anti-gravity machines. . . Chariots of the Gods.

All I can say is that, knowing that these pitfalls lurk behind every loaded word and image, I must enter the territory regardless—for the *data* cannot be denied . . .

Perhaps Von Daniken's hold on us is so powerful because he has reawakened an aspect of an epochal truth that lurks hidden in our entire culture; he has foreshadowed *another* millennial event of which he had no real knowledge, hence could exploit into an (apparently) endless series of "B movie" extravaganzas. At least his sense of timing was impeccable; he single-handedly created an entire category of tabloid headlines, to which "Ancient City Found on Mars!" is just the latest . . .

The scientist in Von Daniken failed because he took on too much unprovable material, and imposed too terrestrial (and Germanic) an image on any potential "terrestrial connections" that he found. We must emphasize here that Cydonia is an unclaimed mystery and must remain so— if we are to understand it finally on its own terms. We can surmise connections, in order to make the inexplicable bearable for the time being (and to sate our curiosity to some small degree until we or our descendants make the trip to Mars). But, until such time as we (or our machines . . .) actually walk among the wonders at Cydonia, and explore and open the waiting Monuments with either our remotely-controlled surrogates or with our own hands, the essential explanation, the nature of "the terrestrial connection"—if any—will remain a mystery.

What we will one day find there could be anything . . . absolutely anything . . . and is quite likely beyond our limited imaginations . . .

So, with that forewarning, let us go to Egypt and see if we can find the outlines of a cosmic tale yet untold. Let us examine the (admittedly subjective) connections between Egypt and Cydonia and see if there is perhaps a thin thread of data, of roughly quantifiable links and correlations between these worlds . . . recognizing all the time that Cydonia is very fuzzy, very faraway . . . and Egypt is in ruins, a partial artifact, of something that happened long ago by the chronicles of our own culture. (Since it precedes us at such a remove, it will cast its beams through us, though very dispersely, and we will likely find all too many suggestions . . . and little truth.)

Make no mistake about it: it is the eerie similarity between the pyramidal structures that have come to stand for "Egypt" and those lying—empty and abandoned—at Cydonia, that almost scream of some "connection." (That—and the presence of two "sphinxes" . . . as Avinsky aptly phrased it . . . on *two* worlds.) But "how" and "when?"—for, whatever happened on the Martian surface (according to the implacable geometry of that critical solstitial alignment) had been drifted over by the dessicated Martian dust and sands for *half-a-million-years*—

By the time Egypt spread its magnificence beside the Nile!

There was no way for either of these cultures to have ever touched each other . . . separated as they were by a gulf at least *a hundred times* as wide as that which separates us from the life of ancient Egypt!

*       *       *

The earliest dating of the Egyptian pyramids is in the Third Dynasty, Old Kingdom—estimated variously by Egyptologists as occurring between 5000 B.C. and 2600 B.C. . . . several hundred thousand years *after* Mars somehow expired . . .

[Egyptian dating methodology is an "art" unto itself; the original dates for the Egyptian chronology currently in widest use were taken from one fragmentary papyrus written by an Egyptian priest living around 300 B.C., Manetho—the so-called "Kings Lists." Manetho compiled a list of 30 "dynasties" which ruled Egypt for thousands of years before his efforts to recreate their sequence. His "History of Egypt," in turn, was based on equally fragmentary records—papyri that had only partially survived the ravages, not just of centuries, but of *millennia*.

[Since early Egyptologists attempted the first reconstruction of an Egyptian chronology from Manetho's lists (by attempting to correlate them with the dates and successions of pharaohs left on Egypt's prolific monuments, or other scattered written records—such as the famed "Turin Papyrus"), scholars, using a variety of additional techniques—from celestial alignments of temples to radiocarbon analysis of bits of mummy wrappings—constantly have labored to increase the accuracy of this critical chronology.]

The pyramids—those most enigmatic and stupendous monuments of all—according to this major effort at accurate dating, resulted from a "flurry of construction" that, for the *best* pyramids, only spanned about *a hundred years*! The period of "real" pyramid construction stretched from the Third Dynasty of the Old Kingdom (circa 2600 B.C.) through about the Sixth (circa 2200 B.C.).

The so-called "Great Pyramid"—legendary as one of the Seven Wonders of the Ancient World—was *not* (according to this chronology) the first pyramid constructed! That honor went to a peculiar "stepped pyramid" erected within a magnificent 35-acre "mortuary complex" dedicated to one of the most powerful pharaohs of the Third Dynasty— King Djoser; erected at a site called "Saqqara" (after the Egyptian god of "orientation": Sokar), it was the only complex of this *specific* design in all the rest of the history of Egypt. Currently, after more than 50 centuries of continuous internment, Saqqara is the site of more graves and ancient tombs of Egyptian royalty and nobles than any other archaeological "dig" in Egypt—including, perchance, the tomb of the Chief Architect of the Stepped Pyramid itself!

This peculiar structure (termed by a more recent architect as "man's first skyscraper"), according to an inscription found within the Complex, was designed by a genius "generalist" named Imhotep, working directly under Djoser. Compared to the discovery of the tomb of a relatively minor pharaoh named "Tutankhamen," the discovery of the tomb of this towering figure—who single-handedly laid the literal foundations of Egypt's claim to immortality—would be an archaeological triumph unparalleled in the history of science.

The Great Pyramid—the largest and most perfectly constructed of this "experimental" architecture attributed to this legendary figure (who later in Egyptian history was elevated to the rank of a god!)—is considered currently to have been erected less than a century after Imhotep's first pyramidal triumph. Its location was a high plateau, the "Gizeh Plateau" ("gizeh" means "edge" in Arabic), marking the boundary of the Western Desert and the fertile Nile. The Plateau is a few miles southwest of Cairo— the current capital of Egypt. (Saqqara, about 40 miles south of Cairo, was the site of one of Ancient Egypt's early capitals—when the Stepped Pyramid was built.) The Great Pyramid—almost 500 feet in height and composed of an estimated two *million* limestone blocks!—represented an extraordinary evolution of architectural technique (if not sheer engineering!) beyond Imhotep's first "crude" effort—and ostensibly in less than a hundred years . . .

According to this accepted scholarship, the Great Pyramid was thus erected in the Fourth Dynasty; its intended "occupant" was the legendary pharaoh Khufu (or "Cheops," as the Greeks—from whom we get much of our Egyptian history—pronounced his royal hieroglyphic "cartouche"). Its sole purpose (again—according to accepted Egyptology) was to be his tomb, to protect Khufu's mummified remains, and more important, his immortal "Ka"—as the latter wandered through Eternity . . .

Two other massive pyramids (one only slightly smaller than the Great Pyramid; the other significantly smaller) were also erected on this windswept height above the Nile. According to the present reading of the evidence, these were constructed to function as the royal tombs for the successors to the reign of Cheops—"Chephren" and "Mycerinus." (Several smaller pyramids on the Plateau adjoining these "big three," merely piles of stones, very crudely stacked, were apparently piled up later . . . how much later no one knows.)

In the two lesser "copies" of the Great Pyramid—"Chephren's" and "Mycerinus'"—built theoretically only a few years after the Great Pyramid itself, there seems to be a dramatic—if puzzling—deterioration in the quality of workmanship, compared to that of Cheops' massive monument.

The Sphinx—that mysterious, mythic figure physically associated with these Pyramids on their sandstrewn Plateau, and facing East—is the crouching figure of a lion, with the head of a man, carved out of the "living rock" a few hundred yards from Chephren's pyramid. It is attributed by some Egyptologists to Chephren, sculpted as his image . . . gazing eternally eastward toward the horizon . . . the dawns . . .

According to Egyptologists who've created this chronology of pyramid construction, this soaring spirit—which impelled the creation of the most awesome artifacts the world would ever know—somehow petered out; "pyramids" would continue to be built for several centuries, but they would be shoddy replicas of those erected in Egypt's "early years." And many later ones would be merely piles of brick (!), not "eternal stone," and only tens of feet in height, not "manmade mountains" towering hundreds of feet above the Nile; most of these did not survive the elements beyond the lifetime of the pharaohs for whom they were apparently intended.

In the ensuing thousands of years, later dynasties would turn from time to time to equally prodigious engineering projects—the awesome carvings and temples hewn out of the living rock at Abu Simbel, overseen by the egotistic Pharaoh of the Eighteenth Dynasty, Ramses II, immediately come to mind. But no later pharaoh would ever return (because they were unable?) to that brief and glorious epoch of piling *millions* of limestone blocks on top of one another—represented by that anomalously brief era of true pyramid construction.

Another writer, John Anthony West, confronted by this historical panorama was inspired to remark,

> . . . every aspect of Egyptian knowledge seems to have been complete at the very beginning. The sciences, artistic and architectural techniques and the hieroglyphic system show virtually no signs of a period of "development"; indeed, many of the achievements

*The Monuments of Mars*

of the earliest dynasties were never surpassed, or even equaled later on. This astonishing fact is readily admitted by orthodox Egyptologists, but the magnitude of the mystery it poses is skillfully understated, while its many implications go unmentioned.

How does a complex civilization spring full-blown into being? Look at a 1905 automobile and compare it to a modern one. There is no mistaking the process of "development." But in Egypt there are no parallels. Everything is right there at the start.

The answer to this mystery is of course obvious, but because it is repellent to the prevailing cast of modern thinking, it is seldom seriously considered. *Egyptian civilization was not a "development," it was a legacy* . . .[7]

But Egypt was not the only Near Eastern civilization to have "suddenly" appeared; according to the very inhabitants themselves (as reported first by Berossus, then via surviving cuneiform tablets) there was *another* ancient "high culture" with this curious absence of a precedent: Sumer!

*      *      *

Now, attempted reconstructions of the history of Sumer have had one dramatic strike against them: the lack of an "absolute chronology" to anchor them.

Strictly speaking, in the time period we're most concerned with here—*before* 3000 B.C.—all dates in Sumerology are relative.

Egyptian chronology would suffer from the same crippling constraint, were it not for an almost-fanatical attention paid by the Egyptians to key celestial events—in particular, the "heliacal rising" of the bright star, Sirius (see Chapter IV). Astronomers have been able to determine when Sirius rose over ancient Egypt, just before the dawn, coincident with the flooding of the Nile; this, in turn, has allowed some Egyptologists (R.A. Parker[8]) to publish correlations of these "celestially-anchored dates" with Manetho's Lists, thus deriving an overall *absolute* chronology for Ancient Egypt estimated (across five thousand years!) to be accurate "to within about a century . . ."

In Sumer, before about 2000 B.C., this degree of celestial precision is impossible.

The Sumerians either had little interest in astronomy (contradicted by surviving intimations from far later periods), or their records were kept on fragile media—which simply did not survive into the present. They certainly lacked that wonderful "environmental clock" the Nile—flooding each spring with almost orbital precision. This natural marker of the seasons and the orbit of the Earth formed a rhythmic backdrop—

celestially reconstructable—for the entire culture which "appeared" along its course from "Upper Egypt" to the Delta . . .

In Sumer, archaeologists have had to turn to relative techniques for dating; if a bit of charcoal or pottery is found above another, in a higher layer in an excavation, most likely it is *younger*. Attempts to turn this relative chronology into an absolute one depend, in turn, on dating a few of these artifacts *absolutely*—via such techniques as radiocarbon dating (see Chapter II). Unfortunately, radiocarbon techniques can be applied only to *organic* residues—remains of fires, bits of cloth, etc.— and the overwhelming numbers of archaeological artifacts are *non*-organic: pottery, clay tablets, jewelry, and the like. Thus the ages of these samples must be interpolated, from association with the datable organic fragments.

For the Egyptologist, "absolute" now means an error—in some 50 centuries—of about a hundred years; thus, we *probably* know when Egypt's most famous woman pharaoh—Queen Hatshepsut—ruled; when she built her wondrous temple at Deir el Bahari; when her treaties prevented war with neighboring states on Egypt's borders . . . to within, at best, *several generations*—the equivalent of *half the lifetime of the United States*! A *lot* of history can be swallowed by an error of "a hundred years." (But for reconstructing events across some five millennia, this is considered "phenomenal precision!")

Mesopotamia (the Greek name for the land of Sumer, the Akkadians, etc.) provides no firm astronomical dates until very late in the period we are exploring—an eclipse in 763 B.C. All prior dates—particularly the crucial date of the so-called *beginning* of Sumerian culture—depend on techniques of marginal reliability, for errors (in this period) of "less than several centuries"—essentially, the radiocarbon dates described before.

The importance of this inherent uncertainty (as it applies to Sumer) is just this: there is no current means of establishing, with *absolute* precision, which high culture—Egypt or Sumer—originated first! Though by convention Sumer is estimated to be older (by at least 500 years), in truth archaeologists will ultimately admit (if really pressed!) they just *don't know* . . .

\*   \*   \*

What if Egypt and Sumer—the world's first two "high cultures"—had been *connected*? Suppose each could trace its origins back to a common source . . . the *same* "extraterrestrial connection?"

Imagine . . . if these two most brilliant panoramas of cultural achievement—each eerily reminiscent of "a legacy inherited" and not a civilization developed—had managed to preserve the record of their birth in two unique but equally remarkable forms . . .

      *The Monuments of Mars*

In Sumer, a *literary* record of the "contact"; in Egypt replicas of the "monuments of Mars" themselves: the pyramids!

But, remember, this is only a story, a speculative fiction to allow us to approach Cydonia by the only means available (in lieu of spaceships): its possible, even apocryphal reflection in us. After all, the epistemological implications of the attempt to discern the Face's ultimate nature are significant in their own right. We are forced to ask hard questions about ourselves and our origins, and questions about how we ask questions, and so on. These speculations are the first test of the Face: we are forced to stare into its enigma (and perhaps have been doing so for aeons . . .) until we go there. And, staring, we see many facets of ourselves . . .

*     *     *

Egypt's "birth" is traditionally dated from the unification of "upper" and "lower" Egypt. This occurred (according to traditional reading of Egyptian texts and archaeological reconstructions) well after a series of scattered feudal "states"—nomes, in Egyptian—were united into the two main geographical entities that stretched along the Nile—from its source, northward, to the Delta.

So important was this "moment of unification" in subsequent Egyptian life that, for millennia after, the pharaoh retained as one of his key titles: Pharaoh of Upper and Lower Egypt . . . Lord of the Two Lands"; his crown was a visual symbol for this unity—white for Upper Egypt, red for Lower. The date of this all-important unification, under "King Menes," is now estimated by various scholars as having taken place circa 3100 B.C.

In Sumer, political development took a completely different course . . .

No unification of Sumer's several dozen scattered "city-states" succeeded for almost *a thousand years* after the identification of "Sumer" as a major entity in Mesopotamia (estimated now as circa 3800 B.C.— see above). The only unity in Sumer was *cultural*—a series of artistic, legal, scientific, mathematical, and administrative accomplishments, and the appearance of the first written language to describe them. These achievements would resonate down through Sumer's myriad successors—Akkadians, Babylonians, Assyrians . . . long after Sumer itself, as an identifiable force, disappeared.

The history of Mesopotamia, following the relatively brief high cultural plateau set by Sumer, is one of a long unending fall into oblivion . . . into the bottomless abyss of pillage, rape, and murder—conducted in the name of "empire" by a seemingly unending line of would-be emperors, with names like "Sargon" . . . "Hammurabi" . . . "Alexander."

Sumer's fatalism (judging by subsequent events—including the current happenings in "Mesopotamia," site of present-day Iran and Iraq!) would seem to have cast a fatal "curse" over its myriad successors in the region—

While, less than a thousand miles away, Egypt would preserve (comparatively speaking, uninterrupted) its stunning heritage—even to its official language, hieroglyphs—across 3000 years . . .

Two cultures; two incomparable contrasts.

In Sumer, everything—including the most massive and elaborate temples—was made of brick. Consequently, little or nothing now remains of most of Sumer's outstanding architectural achievements . . . but echoes—

Eroding piles of *dirt*—in which the treasures of an entire vanished culture float: elaborate gold masks, exquisite statuary, delicately crafted bracelets, jewels . . . and *hundreds of thousands* of clay tablets, preserving the first written language of an entire planet . . .

Forgotten . . . like precious raisins in an earthen loaf.

Egypt too built most of its houses, palaces, and even public buildings out of brick. But its guiding philosophy regarding life . . . and what was to come after . . . was immortalized in its choice of the material it used in the construction of its monuments and temples . . .

Stone.

Egypt believed in Eternity itself . . . and used brilliantly the one material likely to survive a reasonable fraction!

*      *      *

The answer to this question—are these two cultures derived from some *common* point of origin?—is the essence of the larger problem presented by the search for cosmic linkage with events on Mars . . .

For, beyond the surface differences separating Egypt and Sumer, there did indeed appear to lurk deeper—and incompletely understood—*parallels* of culture. These tantalizing links, between two civilizations which superficially seemed to have nothing in the way of common culture, have fueled heated debates between respected scholars for almost forty years!

Example:

The earliest writing known in Egypt is termed "hieroglyphic"—*picture writing*, where the images stood not simply for the thing that was being imaged (such as a bird or other specific object) but for certain parts of words, through *sounds*. This is termed "phonetization" by the experts.

Remarkably, the earliest decipherment of Sumerian cuneiform symbols (which look nothing like Egyptian hieroglyphs!) serve an identical purpose: *they are phonetic parts of words.*

*The Monuments of Mars*

This fascinating "parallel development" has caused some scholars to propose a "Sumerian cuneiform influence" on Egypt's hieroglyphic origins,[9] on the natural assumption of cultural diffusion—and the (presumed) fact that Sumer originated first!

If, however, both cultures originated from a *common source* (and evolved two separate language systems under the influence of their respective regions), it would be logical that the *principles* determining the underlying structures of the language would remain the same, while the choice of *symbol-systems* evolved differently.

Interesting . . . but hardly proof of a deep cultural connection.

However, as I read more about these two cultures, I discovered seemingly unending additional examples of these "parallels"—ranging from arcane astrological notations used on Egyptian coffins (*and* in Sumerian/Babylonian celestial records) to architectural designs that seemed (in Egypt) something of an echo from the East . . . from Sumer.

Although the Egyptians (as we do) used a decimal system for "everyday" affairs, their religious activities were governed by the sexagesimal sixty—precisely the Sumerian base mathematical system. [At least four completely independent scholars—Wallis Budge (the translator of Old Egyptian), Henry Frankfort (in linguistics), Alexander Badawy (from an architectural perspective), and Robert Temple (from mythological, theological, and astronomical parallels)—conclude that there existed deep and very old connections between Egypt and Sumer, connections seemingly ignored by the bulk of "recent" mainstream scholars.]

Inside the 35-acre mortuary Complex Imhotep designed for Djoser, that marvelous invention of the Third Dynasty—the first Stepped Pyramid—rose into Egypt's cloudless azure skies. But it wasn't the Pyramid itself which now drew my attention; it was the spectacular ruins of the recessed *wall* which Imhotep had built to ultimately enclose his assemblage of chapels, ritual courtyard . . . and the Stepped Pyramid itself. With its intricate indentations and highly polished facing, it looked remarkably similar to perimeter designs in Sumer!

The only radical departure from Sumerian precedent, introduced by Imhotep, was his selection of *materials* . . . massive granite slabs and polished limestone.

Is it sheer coincidence that there occurs an inscription on a cliff above a well-known trade route—Wadi-el-Hammamat—leading from the Red Sea to a small village on the Nile in Upper Egypt, revealing the apparent immigration into Egypt along that route of Imhotep's *father*? The inscription, enumerating 25 generations of architects extending back to Imhotep himself, testifies to the migration into Egypt of one "Ka-nofer," a man subsequently appointed "chief of works of the South and of the North-land . . ." [10] by an earlier pharaoh, Djoser's father!

*       *       *

Even if there was an early (now essentially obliterated) Sumerian/Egyptian link, it was obvious that several thousand years of essentially separate evolution had all but *totally* obscured it . . .

What would *a hundred times* that long do to any evidence of my purported "Martian influence?"

The only element which seemed capable of bridging an almost unimaginable span of half a *million* years—of retaining its integrity across a gap of time equivalent to the gap of distance between planets—was also the one set of "artifacts" we'd discovered on *both* worlds . . . the pyramids.

Think about it. A gargantuan undertaking, the most awesome monuments (even yet) on Earth, requiring a considerable fraction of ancient Egypt's "gross national product" (if Herodotus' numbers are to be believed); yet this prodigious accomplishment *completely escapes* notation in any ancient text, to say nothing of frescoes or paintings depicting *why* it was accomplished.

They must have been constructed sans the wheel (which the Ancient Egyptians wouldn't "discover" for about 800 years *after* Imhotep's Stepped Pyramid was first erected), without steel tools to quarry limestone blocks or granite slabs (the only tools extant were *soft* copper), and with no machine assistance—other than an array of harnessed human muscles!

Yet, despite these somewhat serious impediments, in the Great Pyramid alone over 6.5 *million* tons of stone were lofted almost 500 feet straight up—and with a precision which is staggering.

With a base the equivalent of 13 *acres* (approximately 7 Manhattan downtown blocks!), the departure of this behemoth Monument from being absolutely level was measured in 1925 (in the "Cole Survey," commissioned by the Egyptian government) as "less than an inch." (And most of that, it is suspected, has come from settling—under the enormous mass of stone!) Further, the even larger surface area represented by the Pyramid's four sides—some 22 acres in extent—was originally covered with a polished limestone casing, composed of over a hundred thousand separately-fitted blocks weighing as much as 15 tons apiece. The combined effect of this impregnable armor against the elements—actually becoming *harder* (as this particular type of limestone does) with the increasing years—was only possible because of the phenomenal precision in the fitting of these blocks together—

*The Monuments of Mars*

How phenomenal would await the careful measurements of the "dean" of modern scientific archaeology: Sir Flinders Petrie.

Petrie, a surveyor-turned-Egyptologist, came to Egypt in the late 1800's and proceeded to bring a whole new standard to the measurements of Egypt's architectural wonders. Among the targets of his instruments: the Great Pyramid itself. Working with devices routinely capable of measurements to within 1/100th inch (and sometimes 1/1000ths!), Petrie surveyed, measured, calculated—and literally crawled through every nook and cranny in the Pyramid.

His exacting measurements revealed that the aforementioned casing stones (all but a fortunate few having disappeared, since the 9th Century A.D.—victim of the invention of gunpowder and the need for building materials in nearby Cairo!) were crafted with superb precision. Stones measuring $5 \times 8 \times 12$ feet on an edge were laid, according to Petrie's careful observations, with a mean variation from "a straight line and a true square" of less than 1/100th inch—over a distance exceeding 75 inches. The cracks *between* the casing stones were literally "hairline thin"—no more than 1/50th inch in width—and filled with an extremely fine cement, distributed evenly across an area (represented by an average 8-foot by 5-foot block) of some 40 square feet!

Petrie, presented with this phenomenal engineering, was impelled to comment:

> Merely to place such stones in exact contact would be careful work, but to do so *with cement in the joints* seems almost impossible: it is to be compared to the finest opticians' work on the scale of acres (italics added).

(Marveling at such accuracy, such engineering in such a "primitive" architectural creation, I compared these errors in the casing stones to some contemporary technological achievements—and discovered, to my amazement, that that "1/50th inch-crack between adjoining casing blocks" is *less* than the current allowable errors in the placement of those notorious individual *tiles* . . . on the space shuttle!)

A much more noted (and far more controversial!) aspect of the Great Pyramid concerned geometry . . . and the purported *meaning* of these measurements.

This controversy started with an otherwise obscure Nineteenth-Century editor of the *London Observer:* John Taylor. Taylor, who began collecting "traveler's tales" from Egypt in the 1830's (when anything "Egyptian" was the rage), was also an avid amateur astronomer and mathematician. Musing over reports filtering back to London, from such surveys of the Pyramid as that being carried out by one "Howard-Vyse" (some 50 years *before* Petrie's careful measurements), Taylor was struck by a curious relationship between the reported height of the Pyramid and its circumference.

He discovered that, if he divided the perimeter of the base (as measured by the recent Howard-Vyse Expedition) by *twice* the height—the result was equal to 3.144—remarkably close to the universal constant: Pi, or 3.14159 – !

Taylor couldn't believe this result was sheer coincidence, but concluded that the Pyramid had been *designed* to embody this unique mathematical expression: the ratio of the circumference of a circle to its diameter. In other words, the Pyramid seemed an expression—in stone—of a fundamental geometric relationship, a universal constant supposedly unknown to the ancient Egyptians for another 1000 years!

John Taylor became captivated by the Great Pyramid. Over the succeeding years, after countless hours spent poring over each new reported measurement of the massive architectural enigma by the Nile, Taylor would add to his list of relationships embodied in the structure of the Pyramid—ultimately claiming he had found even such fundamental astronomical information as *the number of solar days in one Earth year*! Taylor's "discoveries" would begin an entirely new field: "pyramidology." With it would be launched over a century of heated (and at times downright vicious) controversy over the foundations of this "science": the supposed discovery of the "Pi relationship" embodied in the morphology of one of the oldest—if not most mysterious—constructions known on Earth . . .

Was the implied "circle"—a relationship underlying several centuries of increasingly fundamental physics, progressing from simple doodles in the sands of Egypt to the quantum forces lurking in the "singularities" of Twentieth Century "black holes"—an occulted but specifically intentional reference to cosmic truth . . . or merely a "coincidence," an accident of how the laboring Egyptian gangs hauled and heaved their piles of blocks above the desert floor of Gizeh?

Not surprisingly, this was our original question of the Face . . . and the same answer . . . intentional *design*—would be circumstantial evidence of a marvelous and primordial event. For how else could a universal constant be *deliberately* buried in the Great Pyramid (and a humanoid replica *deliberately* created on another planet in the company of even more amazing *pyramidal* forms)—were there not some ultimate relationship . . ?

Said Taylor,

*The Monuments of Mars*

It is probable that to some human beings in the earliest ages of society, a degree of intellectual power was given by the Creator, which raised them far above the level of succeeding inhabitants of the earth . . . (The builders of the Pyramid were, probably) the *chosen race* [one of the Lost Tribes of Israel!] in the line of, though preceding Abraham; so early indeed as to be closer to Noah than to Abraham." [11]

\*   \*   \*

Taylor's extraordinary misfortune, as I see it, was to discover his "profound mathematical relationships" (and then attribute them to Divine Intervention!) at precisely *the wrong period* in modern history; by ascribing his discoveries of "scientific knowledge" in the Pyramid to Biblical sources, Taylor was directly undermining the critical attempt of Science at this point in history to become its own authority on the natural world, in every way equal to the Church. (Ironically, Taylor would have probably fared much better if he'd attributed the Pyramid's peculiar measurements to *Martians*!)

For these reasons, it should not be too surprising to discover that—for well over a hundred years now—anyone even *hinting* at support of Taylor's claims of arcane "knowledge" hidden in the architecture of the Pyramid is instantly dismissed by "mainstream" science; Egyptologists in particular will insist with vehemence that "any 'mathematical' relationships discovered in this structure are strictly due to chance . . . and wishful thinking!" (An argument curiously reminiscent . . .)

The next man who would take up this intriguing search (in the wake of Taylor's death, in 1864) would actually appear to contradict this blanket statement—for a time; his name was Piazzi Smyth, and he had one immense asset in his pursuit of the truth behind Taylor's increasingly extravagant claims: he was the Astronomer Royal of Scotland. He was also a recognized expert in "spectroscopy"—the fledgling science in determining the chemical composition of the stars (or other celestial objects) by analysis of their emitted or reflected light. At a time when such scientists *built* many of their instruments, the interest of this pioneer in an entirely new field would prove a quantum leap in bringing scientific measurements to Taylor's claims . . .

In December, 1864, Smyth began an exhausting 5-month trip to Egypt. The journey included, at that time, travel from Scotland by ship; overland by rail from Alexandria to Cairo; and then by camel to the Pyramid itself—accompanied by his wife and bearing a great number of "state-of-the-art" (for 1864) scientific instruments.

Using elaborate metal poles, tapes, special cameras, and a variety of other instruments—including those specifically designed by Smyth himself—this world-renowned astronomer proceeded to conduct measurements of everything he could: height, slope-angle, baseline—and the Pyramid's 600+ feet of *interior* tunneling and passageways. In the end, upon his return to Edinburgh, the Royal Society awarded him a gold medal for his "meticulous and detailed measurements at Gizeh." His work would be published in an exhaustive 3-volume "opus,"[12] describing in copious detail his procedures and results.

His conclusion from this exhaustive survey was that there was, indeed, embodied in the very structure and dimensions of this awesome monument, the "Pi relationship"; data on the casing stones (slope-angles measured on the few surviving stones, from which an *original* height for the Pyramid could be computed) coupled with Howard-Vyse's perimeter measurements (Smyth was unable satisfactorily to conduct his own), produced an astounding value for this ratio, of 3.1416—*exactly* the value for the constant (within the errors of Smyth's instruments)!

Not content with just the exterior, Smyth had dragged his cameras, measuring tapes, and temperature-controlled rulers into the dusty, overheated, bat-infested passages within the Pyramid itself. His careful measurements of the dimensions of the "Queen's Chamber," the "Grand Gallery," and the reputed final resting place of Cheops himself—the "King's Chamber"—would stand as the epitome of Nineteenth-Century science—careful mensuration—for half a century—until the work of Petrie.

But Smyth's special contribution to measuring the interior layout of the Pyramid, was his attention to the "Descending" and "Ascending" passages—two very long, very narrow (3.5 ft. wide by 4 ft. high) tunnels, deep into the interior of this gargantuan structure.

The Descending Passage is the means by which all who wish to explore the interior of the Great Pyramid must enter—a long (~350 ft.), sloping tunnel descending at a steep (26°) angle, to a small chamber cut in the bedrock of the Plateau . . . directly beneath the apex of the 6.5 million tons of Pyramid, 600 feet above.

Decades earlier, Howard-Vyse had asked another well-known astronomer (Sir John Herschel) if this long, narrow passage, entering the Pyramid from *the north side*, could point in the direction of the Earth's "pole star." Herschel replied, after a moment's calculation, that though the Earth's *current* pole star (Polaris) couldn't have been seen shining down this 350 foot-long tunnel (because of Earth's *precession* over the previous 4000 years), another pole star could have served—a dimmer star called

"Thuban," in the constellation Draco.

Smyth, expanding on this notion, painstakingly measured the exact angle of this "Descending Passage." This, combined with his equally-careful measurement of the *latitiude* of the Pyramid, allowed him to plot precisely which tiny ($\sim 1°$) patch of sky a hypothetical observer crouching in the chamber deep beneath the Pyramid would see, looking up the long, dark tunnel . . .

By combining this calculation with the 26,000 year precession-period of the entire Earth, Smyth then proceeded to derive the *dates* when the pole star Herschel noted (Thuban) could have glimmered down this long Descending Passage. This alignment date Smyth concluded (very logically) was quite likely the *construction* date for the Pyramid itself.

Using these measurements and calculations, Smyth arrived at *two* dates (two, because of subtle factors inherent within this "nodding" motion—precession—of the Earth) for when the Pyramid could have been aligned with Thuban:

2123 B.C., and about a thousand years earlier—3440 B.C.

Smyth, like Taylor, could not believe that any "primitive" society (such as the ancient Egyptians!) could have designed this precise alignment, the "Pi relationship," and so many other sophisticated features—all in one extraordinary structure; an opinion he phrased thus:

> [The Pyramid] revealed a most surprisingly accurate knowledge of high astronomical and geographic physics . . . nearly 1500 years earlier than the extremely infantile beginnings of such things among the ancient Greeks.

As to "who" incorporated this extraordinary knowledge into this extraordinary structure, Smyth (as much a religious "fundamentalist" as Taylor) had one unfailing answer—and expressed his inevitable conclusions (given the era) all-too-clearly:

> The Bible tells us that in very early historic days, wisdom, and metrical instructions for buildings, were occasionally imparted perfect and complete, for some special and unknown purpose, to chosen men, by the Author of all wisdom. [13]

Which, of course, was the end of Piazzi Smyth's scientific and astronomical career!

Smyth was pilloried by academics and press alike, called by later critics "the world's first 'pyramidiot!'" Even sympathizers lamented that "such a first-class mathematical brain should have wasted its energies in so unprofitable a field." [14] In all the furor—over his attribution of the

Pyramid's inexplicable measurements to "Divine Inspiration"—the truly careful basis of those measurements got lost . . .

Which brings us back to the work of Sir William Flinders Petrie.

Petrie, as a boy of 13, became fascinated by the work of Piazzi Smyth through one of the many books that Smyth's eventual life-long obsession with solving the "riddle" of the Pyramid produced. That work was to infuse in Petrie the inklings of a long-term goal: to prove someday—one way or the other—whether Smyth (and Taylor before him) had been right or wrong in their assertions.

His eventual choice of a career as a surveyor provided exactly the right background for an older Petrie—now 26—who would use this engineering knowledge as the means finally to carry out his plan . . .

In November, 1880, by ship from Liverpool Petrie departed (literally on "a dark and stormy night"—experiencing agonizing bouts of seasickness) on the first of many visits to the ancient sites of Egypt. With him was an even greater arsenal of instrumentation (much of it designed by Petrie's father, himself a mechanical engineer enthralled for over twenty years by the problems presented by the Pyramid).

Petrie's methodology for investigating Smyth and Taylor's claims was as original as the instruments he used; equipped with a ten-inch French "theodolite" (a precision telescope, and a "Cadillac" of this tool of all surveyors), Petrie planned to survey *optically* the entire Gizeh site—establishing the placement and orientation of the Pyramid and its adjoining structures to within *fractions* of an inch. Petrie's survey, with individual readings sometimes conducted *50 times*, pinned down the location of the Pyramid to within a quarter of an inch (and usually within a *tenth* of an inch in most summed observations).

(This accuracy would not be equalled or exceeded until 1925—when Cole established the location of the corners of the Pyramid to within a few *hundredths* of an inch!)

Petrie's survey of the orientation of the Great Pyramid revealed another level of sophistication for this ineffable monument: an alignment with True North (that is, along the terrestrial polar axis) deviating less than 1/400th—within 5 arc minutes of perfection. In Nineteenth-Century architectural terms, this was an accuracy which surpassed any contemporary structure (and *still* does!).

His measurements of the Descending Passage revealed an equally incredible precision: a deviation of less than 1/50th of an inch along the entire masonry part—a length of 150 feet; the overall error in the entire 350 feet of this Descending Passage (including the 200 feet bored through

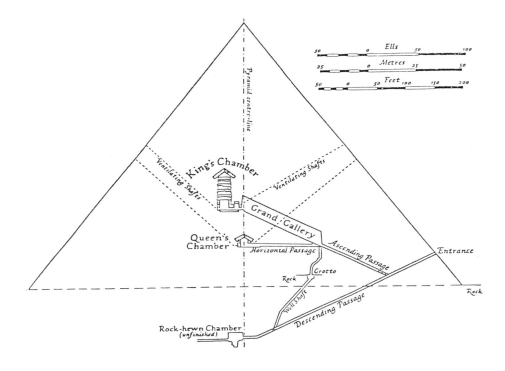

PYRAMID OF KHUFU
Chambers and Passage System

solid rock) was within ¼ inch—comparable to the best *laser-controlled* drilling being done at present!

An equally exacting survey of the Pyramid's interior chambers, including the King's Chamber, established to Petrie's satisfaction that the walls had been constructed *on the same "Pi" proportions* Smyth and Taylor had originally attributed to the exterior dimensions of the entire Pyramid!; the length around the Chamber was to the circuit of one wall as 1 to Pi.

In these extremely detailed measurements—which included versions of the devices Smyth had used (special "measuring chains" and "temperature-controlled rulers"), but advanced over those Smyth had made by 50 *years*—Petrie found many additional examples of the incorporation of this ratio into the fundamental architecture of the Pyramid. He also found *another* crucial ratio—the so-called "Golden Section"—used now and by architects throughout the Ancient World (and made famous by da Vinci, as the proportions of the human form . . .).*

*      *      *

The measurements of the Great Pyramid were provocative, but they certainly didn't amount to a smoking gun. The only obvious place to turn for any further clues was the one place where I felt the most uncertain: the almost impenetrable Sumerian and Egyptian texts—the oldest documents extant on Earth. The problem, as Sagan had made clear, was that with textual evidence alone—without an "outside" referent—a date, an artifact, a verifiable *specific*—one could prove nothing. The texts were subject to the same multiple levels of interpretation reserved (in our experience) for myths and dreams; one could never know for sure whether we were dealing with some internal "alien landscape" or an external reality verifiable through astronomical investigation, with visions of "angelic" entities and dreams of "aliens"—or with actual historic entries of Earth's sphere by creatures from some other concrete (if slightly distant!) world . . .

The images alone could not betray their level. Further, the texts were written in languages which were last spoken on this planet in a time removed from our experience by a minimum *several thousand years;* not only were their images open to interpretation, the very translations on which those images were based were also highly suspect! Who finally knows "Sumerian" well enough to discern with certainty the difference

---

*In May, 1986, a French archaeological team, using a sophisticated method of measuring minute gravitational field anomalies, made a startling discovery within the Great Pyramid: three previously unknown 6 × 9 foot chambers behind at least nine feet of limestone blocks—two of them off the level corridor leading to the Queen's Chamber from the Ascending Passage; the other apparently beneath the base of the Pyramid itself! The French have secured permission from the Egyptian government to drill four 1½ inch holes through the limestone, after which they planned to insert an "endoscope"— a device like a periscope, which would allow them to, among other things, take high-resolution color photographs of whatever still remained within the chambers . . . The chambers might be empty, having only been designed as part of a structural device to relieve the enormous stresses of 6.5 million tons of limestone. But they have surely been preserved from vandals . . . and thus have been untouched for millennia—*since the pyramid was built.* If extraordinary artifacts are found within them, then the question will arise quite logically: *from where*? Thus, a "terrestrial connection" to Cydonia . . . and this extraordinary monument . . . could well be proven before we ever go to Mars. Provided, that they tell us what they find . . .

*The Monuments of Mars*

between a "concrete object" and a metaphor (or an ideogram), or the difference between one metaphor and another?

All we could attempt to do—to avoid the pseudo-scientific babblings of the Von Danikenites—is put these bits and fragments, these textual clues, into a new context . . . and see what happened. For, if this book has any central meaning, it is this: *nothing can be properly interpreted without the context!* There are no "pure" spacecraft or aliens, only hints—as fuzzy as the photos of the Face itself, and as subject to our own fantasies, prejudices, and psychological "projections" as any excursion into poetical interpretation.

What could "save" us would be the discovery of even one "outside" anchor point . . .

*      *      *

In my quest to penetrate the maze of textual material confronting me from Egyptian and Sumerian sources, I was extraordinarily fortunate to have discovered two outstanding "guides": Robert Temple and Zecharia Sitchin.

Temple's work I'd been familiar with since 1976—"the Viking Summer." His book, *The Sirius Mystery*[15] (appearing as Viking was making "landfall" at the Red Planet) was an exploration of the possibility that an emissary *from a planet orbiting Sirius* had visited the Earth "sometime in the last several thousand years," and had left a legacy—

Which resulted in the appearance of those first two "high cultures": Egypt and Sumer!

While one could endlessly debate the thesis behind Temple's work (without "outside" corroborating "anchor points"—other than the textual references to "Sirius" throughout both cultures, and an apparent deep obsession with, and surprisingly accurate knowledge concerning, *this one star*), what impressed me at the time was. Temple's scholarship. Or, as Isaac Asimov (who is a scathing critic of anything even hinting of Von Daniken) felt impelled to comment:

"I couldn't find any mistakes in this book. That in itself is extraordinary!"[16]

Temple, of course, knew nothing of a purported "Mars connection" to the two cultures he was so thoroughly investigating—Sumer and Egypt. Rather, his focus—through a meticulous cross-comparison between Egyptian and Sumerian myths, texts, epic poems, hieroglyphic and cuneiform dictionaries, and literally hundreds of authoritative analyses—was "Sirius" itself—and its fundamental shaping of much of Egyptian cosmology/ theology, if not (surprisingly) that of Sumer as well. He too felt there was a fundamental link between these cultures, not apparently appreciated by scholarship carried on in recent years.

Temple's major contribution to my quest was a detailed cross-cultural comparison of terms and "characters" common to both cultures, with constant reference to the original textual material and original analyses (carried out by those who first discovered many of these texts), which more recent scholarship seems to have forgotten. A perfect example presents itself in the portrayal of one of the central figures in the Egyptian religious pantheon:

The "hawk-god" Horus.

Horus, as a sacred falcon, was one of the most revered symbols to the ancient Egyptians, immortalized in countless temple carvings and three-dimensional representations; one of the most famous is the exquisite statue, carved out of a solid block of black diorite, of Chephren—Horus protectively encircling the Pharaoh's head with outstretched wings.

Horus is the son of Isis and Osiris—goddess and god, respectively, most widely worshipped throughout Egyptian history.

Isis, of course, is identified with Sirius, "one of the rare certainties in Egyptian astronomy." [17] She is also, at once, the wife of Osiris (who is correspondingly identified with the constellation Orion in the Egyptian texts), and the daughter of the "sun-god"—Ra.

Osiris, according to Temple (reporting thoughts by Wallis Budge) "is referred to simply as 'god,' without the addition of any name. No other god of the Egyptians was ever mentioned or alluded to in this manner, and no other god at any time in Egypt ever occupied exactly the same exalted position in their minds, or was thought to possess his peculiar attributes." [18]

One of the first things you learn, regarding the appropriate ways to understand these complex religious tapestries of 'gods' and legendary relationships and deeds, is that the "cast of characters" is *interchangeable;* gods are constantly appearing, disappearing, and *reappearing* in changed form, in different guises and in different stories—which must be simultaneously interpreted (if they are to be understood) on several levels. In addition, according to Temple, the priesthood (particularly the *Egyptian* priesthood) was constantly encoding information in the form of "sacred puns!" This makes doubly (or triply!) difficult the attempted discernment of literal references to visitors from space or other planets . . .

It was not surprising, therefore, to discover that Horus (whose real Egyptian hieroglyphic designation was "Heru") was "the ancient Sun-god"—according to Budge's definitive volume on the subject. [19] This long-preceded the later appearance of "Ra," his siring of Isis (who, in turn, became the mother of "Heru"), or Heru's evolution into the "falcon of Horus"—the specifically-Egyptian designation of the *rising* or *setting* sun, also the Egyptian symbol for resurrection (like his "father"—Osiris).

If you're a bit confused, you're not alone. But follow . . .

One evident connection between "Heru" the sun-god, and "Horus" the god of the rising sun as represented by a *falcon*, was in the role of "Horus of the Horizon." The obvious association of this daily reappearance of the sun, after the preceding sunset, with "death . . . followed by rebirth" seems fairly "safe" to assume; the association of a "falcon" with this activity is far less clear, though one strong possibility might be the tendency of certain hawks and falcons to hover before striking. The apparent "hovering" of the rising or setting sun on the horizon may have led to this association.

This alliteration in countless Egyptian texts—"Horus of the Horizon"—struck a resonant chord. More than once, I considered the structures at Cydonia, their apparently deliberate layout in striking reinforcement of one critical rising of the sun—on the *Summer Solstice*, directly over the Face. Egypt's rapt fascination with the "rising of the sun"—as "Heru"—"Horus of the Horizon" kept recurring . . .

It was Temple, in recounting Budge's many translations of the term "heru," who provided the first breakthrough:

"The word *heru* also has the meaning of 'face.'"

"*Horus* of the Horizon" . . . "*Face* of the Horizon" . . . (note the cover of this book!).

There is more.

Temple:

> In Egyptian the letter "l" and the letter "r" are entirely interchangeable and have the same hieroglyph. Consequently, Heru could just as reliably be Helu. If one takes Helu and puts a Greek ending on it one gets Helios [the Greek god of the sun, in *much* later mythology] . . .
>
> It is interesting to note in the account of the word Helios as given by Liddell and Scott,[20] Homer used the term [in the *Odyssey*] in reference to "the rising and setting, light and darkness, morning and evening" . . . Homer has thus used the *heru*-derived Helios in precisely the manner which we might have expected of an Egyptian, rather than a Greek, poet.

The Greek name "Heliopolis"—"City of the Sun"—for the temple-site of the Ancient Egyptian sun-god, Atum-Ra, therefore, could just as well be termed "The City of the *Face on the Horizon*" . . . (One of Imhotep's titles, discovered on the inscription carved at Saqqara, was "chief priest . . . of Atum-Ra."[21]) Incidentally, Heliopolis is a suburb of modern-day Cairo.

Temple, continuing his explorations of the Egyptian term *heru:*

> It seems that the curious Greek word for *hero* comes also from *heru*, though a word similar to *hero* exists in Sanskrit, the language of ancient India after 1200 B.C. The word in Sanskrit which has the meaning of "hero" is the related *Vira*. It is used in the precise sense of "hero" (as opposed to a god) in the early *Rig-veda* and is thus attested at the time of the first migrations of Aryans into India. There is no question that the two words are cognates of each other [related]. However, I propose for them . . . a common derivation: from the Egyptian *heru*."

Temple then went back to Budge's earliest translations of the term, finding another meaning for *heru* (in addition to "sun on the horizon" and "face") almost identical to the Greek *hero* and the Indian *vira:* "applied to the king [pharaoh] as the representative of the sun-god on Earth."

Temple:

> This is a precise meaning applying to a human being on earth who is neither god nor daemon, but *hero* . . . In Homer "the heroes were exhalted above the race of commen men," but particularly in Pindar the poet, we find the word used to describe a race "between gods and men," in precisely the sense that we should expect the word *heru* to survive in another language. This Egyptian application of the word to their Pharaohs survived almost without change in Greek and Sanskrit and later in Latin and the later Indo-European languages.

Thus, by metonymy (use of the name of one thing for that of another), *heru* can equally apply to "the sun on the horizon," a race "between gods and men," a "face" (on the horizon?) . . . and the "king [pharaoh] as the *representative* of the sun-god . . ." And where, one might reasonably ask, was the one place where all these descriptions came together *in one object*—

If not Cydonia . . .

Which was fascinating . . . if totally unprovable.

This was the trap of such "mythological analyses": one could all-too-easily read *any* meaning into texts and terms—if one went looking hard enough. Without a *specific* reference to the site we call "Cydonia," or at the very least a reference to Mars, any connections of "heru" with the Face remained conjectural.

Right about that time I came upon an anthology of articles edited

by Ian Ridpath, formerly published in the respected *Journal of the British Interplanetary Society* (the organization founded many years ago by Eric Burgess and Arthur Clarke). It contained a provocative chapter heading:
    *"Signpost to Mars"*[22]
    I turned to it with curiosity and apprehension, to find the following:

> One of the weirdest coincidences of the whole affair is that Cairo, the site of the (two greatest) pyramids, was originally named El-Kahira, from the Arabic El-Kahir—
> "Meaning Mars . . . "

I blinked . . . several times.
Cairo—*Mars*?!
I fired off a note to Lambert Dolphin, and he confirmed this fascinating tidbit: when Cairo was renamed (sometime in the 10th century A.D.), from something loosely translated as "the camp" (!), its new designation was taken from the same Arabic root-stem as "Mars"—*

Again—what were the *random* probabilities that there would exist two isolated worlds, both with "pyramids" and "sphinxes," and now, that the *one* site on this planet where the most perfect, most archtypal forms still stand—Cairo—would also form the key linguistic bridge that *links* those worlds . . !?

But even this serendipitous discovery would pale beside developments to come . . .

Throughout this search for some fundamental linkages with Mars, one considerable problem always loomed: how such a "Cydonia connection" with the splendor that was Egypt (and Sumer) was possible at all—given their irrevocable quarantine from whatever had taken place on Mars by that "half a million years."

Temple was not the only scholar I turned to in an effort to understand these potential extraterrestrial-mythological references. Remember, my other guide: Zecharia Sitchin? The blurbs on his book, *The Stairway to Heaven,* termed him a man "with a profound knowledge of modern and ancient Hebrew . . . the Old Testament . . . history and archaeology of the Near East . . . [who] attended the London School of Economics."[23]

I soon discovered that it was Sitchin's thesis that the two "high cultures" of the Near East—Egypt and Sumer—owed their existence to "advanced visitors from another planet."

---

*Aldridge, J., *Cairo: Biography of a City,* New York: Little Brown & Co., 1969

Another Von Danikenite!

Something, however, stopped me from immediately consigning the book back into oblivion . . . Instead, I turned to the notes section—and confronted scores of dense academic references and texts, with titles like *Journal of Near Eastern Studies* (Chicago) and *Revue d'Assyrologie et archeologie orientale* (Paris). Kramer's books (all *six* on Sumer) were there, as were Jacobsen's, Ebling's, Frankfort's . . . *all* the major scholars in the world of Near East studies. With growing excitement, I read further in the jacket notes, ". . . Mr. Sitchin had devoted thirty years to gathering and synthesizing the data for this remarkable book." (Later, I learned that this volume was the second of a comprehensive *trilogy* on the subject of possible Near Eastern extraterrestrial contact, which Sitchin termed— somewhat ironically—*The Earth Chronicles* . . .)

Here (like Temple) might be someone with some of the same questions I was now forming, regarding the inexplicable genesis of two of the world's earliest and most "splendiferous" high cultures—but with a depth in cuneiform, hieroglyphs and all the other nuances of Near Eastern studies that I obviously lacked.

My initial interest, naturally, focused on Sitchin's treatment of Egyptian myths and texts. Very quickly, with scores of references to scholarship extending back over a century (but sidetracked in more recent years by other, more "fashionable" interpretations) he too confirmed my own suspicions regarding the possibilities of strong, fundamental links between these two great cultures—Sumer and Egypt.

In closing this section of his treatment of these two cultures, Sitchin noted:

> Hieroglyphically, the sign for Ur [a great city in Mesopotamia] meant "the far-foreign [land] in the east"; that it may have referred to the *Sumerian* Ur, lying in that very direction, cannot be ruled out.
>
> The Egyptian word for "divine being" or "god" was NTR, which meant "one who watches." Significantly, that is exactly the meaning of the name Shumer [the Old Testament name for "Sumer"]: the land of the "ones who watch."

Having clearly intimated some deep connection between Egypt and Sumer (and with a great deal of additional material, impossible to include here), Sitchin turned his attention to identifying the Sumerian counterparts to Egypt's "olden gods"—if not when they may have had a connection with "the mountain land and the far-foreign land" (which, from an Egyptian perspective, Sitchin identified with Sumer).

Sitchin's argument here was that, when properly decoded, the vast array of gods and goddessees strewn throughout the pantheons of all the Near East cultures could ultimately be traced back to an "original twelve gods" of ancient Sumer. Using loan words, Sitchin demonstrated the gradual transformation of these primordial twelve gods into regional deities throughout the world. Even the Greek gods and goddesses, thousands of years after, replicated their counterparts in *Sumerian* texts. The gods and goddesses of the Indus Valley, thousands of miles to the East, were "proto-Sumerian." In Sitchin's words,

"We are all, ultimately, Sumerians."

That is, we all appeared suddenly, simultaneously, mysteriously . . .

Others have cited the historical importance of Sumer; it was Sitchin's underlying premise—in reassuring consonance with Temple's, and drawing on *identical* source material but the product of completely independent research—that was so extraordinary: that the Sumerian civilization—which ultimately diffused throughout the world, in remarkably pure form in many instances*—was the result of "outside" interference.

Sitchin made no bones about identifying his original twelve "gods" with his purported visitors—in principle, no different from Von Daniken's contentions. The striking difference, as I saw it, was that at each point in the construction of his thesis, Sitchin (unlike Von Daniken) backed up his chain of logic with truly first-rate *scholarship*—with meticulous reference to the original Egyptian, Hittite, Assyrian, or Sumerian texts, and a "weighting" of sometimes half a dozen translations and transliterations that have taken place in the hundred years or so since much of this material was first discovered, before deciding on a specific meaning for a text . . . or even a *term* used within a text.

A case in point:

The Mesopotamian texts that refer to the inner enclosure of temples, or to the heavenly journeys of the gods, or even to instances where mortals ascended to the heavens, employ the Sumerian term *mu* or its Semitic derivatives *Shu-mu* ("that which is a *mu*"), *sham*,

---

*In 1976 it was announced that a clay tablet dug up from Ugarit (on the coast of present-day Syria), from the Assyro-Babylonian culture, bore a musical text based on *our* familiar "octave." Said Dr. Richard L. Crocker, Professor of Music History at the University of California, Berkeley: "We always knew there was music . . . but until this, we did not know that it had the same heptatonic diatonic scale [as] contemporary Western music and Greek music of the first millennium B.C."

or *shem*. Because the term also connoted "that by which one is re-membered," the word has come to be taken as meaning "name." But the universal application of "name" to early texts that spoke of an object used in *flying* [a "skychamber"] has obscured the true meaning of the ancient records.

Thus G.A. Barton (*The Royal Inscriptions of Sumer and Akkad*) established the [current] unchallenged translation of Gudea's temple inscription—that "Its MU shall hug the lands from horizon to horizon"as "Its *name* shall fill the lands" . . .

Sitchin's scholarship is evident where he buttresses his contention that archaic Sumerian "skychamber" (MU) evolved into more "modern" Akkadian and Babylonian "name," by noting:

> That the [original] purpose of the commemorative stone pillars was to simulate a *fiery* skyship can further be gleaned from the term by which such stone stelae were known in antiquity. The Sumerians called them NA.RU ("stones that rise"). The Akkadians, Babylonians, and Assyrians called them *naru* ("objects that give off light"). The Amurru called them *nuras* ("fiery objects"—in Hebrew *ner* still means a pillar that emits light, thus today's "candle"). In the Indo-Euorpean tongues of the Hurrians and the Hittites, the stelae were called *hu-u-ashi* ("firebird of stone").

Sitchin concluded his fascinating lesson in etymology, and its application to this almost impenetrable problem, by saying,

> The persistence of biblical translators to employ "name" wherever they encounter *shem* has ignored a farsighted study published more than a century ago by G.M. Redslob (in *Zeitschrift der Deutschen Morgenlandischen Gesellschaft*) in which he correctly pointed out that the term *shem* and the term *shamaim* ("heaven") stem from the root word *shamah*, meaning "that which is highward." When the Old Testament reports that King David "made a *shem*" to mark his victory over the Armaeans, Redslob said, he did not "make a name" but set up a monument pointing skyward.
>
> The realization that *mu* or *shem* in many Mesopotamian texts should be read not as "name" but as "sky vehicle" opens the way to the understanding of the true meaning of many ancient tales, including the biblical story of the Tower of Babel . . .

The simple difference between alternate translations of these texts by Kramer, Jacobsen, and all the other "mainstream" scholars, and Zecharia Sitchin, comes down to *context*. If you assume (as Kramer and, to a lesser extent, Jacobsen) a simple, agarian people, slowly being urbanized in cities (Sumer's "big invention"), then you are only going to discover terrestrial objects in these arcane texts and legends.

If, however, from delving into the earliest interpretations of key words, you (like Sitchin) glimpse an extraordinary *technological* backdrop for these most vivid literary texts, and use that as your *assumed* context—you might quite naturally arrive where Sitchin has. You might come to drastically differing interpretations of these so-called "myths," including the idea that they, in fact, represent a distorted *history* . . . events completely outside the *context* of a simple agrarian society en route to city-states and taxes . . .*

For instance, "the black-headed people"—a consistently puzzling term used by the Sumerians in referring to themselves—if referred to the earliest-known ideogram translated as "black," could just as legitimately be read "the people associated with the vault of heaven," or even "people *descended* from the vault of heaven . . . or, the *stars!*" (According to Temple, the Greeks for some reason referred to the Egyptians as "the black-*footed* people" . . .)

One etymological indication that this hypothesis might, in fact, have substance comes from the mainstream scholars themselves; both Kramer and Jacobsen, for example, admit to puzzlement, and even outright bafflement, when confronted by thousands of early Sumerian words. [24] Significantly, the problem becomes worse *the more ancient* the translation that's attempted—a tell-tale indication of a change in *context* of these terms: either increasing distortion by the sheer weight of millennia, or—

A context so *removed* from "conventional interpretation" that it might as well have originated on another planet . . . which is exactly what Zecharia Sitchin is proposing!

Which brings us to an overview (now that I hope we have established Sitchin as a serious researcher) of exactly what he *is* proposing.

\*         \*         \*

From translations of pieced-together fragments of innumerable fragile tablets of baked clay, and from cylinder seals and inscriptions buried for millennia amid the rubble of Sumer's once-great palaces and urban centers, Sitchin wove his extraordinary tale . . .

Of a race of "beings" from another member of this solar system—the NE.BI.RU—who come to Earth on something of an "emergency mission." These beings, for many years after their arrival, labor by themselves

---

*Take Schliemann's example of the *Odyssey*—and the eventual discovery of Troy!

on Earth, at some "project" apparently connected with the emergency that's brought them here. But eventually, there is a "mutiny" among the "crew" (the "Anunnaki")—and to save the "mission," one leader is inspired to create—out of the available proto-human material—"workers" to assist them—

And the human race—"the black-headed people"—are thus born.

Eventually among these "beings," even greater dissensions arise and one faction (led by "Enlil") demands that the "workers" be exterminated—by means of a "great flood!" However the expedition's *other* leader ("Enki"), who created these workers in the first place, manages to save the fruits of his "experiment"—by instructing a few (in particular, an individual named "Ziusudra") in how to build a boat . . .

Ultimately, Enki becomes something of the workers' protector—teaching "the black-headed people" the arts and culture of civilization, "lowering kingship from heaven," and leaving a legacy of knowledge—from metalwork to legal codes—to this fledgling race, the inheritors of a new world . . .

*      *      *

One doesn't have to accept this whole science-fiction melodrama to take Sitchin's meticulous reconstruction seriously. I myself would have dismissed it with a laugh only a few years ago.

But the monuments on Mars cry out for an explanation. The Face—if it is artificial—absolutely demands that something in the conventional time-honored history of Earth be changed.

It is impossible, in the space allotted here, to do justice to Sitchin's painstaking reconstruction of this central "story line," from hundreds of quoted (and often reproduced) cylinder seals and cuneiform texts brought to light by archaeology across the Near East in the last one hundred or so years. Suffice to say, there is enough detail in what I have presented for the discerning reader to identify a disturbing parallel—

With key events recounted in the Western *equivalent* of these, Sumer's most treasured (thus copied and recopied, countless times . . .) "holy texts."

That equivalent, of course, is our Old Testament.

As I read Sitchin's recreation of this Sumerian "epic," I couldn't help but remember Soviet ethnologist, M.M. Agrest's contention many years ago, that, "certain Biblical events could be interpreted as evidence of an extraterrestrial visitation . . ." (see Chapter XV). That the *events*—stripped of any contextual interpretations—are essentially identical in the Old Testament and in Sumer's (much more ancient) texts, is clear. What this ultimately means is far less certain . . .

Sitchin focused much of his attention in attempting to discern the ultimate *origin* of "Enil," "Enki," and the other members of the Sumerian pantheon. This, in turn, focused his attention on a term that Temple, in his book, describes as "one of [those] infuriating Sumerian words which we would like to understand.

"Where does it come from? What does it mean . . ?"

The term is "NE.BI.RU."

Much of Sitchin's thesis—that his "visitors" ultimately originated from the "NE.BI.RU"—centered around efforts to decipher the meaning of this extremely cryptic reference, whose symbol (with all its many later connotations . . .) looks like this: +.

The textual sources for the term, including the Babylonian creation myth entitled the *Enuma elish* ("When on high . . ."), are not much help:

> NE.BI.RU shall hold the crossings of heaven and earth . . .
> He who the midst of the sea restlessly crosses,
> Let "Crossing" be his name, who controls the midst, etc. . . .

One major weakness exhibited by Sitchin, in this otherwise impressive work, is his astrophysics—for as ultimate explanation of the mysterious "NE.BI.RU" Sitchin proposed a hitherto *undiscovered* planet (*The 12th Planet*—title of the first volume of his trilogy[25]); further, that it orbits in an extremely elliptical path which *crosses* ("NE.BI.RU . . . Let 'Crossing' be his name . . .") the orbits of the outer planets of the sun!

To me, this whole idea was highly improbable—on grounds ranging from simple celestial mechanics to sheer temperature; the likelihood of life originating—let alone evolving to intelligence!—on a world whose temperature most of the time (in that particular orbit) would hover near *absolute zero*, seemed dubious—at best!

It was Temple, from his completely independent efforts to understand the meaning of the arcane term, who ultimately offered, not only a brilliant means around this impasse—but the means of saving Sitchin's thesis and his otherwise exemplary scholarship. Wrote Temple:

> . . . let us look at the Egyptian language again. We find the word *Neb* is extremely common and is used in many combinations and means "Lord." Without further ado, let me make clear that I believe the Sumerian Nebiru to be derived from the Egyptian Neb-Heru. If we treat Heru in its older Egyptian sense as the sun, then descriptions of Neb-Heru in the Babylonian *Enuma elish* could read as a perfect description of Neb-Heru—"the Lord the sun": "Nebiru shall hold the crossings of heaven and earth, etc."

But if we remember that *another* meaning derived by Temple for "heru" is "hero," then we can also interpret this enigmatic passage as a poetic description of a *physical* "crossing of heaven"—

Perchance, *the actual transit of someone "heroic" on a legendary journey from Mars to Earth?* . . . or maybe, the original generations-long odyssey *from beyond the solar system* . . . to the deserts of Mars itself?!

Extraordinarily, such legends—concerning "ships" and "epic voyages"—resound down through the world's most "epic" literature, and feature "larger-than-life" characters, "a race 'between the gods and man'"—with names like "Gilgamesh" and "Hercules"—on daring voyages pursued in the face of overwhelming obstacles. And, in the case of one of the most familiar versions, a voyage that *repopulates* a decimated Earth . . .

And so we have a series of legendary "arks" appearing throughout history (a word that, according to Temple's ceaseless efforts in exploring the dim "corridors" of Budge's massive 1300-page *Egyptian Hieroglyphic Dictionary*, derives from the Egyption *ārq*—and means "to complete, to finish" . . . in the sense of cycles; also "the last" or "the end of anything."

The most telling (for these purposes) incarnation of a mythic vessel and an "epic" odyssey appears in Greek mythology, as the fabled *Argo* (whose very name derives, according to Temple, from that Egyptian *ārq*). On her initial voyage, the "hero-figure" Briareus assumed command. Later evolutions of the *Argo* tale involve a series of such heroes—including "Hercules" (in early and late form), ending eventually in "Jason"— who captains the most widely-known (though very late) version of this epic. Jason and his fifty "Argonauts" (somewhat reminiscent of the fifty Anunnaki . . .) are on an urgent quest—to return "the golden fleece" from a fabled land called "Colchis."

It is fascinating to learn (via Temple's meticulous research) that the original Hercules was probably derived from Briareus, that he was also captain of the *Argo* (not just making a "cameo" appearance, as in the later version), and that he was acknowledged by the Greeks as having come from Egypt. But most intriguing . . .

> It is well accepted today among scholars that Hercules was in many ways a survival of Gilgamesh, with particular motifs and deeds being identical in both heroes . . . There is a possibility that Herakles ("the glory of Hera") . . . and his protectress the goddess Hera (wife to Zeus and the Queen of the gods) are derived from *heru* . . .

Which introduces all kinds of potential metonymic levels and associa-

*The Monuments of Mars*

tions, for a "hero-figure" associated with an epic voyage . . .

Perhaps the most stunning possibilities center on the *reason* for the *Argo*'s voyage: to find that "golden fleece." Temple effectively demonstrates that this could be but *another* of those endless Egyptian "sacred puns" transliterated into Greek: if you drop the "h" from *heru* and add a Greek ending, you wind up with *erion*—which means "woollen fleece!"

Then, by reference to Herodotus, Temple establishes a firm connection between "Colchis"—the place where the "golden fleece" is being kept—and Ancient Egypt; quite likely, Colchis (on the Black Sea) was an early colonial outpost of the Egyptians—before about 1200 B.C. Thus, myths concerning it, in particular its treatment of the dead, when they became transliterated into Greek, preserve their essential *Egyptian* character (according to Temple). The central example he uses to illustrate this point is his identification of the Greek goddess "Circê" (which means "falcon"), guardian of Colchis' dead, with the Egyptian "Horus . . . as a falcon." Further, in the *Argo* myth the sun-god Helios (which we've already identified with the *rising* or *setting* sun, per Homer) stables his horse in Colchis . . . (More multiple metonymic links!)

So, we have a "golden fleece" (which is really a series of associative links for "heroes," "faces," and the *rising* "sun") held prisoner in Colchis—to be pursued by Hercules (Heru?) and a crew of fifty "Argonauts."

It is here that Temple springs his best surprise, for—writing up to 1976—he knows *nothing* of what lies waiting on the planet Mars . . .

"During the interval of the fleece's stay in Colchis the fleece rest[s] 'in the grove of Ares (Mars) . . .'"

There is a companion legend, where another Greek hero—Cadmus— is instructed to "follow a cow and build a city wherever she should sink down for weariness . . . at last (the cow) sank down where the (Greek— but named after the *Egyptian*) city of Thebes now stands, and here (Cadmus) erected an image of Athene (who originally helped construct the *Argo*) . . . Cadmus, warning his companions that the cow must be sacrificed to Athene without delay, sent them to fetch the lustral water from the *Spring of Ares* . . . now called the Castalian Spring, but did not know that it was guarded by a great serpent . . ."[26] A similar "great serpent" was on guard over the golden fleece in the "*grove* of Ares" in Colchis . . .

My working hypothesis is that some myth could be more the stuff of lost history than spontaneous vision or dream; that is what I am trying to test. The constant ambiguity of any textual interpretation firmly in mind, I consider this potent stuff!

We have here a multi-layered, apparently heavily encrypted, highly metonymic epic tale, coming to us in recursive forms from the dawn of mankind's efforts to record and transmit faithfully "large events" which somehow shaped its past (if you subscribe to "myth as history" . . .). The tale is multiply redundant—involving key root words which stand for "rising sun," "heroes," even *faces* . . . all involved somehow with being kidnapped *to Mars*, pursued by valiant "heroes" in a fabled "voyage," in an equally-fabled craft "crossing" some immense expanse of heaven and earth—

And ultimately *returning* . . .

Oh yes, one more thing . . . *ãrq ur*, which translates from ancient Egyptian as "the Great ending . . . ," is the description for that "mystery of mysteries," the Sphinx—located within a few hundred feet of the most perfect, most ineffable pyramid on Earth, at a place whose very name in Arabic means "Mars" . . .

*       *       *

Turning back to Sitchin; it was at this point in his discussion of the NE.BI.RU—"Let 'Crossing' be his name . . ."—that he introduced a now-familiar name: Berossus.

Unlike Sagan, who denigrated all temporal references in the "Oannes legend" as "unreliable"—and therefore didn't reproduce them—Sitchin cited these specifically—and in so doing, furnished a crucial datum missing from Sagan's treatment of this increasingly-important tale . . . the *time* when contact with Oannes (whom Temple identifies with *Enki* . . .) was first supposed to have occurred—according to the Sumerians *themselves*.

From Apollodorus:

> This is the history which Berossus has transmitted to us . . . that the first king was Alorus of Babylon, a Chaldean (Sumerian) . . . (who) reigned ten *sari* . . .

From Alexander Polyhistor:

> . . . in the first year [of Alorus' reign] there made its appearance, from a part of the Erythraean Sea [the Persian Gulf] which bordered upon Babylonia, an animal endowed with reason, who was called Oannes . . .
>
> This Being in the day-time used to converse with men; but took no food at that season; and he gave them insight into letters and sciences, and every kind of art. He taught them to construct houses, to found temples, to compile laws, and explained to them the principles of geometrical knowledge. He made them distinguish the seeds of the earth, and shewed them how to collect fruits; in short, he instructed them in every thing which could tend to soften manners and

        *The Monuments of Mars*

humanize mankind . . . When the sun set, it was the custom of this Being to plunge again into the sea, and abide all night in the deep; for he was amphibious.

After this there appeared other animals like Oannes, of which Berossus promises to give an account when he comes to the history of the kings . . .

From Abydenus:

There were afterward other kings, and last of all was Sisithrus: so that in the whole, the number amounted to ten kings, and the term of their reigns to an hundred and twenty *sari* . . .

It was also from Abydenus that the crucial datum came: ". . . now a *sarus* is esteemed to be three thousand six hundred years."

Using the sum of the reigns of these "ten kings" (which should not be taken too literally, but probably was a device used to reinforce the *total* span of years), and adjusting his calculations to allow for the uncertainty introduced by the consistent reference to "the Deluge" as the time from which these "reigns" were to be measured, Sitchin arrived at a figure for when "Oannes/Enki" first appeared . . .

Four hundred forty-five thousand years ago.

I had found my "missing" half a million years.

*      *      *

The fact that this extraordinary numerical correlation was the result of studies carried on in total isolation from our Mars inquiry, spanning more than thirty years . . . and published even as Viking was departing for its historic rendezvous—when the existence of the Face and its associated Pyramids was totally unknown—this essential *correspondence* of these two dates—mine and Sitchin's—had to make any reasonable individual reflect on the staggering implications resident in this entire interlocking mythology, built around "Horus/Heru" . . . the "NE.BI.RU" . . . the kidnapping of a "golden fleece" . . . to Mars . . .

And the appearance of a Being named "Oannes."

Which now seemed linked, by an extraordinarily *specific* date, to a series of stupendous artifacts on Mars . . . including an archaic resemblance to the highest form of "hominid" living on the Earth when these dates so coincide.

In these archetypal glimmerings, shadows moving in the mists from a time on Earth lost forever to the written word, there is the suggestion— and only the *suggestion*—of an extraordinary resolution to the ultimate dilemma of finding an image of ourselves . . . on Mars.

That we—or something we once were—were *deliberately transported there* uncounted millennia ago . . . to fulfill some "grand design" architected by those who came before . . .

*. . . a Promised Land these castaways can never hope to enter . . . on their own.*

*With an Earth they were prohibited by stark gravitational realities from ever inhabiting in person, beckoning irresistibly across the last few million miles, these* initial *"Martians"*—non-human *visitors from an unknown star and inheritors of a genetic science and a biological understanding matured across a thousand other worlds, if not a thousand times those years—methodically began to plan for the creation of descendants who could* inherit Earth; *using the genetic legacy of Earth itself—the highest hominid that evolution had produced—these travelers, from a place unimaginably far away, set about creating* a new species *which would combine the best of two immeasurably distant worlds . . . the evolutionarily adapted* form *of four million terrestrial years . . . and the intelligence*—the spirit—*which had dared to leap between the stars themselves . . .*

*       *       *

Might it be possible that we, who are so proud to trace the origins of "civilization" back 6000 years "all the way to Sumer," might have to seriously ponder the awesome possibility that "human" history could be a lot more intricate than we've imagined . . . and might conceivably encompass a span of time a hundred times those "mere" 6000 years . . ? Further, might we have to seriously consider the even more extraordinary possibility that "Mars" played some crucial role in the formative development of our own species, that *we* might ultimately be "the Martians"— who, sometime in the last half million years, returned to Earth . . . and *stayed*?

For, according to the records once again of the Sumerians themselves, Oannes "had under a fish's head another head, and also feet below, similar to those of a man . . . (whose) voice too, and language, was articulate and human . . ."[27]

According to another fragmentary record of (presumably) the same event, preserved in the writings of a Byzantine Patriarch, Photius (c. A.D. 820–c. 893):

> (Helladius) recounts the story of a man named Oe who came
> out of the Red Sea having a fish-like body but the head, feet and

arms of a man, and who taught astronomy and letters. Some accounts say that he came out of a great egg whence his name, and that he was actually a man, but only seemed a fish because he was clothed in "the skin of a sea creature."[28]

Is this the message of the Face: that Bradbury's hauntingly prophetic vision, given voice in the immortal *Martian Chronicles*, was true—

> . . . and he wondered, quietly aloud, how they had built this city to last the ages through, and had they ever come to Earth? Were they ancestors of Earth Men ten thousand years removed . . ?

It is impossible, given this extraordinary correlation of chronologies and terms, to dismiss the possibility that the Sumerian hints of awesome "contact," the resplendent monument which symbolizes Ancient Egypt, and the enigmatic configurations discovered at Cydonia . . . are fundamentally connected. That the very geometric spacing and arrangement of the figures we have termed "the Monuments of Mars" were, indeed, *meant* to be a message—to communicate more eloquently than any words the incomparable nature of Something which occurred here—

And cast an echo across tens of millions of miles . . . and half a *million* years.

Suddenly, I remembered a comment from another author with a tendency for eerie prophetic work—including *Childhood's End*—my old friend, Arthur Clarke.

Said Arthur,

> There can be little reasonable doubt that, ultimately, we will come into contact with races more intelligent than our own. That contact may be one-way, through *the discovery of ruins* . . . it may even be face-to-face. But it *will* occur, and it may be *the most devastating event* in the history of Mankind (italics added).

Unless it *was* . . . and the world we have inherited is proof.


## Notes

1. Sagan, C. and Shklovskii, I.S., *Intelligent Life in the Universe*, New York: Dell, 1967. *Op. cit.*
2. Jacoby, F., *Fragmenta Graec. Hist IIIC*, New York: Leyden, 1958.
3. Kramer, N., *History Begins at Sumer*, New York: Penguin, 1954.
4. Grossinger, R., ed., *Planetary Mysteries*, Berkeley: North Atlantic Books, 1986.

5. Kuhn, T., *The Structure of Scientific Revolutions*, Chicago: University of Chicago, 1962.

6. Sheldrake, R., *A New Science of Life*, Los Angeles: J.P. Tarcher, 1982.

7. West, J.A., *The Serpent in the Sky: The High Wisdom of Ancient Egypt*, New York: Harper & Row, 1979.

8. Parker, R.A., *Calendars of Ancient Egypt*, Chicago, 1950.

9. Frankfort, H., *The Birth of Civilization in the Near East*, London, 1951.

10. Hurry, J.B., *Imhotep, The Vizier and Physician of King Zoser and Afterwards the Egyptian God of Medicine*, Oxford: Oxford University Press, 1926.

11. Taylor, J., *The Great Pyramid: Why it was Built & Who Built It?*, London, 1864.

12. Smyth, P., *Life and Work at the Great Pyramid of Jeezeh during the Months of January, February, March and April, AD., 1865*, Edinburgh, 1865.

13. *op. cit.*

14. Tompkins, P., *Secrets of the Great Pyramid*, New York: Harper & Row, 1971.

15. Temple, R.K., *The Sirius Mystery*, New York: St. Martin's Press, 1976.

16. Asimov's blurb on *The Sirius Mystery*.

17. Neugebauer, O. and Parker, R., *Egyptian Astronomical Texts*, Vol. I, Providence: Brown University Press, 1960–67.

18. Budge, W., *The Gods of the Egyptians*, Vol. II, London, 1904.

19. Budge, W., *Hieroglyphic Vocabulary to the Theban Recension of the Book of the Dead*, London, 1911.

20. Liddell and Scott, *Greek Lexicon*.

21. Hurry, J.B., *Imhotep, The Vizier and Physician of King Zoser and Afterwards the Egyptian God of Medicine*, Oxford: Oxford University Press, 1926. *Ibid.*

22. Ridpath, I., "Signpost to Mars." In *Messages to the Stars*, New York: Harper & Row, 1978; Saunders, M. "Signpost to Mars," *Journal of the British Interplanetary Society*, Vol. 30, 1977.

23. Sitchin, Z., *The Stairway to Heaven*, New York: St. Martin's Press, 1980.

24. Kramer, N., *History Begins at Sumer*, New York: Penguin, 1954. *Op. cit.*

25. Sitchin, Z., *The 12th Planet*, New York: Stein & Day, 1976.

26. Graves, R., *The Greek Myths*, Vols. I & II, London: Penguin Books, 1969.

27. Polyhistor, quoted in Cory, I.P., *The Ancient Fragments*, Edition I, London, 1828.

28. Temple, R.K., *The Sirius Mystery*, New York: St. Martin's Press, 1976. *Op. cit.*, Appendix II, p. 256.

# GOING BACK TO MARS

"To the stars . . ."

—S. Christa McAuliffe

*This can all be tested* . . . if we return to Mars.

Either there *is* a set of artifacts—inexplicable . . . bizarre—lying on the planet Mars . . . or there is not. It's that simple.

Now—in light of recent political developments, here and in the U.S.S.R.—there is no doubt that we *will* return to Mars—and very soon. Two specific unmanned missions—the Soviet "Phobos" probes in 1988, and the U.S. "Mars Observer" spacecraft in 1990—have officially been scheduled. (And at this writing, there has been a fascinating new development vis a vis the U.S. spacecraft: a decision in favor of *a last-minute inclusion of a camera*—one with an astounding resolution capability on the Martian surface of "one or two meters.")

But even more significant, representatives from the U.S. State Department and the Soviet Foreign Ministry, at a mid-November 1986 meeting at the National Academy of Sciences, "concluded a framework agreement for the two countries to undertake cooperative scientific exploration of Mars and other planets . . ."—according to a report in the *Los Angeles Times.* [1] The agreement, expected to be signed by the leaders of the two countries at the next summit—whenever one is held—"specifically [calls] for the Soviet Union to share scientific data from its 1988 Phobos probe, which [among other objectives] is to land on a Martian moon . . . [and] for the United States to share with Soviet scientists data returned from its Magellan probe to Venus, which is to be launched by the space shuttle in the Spring of 1989, *and its Mars Observer* . . . (italics added)."

Not only does this agreement remove the potential for a nasty political "surprise," it also establishes "a framework for negotiations on ambitious joint projects in the future . . ." according to the *Times*—

Such as a *joint manned mission*—once the shared data from the earlier probes have verified the existence of a set of "monuments" . . .

In the wake of the numbing disaster of the *Challenger* at the beginning of the year, and its shattering effect upon the space program, this agreement with the Soviets would in any case be welcome news, part of the new NASA Administrator James Fletcher's "detailed plans for revitalization of the U.S. planetary exploration program." But the new cooperative agreement with the Russians vis a vis Mars has triggered significant additional developments.

Picking up the "let's cooperate and go to Mars" theme, a key U.S. Congressman, Rep. George E. Brown, Jr., Chairman of the House subcommittee on science, aviation, and materials (highly influential in terms of prospective NASA funding), strongly urged "an alternative to SDI, one as bold and unprecedented, but one that will not simply extend U.S.-Soviet rivalries into a new realm. A true Space Cooperation Initiative, involving a joint trip to Mars and other ambitious undertakings could do just that." [2]

This was followed a week later by Carl Sagan, writing in *Aviation Week & Space Technology*, reiterating his call for a joint U.S.-Soviet mission—but with even stronger oblique reference to reasons not too different from our own . . .

> America urgently needs a technological goal appropriate to carry us into the Third Millennium with a scope and depth that recaptures both domestic and worldwide admiration.
>
> Fortunately such a goal is within reach . . . a systematic program of exploration and discovery on the planet Mars . . . justified for a variety of reasons [including] as a potential scientific bonanza— for example, on climatic change, on the search for present or past life, [and] *on the understanding of enigmatic Martian landforms . . .* (italics added). [3]

The climax of this sudden focus at the close of 1986—on a planet which had been so remote in space and psychological perceptions just a few short months before—came the first week in December, in a lead editorial in the *Los Angeles Times*. In as bold a statement concerning space as has been heard since John Kennedy's sweeping call for a landing on the Moon, the editors of the *Times* dramatically urged,

> We have gone to the moon, and now it is time to go beyond. Mars is the only planet in the solar system that is remotely like the

Earth, and it gives evidence of having been more like the Earth in the past . . . A round trip to Mars would take two years . . . there would be problems, but they can be surmounted, and the value of a joint mission [with the Soviets] would be so great that it is worth doing the work to make it happen. But if it cannot be done jointly, it is worth doing alone.

The United States should be making plans for a trip to Mars. President Reagan should declare this a national goal, and NASA should set to work. [4]

*    *    *

While we wait for better photographs from Mars, and especially of Cydonia, there is a lot of data still remaining on the Viking tapes themselves, which the pioneering work of Dr. Mark Carlotto, of The Analytic Sciences Corporation in Reading, Massachusetts, is just beginning to reveal—stunning new images (some of them reproduced elsewhere in this book) of the Fort, the D&M Pyramid, and of the Face itself.

Carlotto's improved algorithms, brought to bear for the first time in ten years on the "old" Viking data tapes, have added significant, if not remarkable, detail to what we've seen before—especially in the Fort. In looking at Carlotto's versions of this structure, I am overwhelmed; there is almost no remaining doubt in my own mind—after having seen these pictures—that we are looking deep into a once-magnificent and comparatively fragile structure—which some awesome, if unimaginable, force . . . in the millennia its been exposed on Mars . . . has somehow opened up, revealing the deep, mysterious "courtyard" which initially captured our attention from a thousand miles above.

Mark's processing has brought out myriad "decks" and multiple "overlying and descending levels," whose existence is as out of place— in terms of any rational geological "genesis" for such a "morph"—as their presence is confirmation that this *was* a vast "arcology," the painfully exposed skeleton of which lies open to the stars, underneath the reddened Martian sky. Its method of destruction is as much a mystery as its reason for existence in the first place.

> . . . nothing beside remains. Round the decay
> Of that colossal wreck, boundless and bare
> The lone and level sands stretch far away.

*    *    *

In another recent development, an astronomer named Richard Walker, attached to the US Naval Observatory in Flagstaff, Arizona, published the results of a several-year analysis of the *astronomical* aspects of the Great Pyramid. [5] Among many things discovered in this effort (including a reinforcement of the precision architectural aspects, attested to by Smyth and Petrie), was the fact that the famed Descending passage no longer provides any estimate of the *date* of the Pyramid's construction. According to a sophisticated celestial computer program employed by Walker to reconstruct the motions of the Earth (precession) in the last several thousand years, Thuban comes no closer to an alignment with the Descending passage than 1 degree—out of the "field of view" of anyone looking up the ~350-ft tunnel at *any time* in the last several thousand years. With this anlysis, one key *astronomical* means used by Egyptologists to "confirm" their dating of the Pyramid—its construction when the Descending Passage last aligned with Thuban—has disappeared . . .

*      *      *

Improbable, at best (when we began this odyssey—now almost four years ago), there is every reason to believe that in the next several hundred *days* spectacular close-ups of the Face and its associated pyramids will be on television all around the world (after all, what is two years . . . but 24 short *months*!).

And the Final Act . . . of a "play" with a "run" of perchance, half a million years . . . will have begun.

The most fundamental questions raised by the presence of the "monuments of Mars," particularly in light of the legendary whisperings around the world, are:

"What are their connections with ourselves . . . with the very *existence* of the human species?"

The answers—if they exist at all—are waiting for us on the Martian surface. We must *at all costs* guard that surface against terrestrial biological contamination—lest we destroy forever the very evidence we will have come so far to find . . . a *record* which—if it exists—would tell us instantly if we are related to "the Martians"—

Their genetic code.

*      *      *

The scientific theory behind this possibility is as simple as the thought is dazzling: if we can find so much as *one* (even partially-preserved!) body of "a Martian"—then, via modern techniques of gene splicing, recombinant DNA, amino acid sequencing, etc., we will be able to determine exactly how much "its" genetic structure (whatever its ultimate composition) differs from our own—

In other words, if it is even DNA!

If it is, we will then be able to determine if any part of that genetic code appears in *ours*[6]—in the same fashion that biologists are now able to determine with precision that over 99% of chimpanzee DNA is *identical* to that of *Homo sapiens.*

If "the Martians" did in fact somehow "tinker" with our own genetic code—in ways that are hinted at by the Sumerians, and echoed in those four immortal words . . . "In our own image . . . "—then we will *know.*

The odds against replicating—randomly!—even one small sequence of human DNA, is *billions to one.* Thus, if such genetic "identity" can be confirmed, it will provide compelling confirmation that we and "the Martians" are—were—related . . . throwing back the lid to the Pandora's box containing "how" . . . and "why!?"

*      *      *

In a related development that took place early in October, 1985, a group of planetary scientists, convened for a three-day "Mars Water Conference" at NASA's Ames Research Center, announced in a NASA news release that:

> Ice, snow, flowing rivers and *vast lakes* may have played a *major* role in shaping the ancient Martian surface and climate . . . Huge ice-covered lakes may have formed in the canyons near the Martian equator early in the planet's history . . . Primordial Mars may have been warm enough [to have supported these] flowing rivers and lakes on its surface . . .
>
> According to James Pollack . . . the early Martian atmosphere may have been much thicker, with more carbon dioxide to hold in the Sun's warmth . . . A complex geochemical cycle may have maintained this warm climate for as long as *half a billion* years . . . the liquid water then present would have speeded up [over present Martian climate] weathering of rocks, enhancing chemical reactions that take carbon dioxide out of the atmosphere and incorporate it into minerals. But, heat from lava flows coming up from the interior would have decomposed the carbonate rock, returning $CO_2$ to the atmosphere, Pollack says. (On early Mars, whose crust was relatively thin, lava could come up almost anywhere on the planet.) In certain conditions, Pollack says, the flowing lava would have buried carbonate rocks, bringing them to a depth where they would have been decomposed by the planet's internal heat. The lava action would have been great enough to release sufficient carbon dioxide to keep the cycle going in early times, according to studies by Pollack. Eventually, however, Pollack says, the lava flow rate on the small planet

dropped, and the $CO_2$ became locked up in the rocks.

With the loss of carbon dioxide from the Martian atmosphere, heat would have escaped the planet's surface, cooling the planet and freezing its water . . . (italics added).

*       *       *

Half a *billion* years for a "warm, wet epoch" in the beginning of the planet's life increased enormously the possibilities for the *indigenous* origin of simple one-celled microorganisms . . . but did nothing to alleviate the problems for the subsequent evolution of *intelligence*; if the "monuments of Mars" are artificial, they were still the product of "visitors" to Mars— and not the creation of the subsequent "descendants" of those Martian microorganisms!

But with NASA's own acknowledgement of *significant* quantities of water still resident on Mars (if in frozen form), the prospects were increased for another "wild idea" which had initially been born in the crucible of "Chronicles":

That Mars had "recently" been *terraformed*.

> Terraforming—the creation of a different planetary environment, through "planetary engineering" . . . warming up the planet . . . melting reservoirs of frozen carbon dioxide and liquid water . . . importing photosynthetic microorganisms to create a *free* oxygen atmosphere . . .

What Mars had once been able to accomplish on its own (if the planetary scientists were now correct), it could do *again*—with a bit of "outside" help. *If* the materials were there . . . which now seemed certain. NASA itself several years ago had held another conference[7]—to consider ways in which the presently inhospitable environment of Mars could be transformed . . . with *present-day* technology!

What NASA could envision—so my thinking went—so could advanced beings from another star . . . whose choice of Mars—when the Earth was so extraordinarily close at hand, and so benign—must remain a *major* enigma of the Mars story . . . until the day we land.

For the evidence is there . . . that, possibly, "Martians" attempted to create an "analog" of Earth—if only for a time . . .

*       *       *

The only means by which human beings "safely" can return to Mars— the mission which will not irrevocably destroy the answers to these fundamental questions—is *not* to go to Mars—

But to the Martian *moons*. To use the moons as "steppingstones" to exploration of a veritable "planetary Smithsonian!"

Only when the major biological questions have thoroughly been answered, should we take the final step . . . and land on Mars in person. Human beings are very "messy" organisms, carrying a plethora of micro-organisms; even with major technological advances, spacesuit designs—for years to come—will continue to inevitably leak, allowing countless numbers of these microorganisms to escape . . .

The resulting microbiological contamination of an entire world—within *a week* of the first manned Martian expedition—would be a tragic climax to a saga that's been waiting half a million years . . . or even longer.

The answer to this quandary—how do you explore the surface of a world . . . if not an incalculable set of *ruins* . . . if you are effectively pro-hibited from doing so in person—is again (coincidentally?) paralleled by current headlines—

The Soviet Union's impending "Phobos" mission.

Intriguingly, this unprecedented probe, especially for the Soviets, is *precisely* the kind of "pathfinder mission" required to explore the details of exploring Mars with something other than with men . . .

With sophisticated, remote-controlled tele-operated *robots*.

Unfortunately, for robotic exploration directed from the Earth, there is a time-lag—introduced by the finite velocity of light—which is an *inherent* aspect of the system, and which can never be "engineered" out . . . no matter how advanced the eventual techology. Such delays for signals present some fairly obvious (if major!) problems for "cheap, unmanned robotic exploration of the planets . . ." Not the least of these is embodied in a "worst-case scenario"—which envisions a Martian "rover" one day cheerfully rolling up to the edge of a Martian cliff or crater—and promptly falling over it!

(Because of the time-lag between planets, the frantic "stop!" command would obviously arrive at the rover's on-board computer long *after* the impending disaster had been transmitted back by television . . . when a billion-dollar robot had become a a billion-dollar pile of scrap.)

So, how can we seriously recommend attempting the most impor-tant, most complicated, archaeologically-sophisticated exploration of our first true alien culture . . . with robots?! The answer is: we aren't. What we are about to outline is a remote-controlled "Mars exploration pro-gram" conducted through the use of *tele-operated* robots—which are a totally different "breed of cat!"

For what will really be occurring will be the exploration of the "monuments of Mars" with *humans*—transported (Scotty . . .) *electro-magnetically* to the surface of the planet—through a system of sophisticated television cameras, multi-spectral analyzers, and even microphones, on specially-equipped robotic rovers with remote manipulator arms, precision drills, laser probes . . . and any other device which scientists can dream up for collecting and analyzing samples on the surface. And those items too complex for remote-controlled analysis—such as those *biological* materials so critically desired—these will be carefully packed in special *sample-return rockets* and fired from the surface to a rendezvous with an entire specially-equipped laboratory—

Waiting expectantly on *Phobos*.

For, if our team of scientists and specialists in ferreting out the half-million-year-old-secrets of this alien culture are waiting in a control center on Phobos—a mere 6000 miles above the planet—the "lag" between electromagnetically sending and receiving images and data will be measured in *fractions of a second*—as opposed to up to half an hour from the Earth!*

Under these conditions, electronically "transporting" humans to the surface to direct the critical analysis of artifacts and samples not only will prove practical—

But inevitable—given the alternatives.

The obvious interest of the Soviets in determining the *composition* of Phobos is a vital clue to their seriousness surrounding the entire Martian question. For, in verification that Phobos contains significant amounts of *water* lies the key to this entire concept . . .

The fact is, that without Phobos—and the presence of a compound which makes extremely high-performance *rocket fuel* (when split apart, then *recombined* in a liquid hydrogen/liquid oxygen engine)—such a mission as we have just described could not even be contemplated—let alone accomplished—for another *50 years*! Without the presence of the inner Martian moon, the space propulsion systems necessary to achieve this ambitious exploration strategy simply don't exist; to transport the crew, the rovers, the surface landing rockets, the Phobos laboratory, the scientific instruments, the supplies required for the crew (consumables), not to mention the *rocket fuel* required to send and receive artifacts and samples from the surface—to conduct the kind of comprehensive, statistically-significant, remote-controlled sampling of the *entire* planet (that we've en-

---

*Actual timelag can vary from 6 minutes (when Earth and Mars are closest) to 40 minutes, (when Mars and Earth are on almost opposite sides of the sun in their respective orbits). The "average" is approximately half an hour.

*The Monuments of Mars*

visioned here . . .)—would require the development of *nuclear* propulsion.

Because Phobos does exist—and quite likely (in the opinion of an overwhelming majority of planetary scientists) contains enormous quantities of, not only water, but the *other* elements essential for human life *anywhere across the solar system*, nitrogen and carbon—we are talking the conduct of this mission with current (shuttle- and space-station-based!) technology and hardware . . . and within *ten years*.

For it is cheaper to import the vital fuel (and all the other resources needed to achieve true space industrialization) to Earth orbit, to Luna itself, to anywhere (in fact) across the solar system . . . from Phobos and Deimos—

Than to drag these resources up out of the gravity wells of any other body orbiting the sun . . . especially Earth.

In the Martian moons, in all likelihood, we have found an almost limitless supply of the four vital elements—hydrogen, oxygen, nitrogen, and carbon—necessary to insure the eventual spread of human beings all across the solar system . . . and with that, the ultimate preservation of the Earth and the human family against all possible disasters. Jupiter and its moons (and *their* wondrous resources) are next . . . Saturn—when we are ready—from launching points on Ganymede and Europa . . . the outer solar system from our base on the hauntingly "Earthlike" moon of Saturn, Titan . . . each step when we are ready, each rung opening exponentially new resources and treasures, each stage an evolution out into the universe, in the direction of the stars, to which the Monuments of Mars stand as a mute artifact.

In the Great Diaspora—the "inner solar system Renaissance"—to come, in the scattering of the seeds of humanity across the variety of worlds which, even now, are waiting, lies the secret of who we are and where the "Martians" came from.

If we want it, "the First Martian Expedition" can be ready to leave Earth orbit in the 1998 Mars "window" . . .

*      *      *

*The images, ghostly and surreal, arrive on a billion television screens around a breathless world . . .*

*Eerily reminiscent of those faded 1920's half-tones . . . Howard Carter, kneeling, before the still unopened tomb of Tutankhamen . . . peering in, by the light of one lone candle, on a scene no human eye had registered in a "mere" 4000 years . . . a staggering 5000 centuries now flicker spectrally across the screens from the electronic "Carters" on the Martian surface—as the robot rovers crawl slowly towards the City . . .*

*Going Back to Mars*

*and its haunting image of "ourselves."*

*Mixed with these ancient images—staggering silhouettes of strangely flattened pyramids against a crimson twilight—are memories of other vivid scenes, of another "mythic" artifact finally found by the ubiquitous tools of Twentieth-Century technology—*

*The* RMS Titanic.

*A sea-wreathed chandelier swinging gently in an unseen ocean current . . . a bright brass bell glistening from the first light to play across its surface in three quarters of a century . . . dishes of exquisite china, amazingly unbroken on the ripped-out remains of a massive boiler resting dormant on the ocean floor . . . the ship herself, upright and serene in total darkness . . .*

*A once-invincible symbol of technology with great crimson rust stalactites streaming down her flanks . . . which, in a thousand years, will be little more than a swirl of reddish ooze on the timeless ocean floor—*

*While the Martian pyramids shrug off another thousand years of planetary duststorms . . .*

*The weight of Ages is draped here—in the drifts of Martian dust which sift between "the lofty dwelling places . . . " and into the dark and unknown realms which lurk inside . . .*

*Inside . . . a parade of endlessly exotic scenes, each one more alien than the one before . . . countless rooms, some of them rising upward in the gloom until their arching heights are literally lost—even to the electronically-amplified television systems on the rovers . . . A constant parade of the familiar . . . and the unfamiliar: artifacts at once identifiable as "technological" yet carrying an aura of exquisite craftsmanship . . . like another culture . . . far away . . .*

*And overall . . . the scale!*

*The sheer mindbending scale of constructs containing cubic* miles— *towering thousands of feet into the reddened Martian air—not as familiar* skyscraper *slivers of concrete and steel—but as hulking artificial* mountains—*not for the first time evokes images from another time . . . and from another world.*

*     *     *

One way or another Mars is our mirror. If the visitors were wise (as they must have been to come so far), then they may have been able to leave behind the perfect Sphinx, the absolute oracle, in which we could read ourselves exactly as we must, in order to . . . what?—survive?, meet them?, transcend?, evolve? We could not possibly answer this. Even if the architects of Cydonia are ourselves (at whatever level we choose to

*The Monuments of Mars*

take that), the message is still clear: you must look inside your hearts and genes and ask the hard questions you have failed to ask . . . if you want to take your place among the other worlds.

*　　　　　*　　　　　*

*The jerky images being relayed by satellite from the rovers flash upward through the Martian night to receivers intently listening on Phobos, before being redirected onward toward the center of the solar system . . . and an audience of billions . . .*

*The scenes have an aura of total unreality about them—even with the U.N. "logo" overlayed from time to time; these aren't "live" television images (with deference to the distance they've traversed, at the finite velocity of light) coming from the surface of another planet—*

*They're scenes from* Close Encounters: *a remake of that frenzied panorama of equipment, computers, men, and a blazing dome of lights . . . huddled as one brilliant oasis in an otherwise engulfing sea of darkness . . . as the surrounding Time-eroded mesas, rising like giant flat-topped tables in a land of giants . . . are swallowed up by the moonless Martian night . . .*

*Which, even as the thought arises, is broken by the rising of the visibly mis-shapen Phobos—in the* west—*its close orbit taking it completely around the planet in slightly over seven hours . . . the only natural satellite in the entire solar system to orbit* faster *than its parent world revolves . . . unless, of course, Phobos (or, more precisely, its* location) *is not completely "natural."*

*Up there—as its self-image is flashed across the inner solar system by the transmitters brought to Phobos—intense groups of men and women huddle—even as the audience on Earth—around ranks of other screens and banks of overworked computers . . . many marveling internally at scenes they never thought they'd witness in a million lifetimes . . . let alone direct.*

*And over all . . . the massive replica on the horizon looms . . . its remote Presence a constant reminder—if any one was needed!—that this is not "just another planetary mission"*

*Heru-sa-agga . . . the hero of the hill . . .*

*Who . . . and when . . . and—why?*

*Unlike any other culture ever explored by archaeologists, this one is unique—besides the obvious!—in that it was a high-tech culture . . . Somewhere in those "lofty dwelling places," amid the priceless and the trivial—the detritus all cultures unconsciously collect . . . and throw away— there must be records—*

*Only this time, they won't be merely more "inscrutable inscriptions" incised on countless walls . . . or fragile earthen tablets . . .*

*They'll be* images—*capable of being unscrambled and decoded . . .
and then* played back . . . *to be viewed for perhaps the first time in* half
a million years . . .

*Images of daily life, of ceremonies and events both great and small . . .
extending back, perchance, to the construction of the fabled pyramids
themselves—the Monuments of Mars—if not their Guardian . . .*

*The oldest portions of this "Martian chronicle" may be the most
important . . . from* before *the journey that brought these architects to
a collection of scattered specks of dust in an immensity of space and time
we call "the solar system" . . .*

*Somewhere . . . in those cubic* miles *of rooms . . . carefully preserved
for our* specific eyes *by those who built this extraordinary Complex . . .
may be a literally priceless "legacy"—copies of "videos" made millions
of years ago . . . on worlds uncounted light years from this sun . . .*

*A visual panorama of a Galaxy* alive *with other cultures . . . which
by finding and opening this special vault, we may now join . . .*

*An incomparable "Encyclopedia* Galactica*"—a wondrous heritage
of music . . . art . . . and scientific knowledge . . . medical techniques and
alien biological discoveries—including, possibly, the keys to prolonging
life itself—gathered from a million patient years and perhaps as many
races . . . scattered down the cavernous corridors we call the Milky Way . . .*

*       *       *

Imagine . . . if, instead of having to anticipate throughout our lifetimes
(and how many lifetimes to come after . . ?) Christa McAuliffe's dream
of one day going "to the stars . . ."

*The stars have come to us!*

*       *       *

*Morning is coming to Cydonia . . . the dust-filled wisps that pass for air
which (as Viking sadly found) obscures all but the brighter stars, slowly
brightening into a dust-filled dawn. A bluish ellipse forms slowly in the
east, product of forward-scattered sunlight from the nuclear heart of the
solar system, still far below the mesa-limned horizon. And below the blue
ellipse, now tinged with alien hints of pink, a brooding silhouette slowly
is transformed . . .*

*From an inky Presence stretching out on the horizon . . . to a strik-
ing countenance staring upward toward the last remaining stars . . .*

*One bright object in particular—blue-green, bright enough to cast
thin shadows in the rusty sand—hangs above the brooding form. Near-*

314                                      *The Monuments of Mars*

*by, a degree or so away—hangs a dimmer, warmer-looking "star."*

*The Earth and Moon . . . glimmering across the inner solar system.*

*It is an alien sunrise, born on an alien world, looked on by something alien . . . and familiar. Is there truly some connection—between this massive image lying on its back . . . which has witnessed over a hundred million dawns . . . and that brilliant spark of light?*

*Or, are the minds which are, even now, drawing plans to cross the space between that "star" and here, destined to confront merely a reflection of their own obsessive longing not to be alone . . . projected one last time on a rusting pile of sand . . . ?*

A quotation, attributed to the Egyptian god of Wisdom, Thoth, seems an eerie foreshadowing across the dusty millennia of what is to come from that green star . . .

> Men will seek out . . . the inner nature of the holy spaces which no foot may tread, and will chase after them into the height, desiring to observe the nature of the motion of the Heaven.
>
> These are as yet moderate things. For nothing more remains than Earth's remotest realms; nay, in their daring they will track out Night, the farthest Night of all . . .

## Notes

1. *Los Angeles Times*, November 17, 1986.
2. *Op. cit.*, November 29, 1986.
3. Sagan, C. "To Mars," *Aviation Week & Space Technology*, December 8, 1986.
4. *Op. cit. Los Angeles Times*, December 7, 1986.
5. Hodge, C., "Pyramid Buff has New Angle on Old Puzzle," *The Arizona Republic*, January 20, 1985.
6. Miller, J.A., "Mummy DNA Intact After 2400 Years," *Science News*, Vol. 127, No. 17, 1985.
7. Oberg, J., Chapter 8, "Mars—A Closer Look." In *New Earths*, Harrisburg: Stackpole Books, 1981.

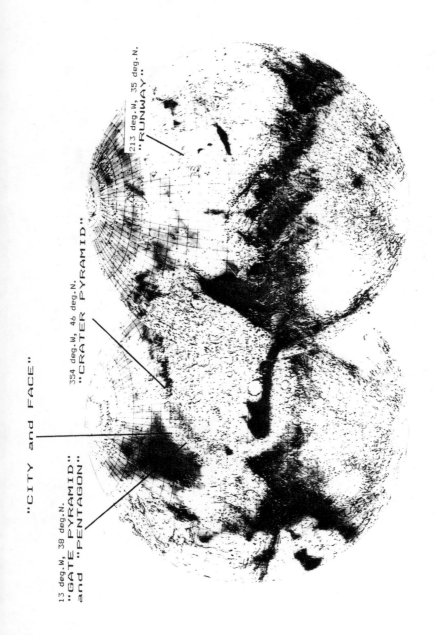

"CITY and FACE"

13 deg.W, 38 deg.N.
"GATE PYRAMID"
and "PENTAGON"

354 deg.W, 46 deg.N.
"CRATER PYRAMID"

213 deg.W, 35 deg.N.
"RUNWAY"

A simplified map of the surface of Mars, indicating four regions under investigation.

# APPENDIX I

During the course of the Independent Mars Investigation, a number of additional "unusual Martian surface features" were discovered. Though the level of analysis performed on these to date does not compare with what has been accomplished for Cydonia, for the sake of completeness—as well as an indication of the kinds of imaging analysis that could (and should) be performed on the additional Viking images—a brief overview of the most puzzling of these additional "anomalies" is given here.

Joe Mosnier, SRI Research Physicist, and Bill Beatty, SRI Senior Geologist, compiled the initial brief descriptions of several of these features. Lambert Dolphin contributed comments, based on the ongoing "Chronicles" discussions. I have added some additional thoughts, where appropriate.

The format will be to cite the Viking frame number, the designation of the feature in the Conference, and its location on the Martian surface (latitude and longitude), and estimated height above or below the "Mean Datum." This will then be followed by the descriptions and comments.

I. Frame 219S16   Second "Cydonia Pyramid"

Lat. 35.0 deg., Long. 13.9 deg.
Mare Acidalium quadrangle (MC-4)
Local elevation: between 0 and 0.5 km, perhaps 0.2 km.

The "pyramid" is roughly square at its base, four-sided, almost N-S in alignment. At the base of the east side the length is approximately 3.8 km; the south side base is about 3.8 km; the west face 3.1; and the north side base 2.7 km. Its height is about 800 meters, based on the shadow length (sun angle 70.08 degrees from vertical in this frame).

The "pyramid" lies in a region of PLD (dissected plateau material), in the Cydonia Mensae area comprising the same geologic structure as the City/Fort/D&M mountains—which lie about 375 kilometers to the northeast.

Beatty's comments:

"Perched on the edge of an erosion plateau, above a valley with several well-developed rilles, [this feature] is a four sided pyramidal structure. The pyramid may have a circular base as indicated by the light and shadow in the photograph. It also has a well-developed longitudinal ridge for a peak."

Comments by Dolphin:

"This pyramid is surrounded by angular forms and blocks which look like chunks of ice broken off a large sheet. This pyramid, more than many others on Mars, suggests the origin [of these features] is entirely natural; yet the regular sides, sharp angles and smooth planar faces are hard to explain naturally. Crystals this size are unheard of [on Earth]. If artificial, what was the purpose and why go to all the trouble and expense?"

Hoagland addendum:

The epistemological problems presented by such an *isolated feature*, when attempting to reach even tentative conclusions regarding its "artificiality" or "naturalness," are clearly demonstrated by the previous comments. My own criteria for "artificiality" rested on demonstrable mathematical *relationships* with equally anomalous surrounding features— as exhibited in the complex some 400 km "up the coast" from this isolated object.

Even if Mars—somehow—has devised a bizarre mechanism for creating natural pyramidal structures (due, for instance, to the extremely long freeze/thaw cycles occasioned by the million year obliquity shifts, and some form of extraordinary "crystal growth"), this cannot explain relationships with surrounding features. The planet, however, might provide a natural "model" on the Martian landscape for potential immigrants to copy—blending an artificial site into the natural terrain.


II. Frame 43A04   "Crater Pyramid"

Lat. 46.1 deg., Long. 353.2 deg.
Ismenius Lacus Quadrangle (MC-5)
Local elevation between 0 and 0.4 km. Estimated at about 0.4 km.

Four-sided "pyramid" oriented NW-SW. Lies on the SE rim of a crater 3.3 km by 4.1 in diameter. Base appears to have equal sides, 1.6

km long. The height is estimated to be 640 meters. The adjacent crater appears to be 75 meters deep (the sun angle in this frame is 84.6 degrees, hence the shadows are very long). Near the adjacent crater, strange features labelled "wormholes" have been noted. They are about 250 meters in diameter.

The local geology is P [plains material]. At about 1.0 deg. latitiude and 2.0 degrees longitude to the S and SE, there is an extended region of PL; in all other directions there is extended P.

Beatty's comments:

"The 'crater pyramid' appears to be a rectilinear, four-sided pyramid. It has the general shape of a cheese-wedge (isosceles), with the peak having a short lineal ridge rather than a pointed apex. Opposite base corners are approximately N-S and E-W. The 'wormholes' on the margins of the adjacent 3.9 km diameter crater to the SW appear to be a series of short, parallel rilles, as though lava had flowed out of a head vent into these channels for a brief period. The subsequent downhill spill and possible later collapses of upper margins caused an interesting but chaotic bit of downsloping terrain. The above assumes the craters are volcanic rather than impact. The geological explanation on the landscape suggests that these valleys may have been formed by water, but it is questionable that water would emerge from the margins of any active volcanic crater."

Comments by Dolphin:

"Brandenburg noted that the crater pyramid is 100 km from any other mountains in the area. I see it as a pyramid 'built' after the impact crater occurred. I do think the craters in the area are of the impact type, with 'splash.' The 'wormholes' look like a system of collapsed tunnels to me.'

Hoagland addendum:

This *complex*—the "crater pyramid" and its associated "tunnels"—incorporates most clearly my own criteria for truly "anomalous objects" on the Martian surface—at existing resolutions. For, rather than a single, isolated feature, the array (like the City/Face association) exhibits fascinating *mathematical* and *geometric* relationships that are difficult to explain with natural models. The "tunnels" are aligned N-S: the same as the corners of the "pyramid." The faces of the "pyramid" are oriented at right angles to this line; they are curiously positioned in relation to sunlight—over the course of a year—as though most efficently to illuminate the structure. This is consistent with the artificial "arcology model" raised earlier.

As Dolphin points out, the most logical analysis suggests that the associated craters are of an impact nature, rather than volcanic. Further, they appear extremely old—judging by their lack of depth and the generally eroded quality of their ejecta blankets. That this sharp, angular feature (the "pyramid") could be a remnant of terrain from *before* the formation of the craters seems most unlikely, both because of its obvious lack of erosion and its need to have survived explosive energies in the "hundreds of megatons range" (if it existed when the craters were formed). Thus, Dolphin's conclusions that this object had to have been "built" *after* the crater formation seem well substantiated. Since "uplift" geological models seem ruled out by the lack of additional distortions in the crater itself, the mere existence of this feature and its associated "rectilinear tunnels" must remain a major mystery.

### III. Frame 86A08 "Runway"

Lat. 34.7 deg., Long. 212.8 deg.
Cebrenia Quadrangle (MC-7)
The local elevation is between 3 and 4 km. Assuming linearity between contours, this feature is at 3.8 km above the Mean Datum.

The main "runway" is 4.9 km long. The height of the long adjacent mesa (5.7 km long) is estimated to be 830 meters (sun angle 42.04 degrees from vertical). The runway lines up E-W (as closely as can be determined from the image).

This site is on the upslope, toward the volcano Hecates Tholus, which is roughly centered 200 km to the SE. The local geology is PH [hummocky plains]; the ancient surface was intensely cratered in the early history of the planet and then degraded.

Beatty's comments:

"The 'runway' appears on the NW lower slopes of the volcano Hecates Tholus. The vicinity has three large hills which have been formed by several long longitudinal faults creating a 'horst-and-graben' effect. The three so-called hangars, located a kilometer or so from the 'runway' are an excellent example of this type of fault pattern. The 'runway' itself (0.9 km wide) is a tilted slab structure, 5 km long, oriented E-W. The south side is occupied by a series of sheared blocks which have the appearance of smaller structures. The western end of this feature is covered with erosional debris from the mountain above. There may reasonably be a parallel fault on the north side of the runway, making it a true graben, but this is entirely covered by debris and rubble. The center portion of the 'runway' is filled with much fine debris, giving it the roadway appearance because of greater reflectivity."

Comments by Dolphin:

*The Monuments of Mars*

"This feature certainly looks like an east-west runway with adjacent taxiway. The regular 'knobs' along the runway suggests an accelerator structure, and the bumps on the mesa suggests large buildings. This is how I visualize the 'runway' area, assuming it is non-natural. I find a natural explanation for this artifact more difficult [to conclude, than Beatty does]."

Hoagland addendum:

Of all the "anomalous surface features" turned up in the course of our deliberations, this one, literally halfway around the planet from the complex at Cydonia, is the most provocative. Its morphology and its association with a series of other remarkable objects cry out for further explanation. The night I first measured its orientation, I discovered to my astonishment that it was *precisely* east-west. One purpose for such a *specifically aligned structure*, with "regular knobs" along its length, might be as some kind of accelerator—to launch spacecraft from the planetary surface. (Proposals to construct such structures on the Moon and other celestial bodies, for the purposes of launching payloads impractical with current rockets, have been seriously suggested for decades—initially by Arthur Clarke.)

The latitude of "the runway"—35 degrees north of the equator—is the same as the maximum obliquity of Mars, during its million-year "nodding" cycles. This would coincidentally (and most strategically) place the orbit of the inner moon, Phobos, in the same plane—given certain assumptions. Further, the elevation of this structure above the "mean datum"—almost 4 km—is consistent with another prime requirement of an accelerator: that on Mars (unlike on the airless Moon) it be built as high as possible, to avoid the effects of the still significant (though ultra-thin) Martian atmosphere.

Whatever the ultimate explanation for this feature, these possibilities hopefully demonstrate how essential appropriate *context* is in any successful analysis; until we have much better imaging of this region of the planet, any artificial conclusions regarding the nature of this unique linear feature depend on the overall likelihood of a high technology civilization inhabiting the planet. If the Cydonia analysis holds up as supporting a case for "artificiality," then the "runway" could become a crucial element in figuring out precisely *what* "the Martians" were doing on the planet . . . or with its fortuitously-placed moons . . .

These represent the best of the half dozen or so "anomalies" discovered in the course of our deliberations. That they were found on only a few hundred images—out of more than 60,000 taken—in the brief months of the Independent Mars Investigation suggests that a dedicated effort—properly funded—could uncover many additional "proto-cultural" features. If a "high tech" civilization, with a penchant for constructing miles-long structures, did exist on Mars, then the artifacts of that inhabitation should exist (even if covered with sand)—all around the planet. Far more extensive searches of the existing Viking data-base, and image-processing of the resulting images, should not be that expensive. The benefits, on the other hand, could be incalculable—if the evidence gathered at Cydonia is any indication.

# EPILOGUE—
# AN UPDATE ON FOUR YEARS
# OF ADDITIONAL RESEARCH . . .
# AND THE "STRANGE" POLITICS
# OF VERIFICATION

> "I've seen the studies and I've seen the photo-
> graphs . . . and there do appear to be formations
> . . . not . . . of natural or normal existence. It looked
> like they had to be fashioned by some intelligent
> beings. . . . For this reason, I have asked NASA to
> provide assurances that the *Mars Observer* Mis-
> sion will include this [set of objects] as one of its
> imaging objectives . . ."[1]
>
> —Chairman Robert A. Roe
> House Committee: Science,
> Space and Technology, 1989

With these words, Congressman Robert Roe in 1989—two years after
*Monuments* initially was published . . . six years after we began . . . thirteen
years after the original, haunting *Viking* images of the "Face on Mars"
unknowingly were taken . . . and then resoundingly ignored—finally
made the Cydonia Investigation *official.*

By expressing his keen interest in an appropriate resolution of this
(now fifteen-year-old) problem (as an elected congressional representative, and as then head of the major science committee in the Congress
with direct responsibility for NASA's budget), Chairman Roe appropriately assumed—on behalf of not only the House Committee, but the
American people themselves—personal responsibility for securing an
answer to the crucial mysteries posed by the "Monuments of Mars."

The sequence of new Cydonia discoveries which led to this appropriate political response are worth recounting; for not only are these new findings explicitly involved in our developing understanding of the meaning now of Cydonia itself, the fact that they convinced Bob Roe that this is an important scientific problem is a vivid reminder of one of our most fundamental rights under the Constitution of the United States . . . the People's Right to Know.

*       *       *

The major research breakthroughs in the Cydonia Investigation, which began this high-level political momentum for NASA verification of Cydonia's uniqueness "post *Monuments*," accelerated dramatically in 1988—with the addition to the Team of Erol Torun.

Torun is a cartographer and systems analyst at the U.S. government's official "mapping" service: the Defense Mapping Agency (DMA), in Washington D.C. Formerly the Air Force Cartographic and Mapping Service, DMA is responsible for, among other things, knowing the locations to within feet (via reconnaissance spacecraft and electronic navigation satellites)—anywhere in the world—of so-called "strategic targets"; when the Pentagon wants to know the exact latitude and longitude of a Russian missile silo, or the location of Saddam Hussein's nuclear facilities or chemical warfare plants—it is the Defense Mapping Agency which provides that crucial geodetic information.

It is also critical to the continued national security of the United States that analysts at DMA be able to distinguish between a "camouflaged hill" on a reconnaissance satellite photograph—and a clandestine nuclear facility; so, recognizing the signatures of artificial structures seen from orbit—as opposed to natural landforms—is a major part of DMA's official mandate.

This, of course, had been the continuing stumbling block to getting widespread NASA recognition of the potential artificial nature of the Cydonia "artifacts": no one in NASA—certainly no one we'd been talking with for over six years in its "planetary science community," which took the original *Viking* images and claimed to have "exhaustively analyzed" them—seemed to have a clue as to how to systematically discriminate between a set of natural hills on Mars . . . and a set of objects with more interesting geometric properties!

You can appreciate then, why the addition to the Cydonia Investigation of a professional from DMA—whose charter in maintaining our national security lies in making *exactly* such determinations, routinely, from Earth orbit—would have caused us considerable excitement.

Torun was very candid, even in his initial comments on our work:

> As a geographer, I found your book to be a fascinating description of an equally fascinating subject. While I was impressed with most of the images presented and your description of them, the object that especially caught my attention was the D&M Pyramid. I have a good background in geomorphology and know of *no mechanism to explain its formation* . . . [emphasis added].[2]

Torun's letter continued:

> [Therefore], I decided to study more closely the geometry of the D&M Pyramid because of its impressive symmetry (and its alignment with the Face). I was not prepared for what I found. The geometry of this object *so surprised me* that I quickly wrote a semi-formal paper so that I could send my results to you and the Mars Project as soon as possible. I feel that these results argue *strongly in favor of Design* as the mechanism for the D&M Pyramid's origin . . . [emphasis added].

That was how the next phase started . . .

Torun's surprising geomorphological conclusions—after several pages of detailed examination of all known natural processes which could have formed a two-mile-long, five-sided mountain on the Martian landscape, were as follows:

> . . . the Geomorphic Hypothesis [to explain the formation of the D&M Pyramid] is thus left with no mechanism [that can explain its existence]. This object's five-sided shape and bilateral symmetry is *unlike any landform seen to date in this solar system,* and even small-scale phenomena such as crystal growth cannot explain its morphology.
>
> All observations to date of the geophysics of Mars, its gravity, meteorology, geomorphology, etc., indicate that Mars is a place where the laws of physics and principles of geomorphology as we understand them apply, with minor variations due to gravity and atmospheric density and content. It is illogical to assume that there is one small place on the surface of Mars where these same principles are being violated. Being thus faced with *no known natural mechanism* to account for the D&M Pyramid's formation, other possible mechanisms need to be explored . . . [emphasis added].[3]

Meaning, of course, the "intelligence hypothesis."

Torun's own subsequent Cydonia measurements, in July of 1988, went far beyond merely supporting our original claim as to the D&M's geomorphological "uniqueness"; Torun had discovered in the D&M the mathematical "smoking gun" behind the "Geometric Relationship Model" originally presented in *Monuments*—

And thereby, the critical proof of the validity of the entire Cydonia "intelligence hypothesis!"

Torun, to his own astonishment, found "coded" in the internal apex angles of this remarkable five-sided structure lying on the Martian surface—the "D&M Pyramid"—a series of highly specific, highly redundant mathematical constants and expressions: "sq rt 3," "sq rt 2," "e/pi," "e/sq rt 5," etc. Further, he discovered these relationships expressed in at least *three* ways: in the **angle ratios** of the D&M internal angles (example: 60 degrees divided by 69.4 degrees = e/pi = 0.865); in the **trigonometric functions** of those same angles (sin 60 degrees = e/pi = 0.865); and in the **radian measure** expressed by a base angle in the object (49.6 degrees = e/pi radians).

The "D&M Pyramid"—through these dramatic and extraordinarily specific findings of a geomorphologist named Erol Torun—was suddenly transformed . . . from being just "a peculiar, bilaterally symmetric hill" lying near an equally peculiar "face-like mesa" on the Martian surface, to what could only have been intended—if these measurements should be confirmed—as nothing less than a literal and demonstrable mathematical "Rosetta Stone" to Cydonia itself (see Figure 15)!

One of the provocative "specific findings" made by Torun, was his discovery within the D&M of that very specific, very redundant mathematical relationship: "e/pi." Focusing only on the D&M—in his own preliminary paper—Torun remarked however, even at this early stage, on the *wider* implications of finding *this* remarkably redundant mathematical expression at Cydonia . . . e/pi.

> . . . as noted by Hoagland [in *Monuments*, Figure 10 original editions, and on the back cover], the tangent of the City's latitude (40.868 degrees N) also equals *e divided by pi*. We thus have a link between the D&M Pyramid, the City, and the Face, united with one number and with the geometry of one angle [sic] of the D&M Pyramid . . . [original emphasis].[4]

It would turn out that the actual relationship of this key mathematical constant—e/pi = 0.865—to both the D&M, and to the geodetic location of the Cydonia Complex on the planet, would become even more provocative . . . and more enlightening . . . as our measurements and discoveries regarding the entire Complex rapidly expanded. (The details of Torun's original analyses, including his original papers on the D&M, will be published in *The Cydonia Papers*.)

But the immediate effect of his electrifying paper was to send me rushing back to my own copies of the orthographically rectified *Viking* mosaics—where, armed with his now precise geodetic scaling factors for the *Viking* images (which we had NOT possessed before!), I promptly proceeded to discover—

Exactly the *same* geometry—the *same* angles, the *same* trigonometric relationships, and the *same* mathematical constants—as Torun had discovered in the D&M—

But connecting ALL the originally identified "enigmatic objects" at Cydonia—the "City," Face," "Cliff," "Tholus," and the "D&M" (see Cydonia Relationship Map, Figure 10)!

This remarkable sequence of discoveries—first Erol's, then my own—is elegant confirmation of the correctness of our original "relationship model," expressed repeatedly here before (Chapter XIII and Appendix I):

> . . . while it is unlikely that Nature could have randomly created a set of unique morphologies in one tiny location on the planet, if we can demonstrate that, in addition, they are *mathematically and geometrically related,* then we have a problem of a whole different order.

Now, between us, Torun and I had overwhelmingly demonstrated exactly those *predicted* mathematical relationships!

Torun's response came on September 8:

> I was very impressed with your discovery that the trigonometric relationships I observed in the D&M Pyramid are expressed throughout the entire Cydonia complex. Your observations attest to the quality of Mark Carlotto's image processing of the D&M and, of course, to the apparent artificiality of the complex itself. I see that you found *another significant angle,* 50.6 degrees, which connects the D&M and the City Square with a vertex at the Face. The tangent of 50.6 = ~1.22 = e/sq rt 5; this number is expressed *two* ways within the D&M—in the radian measure of the back angle of the Pyramid (69.4 degrees = e/sq rt 5 *radians*), and in the angle ratio between the back angle divided into each side angle (85.3/69.4 = ~1.22 = e/sq rt 5) . . . [emphasis added].[5]

Suffice it to say, these startling new findings—in the Summer and Fall of 1988—ushered in an entirely new era for the Cydonia Investigation: the Era of Mathematical Specificity and Geometrical Redundancy.

Redundancy . . .

If only one word could be used to describe what we have now discovered at Cydonia, it would have to be that word—"redundancy." The presence of so many mathematical constants, encoded to a "three-significant-figure level" of precision (in terms of current measurements; their final precision could go much higher when we have new images from *Mars Observer)*, must now place any "random" explanation for this extraordinary phenomenon squarely in the realm of fantasy. These repeating constants simply appear in too many angular relationships, too redundantly communicated throughout the entire Cydonia Complex—both within one key object, and between a highly limited set of other objects in the *immediate vicinity*—to be dismissible as "chance."

Carl Sagan's previously quoted, highly relevant observation (Chapter V), "Intelligent life on Earth first reveals itself through the *geometric* regularity of its constructions,"[6] now seemed *crucially* apt, as did the thinking of the "scientific giants"—Lockyer, Gauss, Lowell, etc.—from whom Carl had obviously "borrowed" this epistemological dictum; for, irony of ironies, the main nineteenth century scientific proposals in the literature, for detecting intelligent ife on *Earth*, had *all* revolved around discussions regarding the feasibility for deliberate construction of "huge *geometric* figures—triangles, circles, squares, etc."[7]—perhaps in places "like the Sahara Desert," signaling (through "fundamental mathematical constants") the presence of intelligent life *here* . . . to their "obvious" nineteenth century equivalents *out there*—

The "Martians!"

Blatant historical ironies aside . . . the sheer redundancy now of key, incredibly specific, *fundamental mathematical constants and geometry* at Cydonia—if not sheer common sense—argues compellingly, at this point, that we cannot be dealing with anything other than a *designed complex* on Mars—if not one with a *highly specific purpose.*

And a *grand, redundant architecture*, apparently overwhelmingly intended to *communicate* that Purpose.

In other words: a "message."

The Message of Cydonia . . .

Critics, of course, will try to argue that we are "merely projecting" our own internal order on a set of "random rocks." That's fine; that's what critics are supposed to do: keep the system honest.

The neat thing is that there is a foolproof process to differentiate between that hypothesis and ours: simply *test* the measurements!

What Torun measured, in exhuming his extraordinarily specific order in the D&M, were only six (count them!) geodetic points—points which, in logical relationship to one another (the five base corners and the apex of the D&M), promptly produced *redundant* sets of identical angles, that *redundantly* communicated highly specific and *redundant* mathematical information: a series of highly specific (and related!), *repeating* mathematical constants. And if that were not enough, the same angles "just happened" to communicate this redundant information *redun-*

*dantly*—in three completely separate ways . . .

What I discovered in the larger Complex—measuring between six additional, but also highly logical, geodetic points (the exact center of the "City Square"; across the "eyes" and past "the chin" of "the Face"; the top and bottom of the linear "Cliff"; the apex of the conically-shaped "Tholus"; plus the apex of the pyramidal "D&M")—were not only *identical* angles, *identical* constants, and the *identical three ways* to derive those constants (angle ratios; trig functions; and radian measure)—I discovered several additional angles . . . which, however, also promptly resolve to precisely the *same* mathematical constants (see Torun's remarks, above). And the constants, far from appearing helter-skelter or at random, appear to have been deliberately assigned to *specific* objects in the Complex; some appear more frequently in connection with one morphology, than with any others.

Redundancy . . . and implied "meaning within meaning."

The statistics against all this happening by chance are eminently testable—if anyone simply *does the measurements,* and applies the same statistical tests that Torun has applied (see Tables I and II).

Yet—in the three years which have elapsed since Torun and I carried out these original measurements [and published them worldwide on CompuServe[8]—in addition to detailing them now in three invited, official NASA Briefings, for several thousand NASA engineers and scientists, at two separate NASA Centers (see below)]—

NOT ONE "critic"—either within NASA or outside—has seen fit to duplicate this most elementary of tests—and then *publish* their results. Especially *not* Carl Sagan. If they're right, if (as Carl once claimed in *Parade* Magazine) we are "merely seeing Jesus Christ on a tortilla chip"— then their measurements should totally *disprove* the relationships we've found.

Could it be that they've not done so . . . because they simply can't!?

Arthur Eddington, the eminent British astronomer who carried out one of the first definitive tests of Einstein's General Theory of Relativity (by verifying Einstein's predicted deflection of stars viewed near the limb of the Sun, due to gravity's literal geometric curvature of space) has been widely quoted in this context. Said Eddington:

"Gentlemen, you do NOT have a science—unless you can express it in numbers!"

With Torun's discovery of precise, repeating *numbers* in the D&M— the now obvious "Rosetta Stone" not only to verifying, but to "decoding" (and perhaps eventually understanding) the reason for Cydonia's existence—the "intelligence hypothesis" suddenly, in the Summer of 1988, assumed its true and rightful character: still a highly controversial . . . but now ultimately reproducible and *testable* . . . pursuit of contemporary science.

*   *   *

Which has made NASA's continuing refusal to carry out *any* of those tests—and now upon "real numbers"—all the more mysterious . . .

It cannot be because they *do not know* what we have found.

For—even as we were experiencing these breakthroughs . . . we began to receive *official invitations* from separate NASA Centers (!) to present these "numbers" to literally thousands of rank and file NASA personnel and scientists across the country; the first breakthrough in this direction came with a formal August, 1988 invitation to address the NASA-Goddard Space Flight Center Engineering Colloquium . . . , scheduled for December 19, 1988.

*   *   *

What happened on this occasion was as follows (in terms of the politics of sorting out whether there is, indeed, some kind of official "coverup" now occurring on the Cydonia data).

After we had received this first official invitation to present the Cydonia Investigation to a technical NASA audience, NASA Headquarters—at the last minute—actually stepped in and *tried to get us "uninvited"* to the Goddard Colloquium! Only concerted "11th hour" action by some members of the press (who have been following this story since *Monuments* was initially published—"What do you think will happen if you suddenly cancel Hoagland's presentation?; do you really want to give credence to the charge of a 'Watergate-style coverup'— which is already floating around on this subject . . . ?") prevented the faint-of-heart on the Goddard Committee, in the face of some pretty severe pressure originating "downtown," from rescinding their August invitation in December—only *days* before I was scheduled to appear!

When that tactic failed, Headquarters abruptly (over a weekend) scheduled *a simultaneous "Mars" news conference* in downtown Washington D.C., on the *same* afternoon, and at the *same* time, as our Goddard presentation—which had the effect of pulling all the national press previously scheduled to come to Goddard to see us, *away* from our event— The first "official" presentation of the "Intelligence Hypothesis" to NASA in all the years since *Viking*.

Only Keith Morgan, from ABC, eventually showed up at Goddard— later to report back to his colleagues at the Washington Bureau how they'd been "suckered."

At the time, we viewed these developments as just "bad timing" and internal NASA politics with Goddard. Later events—specifically, equally suspect occurrences which have recently taken place around *other* NASA Center invitations (a scheduled address in September, 1991, to employees of NASA-Ames, which was initially arranged by a group of Ames' scientists, then abruptly *cancelled* by the NASA-Ames Office of Public Affairs, when the head of the employee group was briefly *out of the country* over a weekend!); National Press Club appearances by our Team which were mysteriously "misscheduled" in the in-house newsletter (for the first time in twenty *years*); an exploratory "feeler" to address the prestigious "Bohemian Club" in Northern California—whose members are current and past high-ranking policy-makers in government—suddenly, abruptly, cut off without explanation; and ultimately, even more flagrantly suppressive actions by NASA Headquarters itself on the subject of "an official NASA television presentation" of our work (see below)—would all cast this string of continuing "unfortunate coincidences" in a very different light . . .

What was NASA Headquarters (or others in this government) afraid of?: that any *specific* discussion, or tests, of the Cydonia Geometry would, in fact (horror of horrors!) publicly *verify* the Intelligence Hypothesis?!

For, against this backdrop of dramatic new Cydonia research results from us, and the sudden beginnings of "front door" NASA Center invitations to show our actual data, another curious event had deftly been arranged by NASA . . . which went all but unnoticed by the press:

The exquisite "Malin Camera" (see Chapter XI) has been officially (though, *very* quietly!) added by NASA Headquarters to the *Mars Observer* spacecraft—a spacecraft which, for overriding cost reasons, was initially NOT supposed to carry—under *any* circumstances (in the wake of the exhaustively thorough mapping, carried out pole to pole by *Viking*)—an imaging system back to Mars . . .

Yet despite these remarkable technical developments—in terms of now being able to actually test the "intelligence hypothesis" on Mars itself—there persisted (and still persists) a practically bizarre reluctance by NASA planetary scientists at JPL to simply putting our, now highly *specific* measurements and theories re Cydonia, to any such quantitative, highly public *tests*—

Or . . . to agree (even in principle) to carry out the most effective test of all—new images of Cydonia, *fifty times sharper* than *Viking's* 50-meter resolution data, to be taken by the upcoming, unmanned *Mars Observer* spacecraft, returning to Mars in 1993.

The epitome of this absurdity was reached late in 1988, during an "official" response to the (now increasingly inevitable) "Cydonia question"—which seems to be coming up more and more during these events . . .

*          *          *

The "revealing" incident occurred at a technical college in northern Michigan, during a lengthy *Mars Observer* briefing—which had repeatedly extolled the scientific and technical "virtues" of the Malin Camera; the speaker—a senior member of the Smithsonian Institution's "geology group" (which is formally a part of NASA's *Mars Observer* Team), a Dr. Zimbelman—had been going on at some length, re the potentials for acquiring unprecedented, exquisitely detailed *geological* images of Mars with this "last-minute technical miracle."

The questioner in the audience simply asked of Dr. Zimbelman during a pause: "Given, as you say, the superb nature of the new camera on the [*Mars Observer*] spacecraft—are you planning to take new pictures of 'the Face and Pyramids'?"

To which Dr. Zimbelman officially answered: "We'd like to . . . but we can't *focus* the camera."[9]

*          *          *

It was this increasingly obvious "stonewalling" on the entire subject of Cydonia—by *anyone* officially associated with the *Mars Observer* Mission—that finally prompted us to "go for broke" in 1989—

And seek a *political* solution to this continuing "scientific" impasse: Through a Washington lobbyist we specifically hired for the task (that's where some of the royalties from *Monuments* have gone, by the way . . . ) we wrote to Robert Roe, Chairman of the House Committee on Science, Space and Technology, in early April, 1989, and asked for a face-to-face meeting: to simply show him—as Chairman of the key congressional Committee with oversight responsibility for all of NASA's plans and budgets—our spectacular, now *reproducible,* data and analysis. We included, almost as an afterthought, a video of our December NASA-Goddard presentation . . .

Bob Roe agreed to see us three days later.

Participants in the meeting included Erol Torun, our colleague at Defense Mapping in Washington D.C.; Lois Lindstrom, the Washington lobbyist who helped arrange the meeting; and Dr. Mark Carlotto, our eminently able *Mars Mission* image processing expert, with The Analytic Sciences Corporation (TASC) in Reading, Massachusetts.

    \*        \*        \*

A word about Carlotto: Mark had come to us in 1985, in the waning days of the "Mars Investigation Group" at the University of California at Berkeley (see Chapter XI). Tom Rautenberg originally "found" him— after I'd directed Tom to ask Marvin Minsky [an old friend, and founder of the Artificial Intelligence (AI) Laboratory at MIT] "who would be the best at image processing." Minsky had directed us to Mark, an MIT graduate and former member of the AI Lab, who was now working at TASC, a Boston-based corporation heavily involved in reconnaissance satellite imaging enhancements and interpretation for various "official agencies." When Rautenberg approached him on our behalf, Carlotto was already familiar with our work, having seen a wire-service story earlier that year on the U.C. Berkeley research group, with quotes by myself and Churchman.

From the beginning (1985), Carlotto's personal contributions to the Cydonia Investigation were extremely significant: starting with the first new computer imaging enhancements of the original *Viking* data in the more than four years since DiPietro and Molenaar's initial efforts. It was on these new images (processed with state-of-the-art 1985 computer algorithms, developed by Carlotto and his former colleagues in MIT's AI Laboratory[10]) that we first saw "teeth" in the "mouth" of the Face— and on *both* sun angles (35A72 and 70A13—see Figures 4 and 5)!

(This discovery, of course, immediately became an additional subject of controversy—regarding the "objective" reality of such features within this already too anthropomorphic "face-like" structure; for, apart from technical questions relating to the veracity of the image-processing algorithm which had revealed the features, the presence of "teeth" raised fascinating questions regarding possible *reasons* for "building in" such "anthropological detail"—if "the teeth" were real, of course. The problem will quite likely not be resolved until we get the much higher resolution images from *Mars Observer*—if (given the serious anthropological implications) even then . . . !)

Mark's second major contribution hearkened back to the urgings of Dr. Robert Jastrow, founder of NASA's Goddard Institute for Space Studies in New York, and well-known author of many best sellers on astronomy and the search for life beyond the Earth. Bob and I had been friends and colleagues since my "Cronkite" days, in the late sixties.

In one Independent Mars Investigation strategy meeting in San Francisco in 1984, before we made the first "formal presentation" of our work at the Boulder "Case for Mars" Conference that Summer (Chapter XI), Jastrow had stressed that "if you are to seriously interest the NASA community [in the Cydonia 'enigmas'], you must for starters do detailed *three-dimensional modeling* of 'the face,' to discover if the actual morphology conforms to 'a face-like object'—to eliminate the 'trick of lighting' accusation."

The Independent Mars Investigation had never developed the necessary funding to make this crucial computer analysis possible; however, when Mark Carlotto came on board in 1985, one of his first objectives was to achieve precisely such a *vital* 3-D rendition of the "Face."

Remarkably, when this was accomplished, in 1986, the result matched, to an amazing degree, a *completely different approach* to such a crucial three-dimensional analysis: our original Independent Mars Investigation project artist—Kynthia's—laborious "analog clay model reconstruction," part of which formed the cover of the original editions of *Monuments*.

Carlotto's first published work on this highly controversial subject— in a formal paper specifically written to describe his three-dimensional computerized analysis of the "enigmatic objects" at Cydonia—has now effectively *eliminated* "tricks of lighting" as a valid scientific explanation for the existence and appearance of "the Face." The paper containing this immensely significant conclusion, and the detailed mathematical and topological analysis of the 3-D substructure of "the Face" supporting it, made the *cover* of the May, 1988 issue of *Applied Optics*—a refereed scientific journal devoted to the state-of-the-art of current image processing technology and problems.

In the paper, Carlotto simply concludes:

> . . . the results of the 3-D analysis [of the Face] show that the impression of facial features is not a transient phenomenon. Facial features are evident in the underlying topography and are shown to induce the visual impression of a face over *a wide range of illumination conditions and perspectives* . . . [emphasis added].[11]

*Applied Optics* editor Bill Rhodes, an engineering professor at Georgia Institute of Technology, in Atlanta, replying to reporters' questions regarding the level of "science" reflected in Carlotto's paper, stated:

> Carlotto has been impartial and objective in the methods he chose to use; they were classical [imaging] techniques.[12]

So, finally—after *five years* of waiting patiently for this crucial state-of-the-art 3-D analysis of the most controversial object at Cydonia—we had the critical, objective proof that it was NOT all just a "trick of light and shadow."

*The Monuments of Mars*

A year later . . .

Armed with our own quantitative geometrical Cydonia analyses, as well as these vital shape-from-shading 3-D topographical analyses of the "enigmas of Cydonia" from Mark Carlotto, we at last sit down that April morning, in 1989, with the Chairman of the most powerful science committee in the Congress of the United States, Robert A. Roe, flanked by his chief administrative aid, Bob Maitlin—to discuss what must be done *politically—*

To now *force* replication of these crucial findings within the space agency itself . . . if not insure the acquisition of those vital new Cydonia images from *Mars Observer.*

After introductions and the usual pleasantries, Carlotto led off with a recapitulation of his striking 1985 3-D findings on "the Face." He then discussed some additional (then unpublished) Cydonia computer work [since presented in the *Journal of the British Interplanetary Society*[13]]— regarding the application of a *fractal* mathematical modeling technique as a means to "searching for artificial objects on planetary surfaces"; Carlotto's literally only days-old new results had gratifyingly indicated— and by a technique totally "decoupled" from all previous approaches—that the two *most* "unnatural and unfractal" sets of objects at Cydonia are (surprise . . . surprise) the "City" and the "Face."

Again, this *objective* result was strikingly at odds with NASA's endlessly repeated claims across the years, that "the face is merely an average Martian mesa, which 'happens' to resemble a familiar-looking subject . . ." Carlotto's mathematical technique, when applied to this supposedly "completely average Martian mesa," resoundingly revealed that—irrespective of what the Face even looks like—to a computer, its very surface was mathematically *anything* but "average!"

And still—with all its vast resources and the original *Viking* data tapes—NASA refused to simply duplicate these provocative results . . .

We next presented to the Chairman the latest Torun/Hoagland geometric findings on the "relationship model": the extreme specificity of the mathematical constants now uncovered, and their sheer *redundancy* throughout Cydonia. At one point I interjected:

"There's nothing about this [the "intelligence hypothesis"] that isn't 'scientific'—in other words, that isn't testable!"

To which the Congressman instantly shot back—

"Who says it isn't?"

"NASA," I replied.

"Well, they have an agenda ... don't they?" the Chairman immediately responded.

After we had finished the formal "briefing" segment of the meeting, Roe turned to me and pointedly asked:

"Why are you here? Something like 17,000 people are trying to get in that door," he waved meaningfully toward the front of his corner office in the Rayburn Office Building, just across the street from the U.S. Capitol itself, "Why do you think *you're* here this morning?"

"Because ..." I paused—a part of me remembering again that it had taken only *three days* to set up this major meeting, with the committee chairman of *the* major science committee in the U.S. Congress, and on a subject as "out-of-bounds" in Washington (at least at NASA!) as "possible ET ruins on the planet Mars ..."

"Because," I continued, "we have a scientific problem of *major* importance—at least, the evidence which we've presented here would strongly indicate that that's the case—which requires now the action of a federal agency—hopefully several separate agencies—to *test* its major premises and evidence. Yet, apparently, we also have a major and continuing *political* problem within *one* agency, NASA, and have had it for over thirteen years ... a strange resistance to simply carrying out those rather simple scientific tests at this point—including getting any kind of a commitment, *on-the-record* now—"

I handed him the letter from Michigan, containing the NASA/Smithsonian double-talk about "not being able to focus the camera on Cydonia."

"—to simply rephotograph the Face and Pyramids, during the upcoming *Mars Observer* Mission."

"OK," he said, standing up. The meeting was obviously coming to an end—

"Here's what I'd like you to do: write me a letter, briefly going over the points we've discussed this morning—with *specific* recommendations for what you would like me to do. Mark it "personal," and to my attention on the envelope, so it gets right to me. And I'll pursue it."

I handed a copy of *Monuments*—and various papers, diagrams and photographs we'd brought specifically for the Chairman's later "homework"—to his aide, Bob Maitlin, and the five of us began moving toward the door. The Congressman fell in step alongside Torun, putting one arm around his shoulder as they walked, as he said collegially, "By the way, don't listen to those people [at NASA] who say this isn't 'scientific.' You just keep up the good work!"

That was it—our "breakthrough meeting on the Hill" with the Chairman of the House Committee on Science, Space and Technology, Robert Roe—at which we were able for the first time (or, for that matter, *anyone* was able in the 13 years since *Viking)*, to effectively marshall and present the *evidence* painstakingly accumulated over the length of our own, at that time, six-year investigation—and to a key policymaker in the U.S. government—regarding possible *extraterrestrial ruins* lying in a northern desert called "Cydonia". . . on the planet Mars.

Only later—in continuing to wonder at the rapidity with which we got this crucial meeting—did I discover that Congressman Robert A. Roe, in addition to his (then) Chairmanship of this key congressional science committee in the Congress . . . was (and still is) a ranking member of the "Permanent Select Committee on Intelligence."

*               *               *

Six months went by . . .

And then: first in a series of letters to constituents; then in print, in a major story in the *Wall Street Journal* (and another in *USA Today*); and finally, even during a live radio interview on which I was also a guest (we've got the tape!), Representative Robert Roe stated his courageous— and frankly, astonishing (in view of NASA's long history of "debunking" the entire issue)—"position" on Cydonia:

> I've seen the studies and I've seen the photographs . . . and there do appear to be formations of a face and pyramids [on Mars] that do not appear to be of natural or normal existence. It looked like they had to be fashioned by some intelligent beings . . . For this reason, I have asked NASA to provide assurances that the *Mars Observer* Mission will include this [set of objects] as one of its imaging objectives . . . [13]

On the critical question of NASA verification of Cydonia via *Mars Observer,* Roe went on:

> I'm under the impression, and the Committee's under the impression that, yes, NASA's going to rephotograph this area like they said they were going to. I'll ask for reaffirmation. *We're interested enough to see that it gets the proper attention* [emphasis added].[14]

But apparently, NASA wasn't.

In late 1990—a year after Roe's inquiries, and NASA's abrupt reversal of its former plans NOT to rephotograph Cydonia—a free-lance journalist, Walter Gelles, proposed a story on our continuing work to the *Washington Times,* the second major newspaper (besides the *Post*) read by policy-makers in the Nation's Capitol, Washington D.C. James G. Osborn, natural science editor of The World and I, the *Times* magazine section, replied to Gelles' proposal:

Dear Mr. Gelles:

. . . your lead paragraph stating NASA's plans to include the [Cydonia] area on the agenda of the *Mars Observer* lends credibility to a topic that has heretofore been grouped with UFOs with regard to scientific credibility.

To verify your account I called the NASA Newsroom here [in Washington]. Vera Hirshberg, director of the Space Exploration section, said that she is well aware of the claims of some people to products of [a] civilization on the Martian surface, including a face and a pyramid [sic], but stated that NASA still claims that the features are merely products of light and shadow. More importantly, *she denied that the Cydonia region is a scheduled target of the Mars Observer,* and wanted to know what official was spreading this rumor (emphasis added).

I will not even think of pursuing this manuscript unless you can quote a NASA or Congressional official linking the monuments to the *Mars Observer* mission . . . [15]

Gelles immediately responded:

Dear Mr. Osborn:

Thank you for your letter of September 26th concerning my article about the Mars enigmas.

. . . I have enclosed two letters from Congressman Robert Roe . . . In Rep. Roe's letter to Keith Morgan of ABC-TV News, he says: 'It is my understanding that NASA does intend to try to capture, with the narrow-angle camera, the Cydonia region, including the unique features you have referred to as the "pyramids" and the "face". . .' In Rep. Roe's letter to [constituent] Larry Caldwell, he says essentially the same thing . . .

Congressman Roe has been assured by NASA liaison Heninger and other NASA officials that NASA *will* rephotograph the Cydonia region of Mars on the *Mars Observer* mission. NASA is now sponsoring a film mini-series, 'Hoagland's Mars,' which will *report in depth on Richard Hoagland's scientific work on the Mars enigmas* . . . and there are plans [by Lewis—the producing NASA Center] to make it [this production] available to the public through the PBS television network. Rep. Roe has assured the public, based on NASA's promises, that his Congressional committee will closely monitor the *Mars Observer* re-photographing of the Cydonia region [emphasis added].

*The Monuments of Mars*

... your letter is a very disturbing one, for it suggests either that Vera Hirshberg (whose first name means 'truth') is lying, or else that NASA has again abruptly reversed its policy, and [it] has been lying for the past year to the most powerful scientific and technical committee in the Congress, deliberately misleading Congressman Roe and the American people . . . [16]

Then, Gelles copied the *Times* letter, and his response to Osborn, to Chairman Bob Roe—with a separate note to Roe himself, reiterating his concern at this new Cydonia development:

Dear Congressman Roe:

I am very disturbed by the enclosed letter from the *Washington Times'* monthly magazine . . . I wrote a serious, scientific article about Richard Hoagland's work. Now it appears that NASA is lying (or has completely changed its tune) in order to block the publication of my piece.

I want to know what is going on. I urge you to look into this matter immediately, and look forward to your reply.[17]

Within days (Oct 9), Roe replied:

Dear Mr. Gelles:

Thank you for your recent letter regarding NASA's plans to rephotograph the Cydonia region of Mars.

To ensure there will be *no further mis-communication regarding the plans to rephotograph the 'Monuments of Mars,'* (emphasis added) I have asked NASA to comment on your letter. Once I receive the agency's reply, I will send it to you.

Again, thank you for writing. I will be in touch . . . [18]

That, however, was the last letter on this subject that Gelles officially received from Robert Roe. The promised "NASA letter"—clarifying Hirshberg's strange comments to the *Times*—never arrived. And Roe himself was never again in touch with reporter Walter Gelles on this issue.

But something else did happen . . .

On January 3, 1991—without warning—Robert Roe abruptly resigned the Chairmanship of the House Committee on Science, Space and Technology, to take the reins of another Congressional Committee: Public Works and Transportation. Our "man at the helm"—who, after our key meeting, courageously, publicly and *repeatedly* stated his own impressions of the Cydonia data, and who solidly supported the only democratic means the American people now have to insure that NASA lives up to its commitment to this key Congressional Committee to rephotograph

Cydonia—suddenly was gone—

Leaving very much in doubt NASA's *real* objectives toward Cydonia, during its 1993 return to Mars . . .

*　　　　*　　　　*

There are additional reasons now for seriously doubting NASA's continuing intentions on this increasingly contentious issue of "Cydonia," despite its past assurances to Roe *and* to the Congress. These doubts center on NASA Headquarter's less than candid *actions*—well in advance of the actual *Mars Observer* mission—regarding the 1990 production of a NASA-Lewis television show featuring our work, which Lewis produced for nationwide distribution via PBS:

A program specifically titled by NASA-Lewis, *Hoagland's Mars*.
Some further background . . .

*　　　　*　　　　*

In early 1990, as part of a sudden rash of "official NASA invitations" to present the Cydonia data, I was invited by the Director of NASA's Lewis Research Center, in Cleveland, Ohio, Dr. John Klineberg, to present our Cydonia results to the entire NASA-Lewis Center—in a Center-wide address not only to be attended in the main Lewis Auditorium, but to be carried throughout NASA-Lewis on closed-circuit television.

I was informed that, if I accepted, special "viewing rooms" would be set up—in the Cafeteria; in a second auditorium in a separate building; in various offices and lounges scattered through the complex—and a special "account number" established—X01-2497 (so NASA employees could still *charge their time to the government* during the 2:00 PM presentation!)—to enable as many of the 4000 scientists and engineers who work at Lewis as possible to see and hear for themselves our evidence—accrued during the (then) seven years of the ongoing Cydonia Investigation.

A very special, very dedicated member of Dr. Klineberg's staff, Joyce Bergstrom, head of the NASA-Lewis ALERT program, became the contact person for our presentation. Bergstrom was in charge of making all the subsequent arrangements for the March 20th address—including official NASA videotaping of the event. She was also responsible for the subsequent official distribution of copies of this video to members of the television media: to ABC News (Washington Bureau), and others; as well as untold individual copies eventually sent out in response to unprecedented numbers of personal requests—from all around the country—for "the NASA-Lewis Cydonia Briefing." As part of our original agreement to come and speak to NASA, we also received our own Master copies of this tape—a point that would become somewhat important later on.

A few days before I was scheduled to arrive in Cleveland for the address, Ms. Bergstrom called, and asked "would you object to a special NASA interview for PBS the night *before* the Center briefing?"; the interview was to be conducted by Dr. Lynn Bondurant, Chief of the NASA-Lewis Educational Programs Office.

Things *vis-à-vis* "Cydonia" were suddenly, decidedly, looking up on the NASA front!

We arrived at the NASA-Lewis Research Center Main Gate promptly at 6:00 PM, the evening of March 19—expedited through Security by special orders from Bergstrom's office; to take advantage of our brief, two-day visit to the Lewis Center, the interview had to be scheduled after normal business hours. (NASA centers are not military installations, but some military-related research and defense work is carried out by some NASA installations—so a certain level of "security" is normal in "visitor control.")

When we were taken downstairs in the main administration building to the site of the scheduled interview, Bergstrom had outdone herself (and this "extra interview" wasn't even part of her normally-scheduled duties surrounding my formal "ALERT" presentation the following afternoon); laid out in the official NASA-Lewis TeleConference Center (which Bondurant had commandeered as his TV "interview set") were scrumptious hors d'oeuvres, crackers, dips, fresh fruits, carbonated refreshments—even fresh-brewed coffee!—everything tastefully arranged by Bergstrom herself and her equally-dedicated small staff.

All this for only two guests—Nancy McIntosh-McNey and myself, plus Bondurant and a couple of TV technicians.

Talk about "red carpet!"

The interview was (if anything could be, after that) even more impressive.

Bondurant (unlike many others I've been questioned by, in the last four years since *Monuments* was initially published) had really "done his homework"; he knew *Monuments* from cover to cover, and during the *two-and-one-half-hour* interview, asked perceptive questions over a wide range of topics covered in the book—from the geological history of Mars, to the provocative new measurements that now overwhelmingly confirm the "geometric relationship model," to the eventual (and most controversial) architectural speculations I engage in around the "pyramids" themselves.

Through it all, Bondurant's questions were to the point, low-key, and very serious.

And the setting was as impressive as the interview.

Bondurant placed me at one side of the long, oak-polished conference table surrounded by luxurious high-back padded chairs. He sat at the end, note-pad balanced on his lap. The cameraman aimed the camera over his shoulder, framing his shot so—behind me on the wall, in large red letters—appeared the NASA logo:

**"NASA Lewis Research Center."**

All very authoritative, very detailed . . . very "official". . . for a "trick of light and shadow."

As you can see for yourself—on the full, unedited version of the interview tape, which we were subsequently given, and which we will probably make available in some form someday (through *The Mars Mission,* see "Hoagland's Mars," below).

When we finished, it was after 10:00 PM. The director, helping the cameraman pack up his gear, then asked me, "Do you have the graphics and visuals for this?" I made the mistake of saying "yes"—so we promptly found ourselves in *another* building in the sprawling NASA-Lewis complex, loading our slides (actually prepared for the next day's presentation) into a television "slide chain," while the cameraman, in the studio, simultaneously did "tilts and pans" on the photographic blowups and graphics I had with me.

We finally left the Center well after 12:00 AM—

Only to return early (!) the following morning, for an "official" VIP tour of NASA-Lewis—arranged by Bergstrom. This was followed by a VIP luncheon with the Senior Center Staff and, after that, a private meeting with the Center Director, John Klineberg—followed by my address in the Main Auditorium.

Again, not bad—for research not even acknowledged as "legitimate" by "official NASA spokespersons". . .

The "big event" was scheduled for 2:00 PM.

NASA-Lewis Director, Dr. John Klineberg, stood before the absolutely jammed auditorium (there were even people sitting on the broad, carpeted "steps" up and down the aisles), with unknown additional numbers watching from all over NASA-Lewis on closed-circuit television—
*All*, I kept thinking, *charging their time officially to the Government of the United States!*—and began his introduction.

And with it, came the first of several major questions surrounding this whole "VIP NASA presentation"—

*The Monuments of Mars*

Said Klineberg:

—Richard Hoagland is also *the man who managed to convince the President to state that a return to Mars is one of our major goals* [emphasis added].[19]

Several members of the Cydonia Investigation were present in the Lewis Auditorium when Klineberg made this—to us—literally shocking announcement—

*For it was total news to me!*

Afterward, we compared notes:

Yes, Klineberg seemed perfectly serious when he made the statement; and, yes, he'd made it only *minutes* after we'd discussed, in that private meeting in his office—the Director's Office of the NASA-Lewis Research Center, mind you—the new "political sensitivities and realities" surrounding the entire Space Agency.

So, what was going on?

Klineberg had *specifically* commented in that meeting (attended by myself, Nancy McIntosh-McNey, Dr. John Wilson and his wife Diane, and several senior NASA-Lewis officials) on the increasing scrutiny of NASA by the Congress—because of certain "controversial events" which had taken place in recent years (such as the Challenger disaster). And he had voiced special concern over the care with which all potentially "political statements" must be framed "in this new era of increasing scrutiny [of NASA] by both elected political officials, and members of the press."

Then—within minutes of these cautionary pronouncements—John Klineberg had made, before thousands of NASA employees and running television cameras what, by *anyone's* definition, could only be viewed as a *totally irresponsible statement,* on what most in NASA still publicly consider a totally off-the-wall subject . . . unless—

Unless—it actually were true! -

Later, when asked by a reporter why he had included these remarkable comments on "our 'influence'" in the President's decision "to return to Mars," Klineberg merely demurred, terming his remarks "my own droll sense of humor."

Then he added: "It got people to think, and that's probably what you're after in the first place."[20]

Which is all well and good . . . except for a few nagging inconsistencies that kept cropping up . . . after the NASA-Lewis presentation.

As previously noted, NASA-Lewis had promised, during our discussion of arrangements, to furnish a 3/4-inch format, broadcast-standard videotape copy of my entire presentation—which I eventually received. However, when I viewed my copy (and the two "home video" VHS copies that had also been included), John Klineberg's entire introduction— *including his strange comments vis-à-vis me and the President of the United States*—had all mysteriously *vanished . . .* to be replaced by a typed intro, superimposed over the now well-known "flyover of Mars," created by NASA-JPL from *Viking* 3-D computer imagery enhancements.

There was also something very curious about the *date.*

The "date" of the presentation—March 21, 1990—was clearly marked on this "edited and produced" taped version of my presentation . . . and the date was *wrong;* my briefing had been delivered on *March 20!*

What was going on?

Perhaps related to this question was another, earlier "curious anomaly" that had also taken place around this "NASA-Lewis" tape.

ABC News (through Keith Morgan) had been alerted to my forthcoming appearance and briefing at the NASA-Lewis Center several weeks before my scheduled trip to Cleveland. I had even been asked to meet with a producer for Peter Jennings' "World News Tonight" at ABC News in Washington, to discuss details of the upcoming address—which I did. So, a few *days* before my next NASA presentation, this producer called Bergstrom and *specifically asked* to be "Fed-Exed" a 3/4-inch copy of the tape following the March 20th presentation.

Several days went by (following March 20), and still no tape showed up at ABC. So, eventually, someone else from "World News Tonight" called NASA-Lewis—and was told that Bergstrom's office "had forgotten" to send to ABC News—a *major* national news organization—their *requested* copy of the tape!

When it finally did arrive (over a *week* later, and only after *two more* phone calls . . .), not only had John Klineberg's entire introduction disappeared (just as I would subsequently discover on my copies), but the tape furnished to ABC was *not* a 3/4-inch broadcast-standard copy—it was only a home video VHS cassette.

When I received my own copies (a week or so after ABC) and independently confirmed that Klineberg's entire introduction had vanished, I immediately called NASA-Lewis and inquired as to why the Director's extremely generous remarks had been mysteriously replaced. I was then informed that there had been "simultaneous equipment failure . . . in *two* video tape machines"—that afternoon at NASA-Lewis . . . a failure which, marvelously, had cleared up just as I began to speak!

A footnote to this (to say the least) "suspect" sequence of events.

As noted above, the date on the final version of the tape that NASA eventually did furnish (both to me and ABC), read: March 21, 1990—one day *after* the actual presentation (and taping) date.

*If* the tape was sent somewhere "off lab" from NASA-Lewis, for the specific "editing out" of Klineberg's ill-advised opening remarks—somewhere like *NASA Headquarters,* in Washington D.C., perhaps . . . ?
—and *if* it arrived "overnight," those who edited it *the following day* might well have logically put *that* date on the final product (March 21), rather than the original taping date (March 20).

On the other hand, if the editing had been carried out at NASA-Lewis, where both signs and billboards for several *weeks* around the Center had been reminding NASA employees about my March *20th* appearance—complete with special "billing code"—it is hard to imagine (in the small TV "audio-visual" office at NASA-Lewis, with only three employees) someone putting on the wrong date—of a presentation where major "equipment failure" had *wiped out the Director's entire opening remarks;* a presentation originally considered by Dr. Klineberg's office important enough to be broadcast, via closed-circuit television, to the *entire* NASA-Lewis lab.

The evidence then is "consistent" (as physicists love to say) with the fact that the Master Tape of this controversial presentation was subsequently—for some "reason"—deliberately sent "off lab" for post-presentation editing—and handled by NASA personnel (or contractors) *not* familiar with the actual date of the "Hoagland NASA-Lewis Cydonia presentation."

Further evidence consistent with this hypothesis comes from the fact that the "NASA-Lewis" logo, on the opening of the "edited" tape, was abbreviated "LRC." What was an "LRC?"; only someone deep in NASA (certainly not the general public, or even television types like those at ABC . . . ) would ever recognize that "LRC" referred to "NASA-Lewis Research Center."

Which might have been the point . . .

When I asked Bergstrom specifically about this, she was extremely displeased (she herself had not seen the finished tape at that point), and informed me that the designation of the Center *should* have read "NASA-Lewis Research Center"—*all typed out*—which was "normal" for all such Lewis presentations, and was subsequently (per her specific orders) incorporated in my second copy of the "Hoagland tape" from Lewis.

Finally, ABC's failure to receive its promised copy—and in a timely fashion—is *also* consistent with the tape being actually flown off the NASA-Lewis facility for editing (therefore NOT available to *anyone,* including ABC, for several days)—probably Washington—where "someone" made a major political decision to simply eliminate John Klineberg's entire "embarrassing" introduction—

Including his publicly crediting *this* investigation with being a major factor in President Bush's decision to "someday return to Mars."

Only after this editing out of Klineberg's "politically-sensitive remarks," in this scenario, was the tape then "fit" to send to ABC—and then, *only* in a "home video" version—to further guarantee that it *not* be deemed "newsworthy" by the ABC producers.

Nothing else makes sense.

Fortunately, unknown to anyone at NASA-Lewis, we had someone in the NASA-Lewis Auditorium that afternoon with an *audio* cassette recorder—which thus serendipitously preserved forever John Klineberg's "quirky" comments . . .

For the first time, directly involving the President of the United States . . . with the "monuments of Mars!"

*      *      *

Now, remember Bondurant's elaborately-staged "official NASA interview" the night *before* my NASA-Lewis Center presentation . . . ?

The political importance of the preceding, otherwise inexplicable sequence of events surrounding the simple videotaping of my invited NASA-Lewis presentation (and the apparently less-than-candid NASA "explanations" which subsequently followed), would soon pale, by comparison, with the outright "hanky panky" surrounding Bondurant's intended NASA-Lewis/PBS satellite television production—

*Hoagland's Mars.*

On December 13, 1990, this highly unusual NASA-Lewis Research Center television production—which would have been *the first official NASA overview* of the intensely controversial subject of the "Face on Mars" in the entire fifteen years since *Viking*—was abruptly halted *by direct order of NASA Headquarters,* in Washington D.C. The program, in production for over nine months (since the night I was specifically interviewed by Bondurant), had at this time (December 13) even been officially placed on the schedule, by NASA, for satellite transmission to television stations all across the United States—on January 6, 1991.

But on that December 13th, 1990—a Thursday afternoon—all that effort suddenly came to an abrupt halt—by direct (and unprecedented) order of Bondurant's boss at NASA Headquarters, Dr. Robert Brown, Director of NASA's Washington Educational Affairs Division.

In terms of content, in addition to presenting the results of our own work, Bondurant's thirty-minute production, *Hoagland's Mars*, was to have included summaries of the findings by many of the others who have tackled this extraordinary question over the years—from Vince DiPietro's and Greg Molenaar's early imaging enhancements,[21] to Erol Torun's now-thorough geomorphological analysis of *all* the "anomalous objects" concentrated suspiciously in this one tiny region termed "Cydonia,"[22] to the latest imaging efforts of Carlotto, including his now highly revealing "fractal" (pattern-recognition) analysis of the same "anomalous objects."[11, 13]

It was also to have included the fact that (up to the time of broadcast) I had been invited at least *three* times, by two *separate* NASA Centers, to present the Cydonia evidence favoring the "intelligence hypothesis."

And yet, after eight *hours* of these cumulative private briefings, for some reason, NASA *Headquarters,* at the last minute, apparently *didn't want* the American people to see on "official" NASA television even a thirty-minute summation of that evidence—evidence now significantly supporting that Hypothesis.

The crucial question, which now must pointedly be asked:

"Why not?"

If there's truly "nothing there," if we have all been wasting a lot of scientific time and research on just mere "tricks of light and shadow," why bother to "pull the plug" on "one more television rehash"—and this one under *the complete control* of "NASA?"

Why intrude administratively (in a most heavy-handed way, and over such a "trivial" issue) on the educational prerogatives and functions of a distinguished NASA educator with over thirty years of service to the Agency—if not in the internal affairs of an entire, autonomous NASA Center . . . ?

Why, in terms of press response, risk even the *allegation* of a "cover-up" and all the unwanted attention that would raise to this entire issue— if this subject is indeed "pure nonsense."

On the other hand, suppose it's not . . .

In the immediate firestorm of public and press reaction over the sudden NASA-Lewis program cancellation, secretaries answering the phones in Brown's NASA Headquarters' office naively admitted to reporters in the first few days (before "professional" news management people began to "man" the phones), to receiving "*thousands of angry calls* . . . and from all over the United States" regarding the program's sudden termination.[23] This was due, in part, to my weekly appearances each Friday (fortuitously, one day following NASA's abrupt action) on Chuck Harder's "For the People" nationwide radio network, and a completely separate appearance (previously scheduled) the day after that, on Saturday (December 15), on ABC Radio in New York; the public reaction to my announcement, on these two programs, of the sudden termination of NASA-Lewis's planned television production, *Hoagland's Mars*, was prompt, angry, and almost overwhelming—apparently communicated immediately through thousands of calls to NASA Headquarters in Washington.

In eventually answering some of these angry calls, which for several days completely swamped "Education's" many phones, Mr. Frank Owen, (then) Deputy Director of NASA's Washington Education Office, attempted to defend NASA's actions by saying "on the record":

> Hoagland was attempting to use the [NASA-Lewis TV] program to sell his book; we had to investigate.

Privately, he told colleagues (who promptly told us) something considerably different:

> The planetary scientists [who managed *Viking,* and who took the original *Viking* images at JPL] *threw a fit* when they discovered that Lewis was going to air the 'Hoagland' program . . . they do not want, under ANY circumstances, *to legitimize this subject.* And they have a lot of clout here [at NASA Headquarters].[24]

So, why should a handful of "planetary scientists" at *one* NASA center (who 15 years ago abrogated their responsibilities to NASA and the American people—if not to science and to truth—by *not* seriously investigating the Cydonia anomalies in consonance with *Viking's* stated mission, the "Search for Life on Mars") now be allowed a veto (at a completely *separate* NASA Center!) on what the American people can or cannot be told regarding a completely *independent and privately-funded* investigation of that public-domain data?

Do these few "JPL scientists" believe, in spite of their performing no discernible scientific investigation of the subject, that they somehow still hold a monopoly on *Viking's* 15-year-old data . . . or on scientific truth itself?

Or do they, because of their continuing contracts with the U.S. government and with NASA in particular (which maintains *exactly* such a monopoly on shuttles, rockets, spacecraft and computers required to get us back to Mars at all to even *test* the "intelligence hypothesis") somehow think they literally "own the solar system?!"

The First Amendment issues, raised by this remarkable example of deliberate government censorship of an *idea*—if not access to the *evidence* that now strikingly supports it—at this point are very clear . . . and are of major political, if not Constutional, significance.

*             *             *

Confronted now with what could only be termed *outright, official censorship by the U.S. Government on the subject of Cydonia*, we fought back.

We contacted an Emmy-Award-Winning television producer (Cliff Curley—who had written to us some weeks before). We then produced in thirty days, from our own Master Tape of the NASA-Lewis presentation, our own broadcast version of *Hoagland's Mars*!

But *not* the "limited" 30-minute version that Bondurant had planned; this is the the *full* 83-minute, essentially unedited NASA-Lewis address that I gave in the NASA-Lewis Auditorium, March 20, 1990 to all 4000-plus NASA engineers and scientists in Cleveland. All we did was update it with developing new information, and add some new color graphics and a spectacular 3-D computer "flyover" of the Face and Pyramids themselves which had just been completed by Carlotto.

The result was almost worth the NASA cancellation of Bondurant's "official" program.

Several thousand copies of this "commercial version" of *Hoagland's Mars: The NASA-Cydonia Briefings, Vol. I* have now been distributed across the Nation and around the world; it has even (finally) run on Public Television—the PBS station in Anchorage, Alaska contacted us directly, and asked permission to show the program twice (which they've now done). Unknown numbers of copies have also been distributed in England, France, Japan, Africa, South America, and even central Europe (through shortwave advertising—see below), as well as to the official television networks of many of these countries.

People who are buying it as a home video (through *The Mars Mission;* or through the nationwide "800" number we specifically set up; or through the radio advertising on Chuck Harder's national "For the People" radio network, and its short-wave radio *international* rebroadcast) have in turn been calling in friends, relatives—even total strangers!—to run the tape, to show them what NASA now knows (but apparently doesn't want *you* to know) about the "monuments of Mars."

*             *             *

But active citizen participation notwithstanding, there is now unquestionably a disturbing pattern that is emerging in these events, on the subject of "Cydonia": an inescapable escalation of blatant NASA (government) censorship, if not literal outright lying to the media . . . and Congress.

The latter now includes denying again and again the very existence of a formal NASA agreement with the House Committee on Science, Space and Technology, on the *specific subject of Cydonia,* or with its former Chairman, Robert A. Roe!

Far above the vanity or wounded egos of a few NASA planetary scientists (who, having initially "missed it," now probably believe they have good personal reasons for "hiding" from what's waiting at Cydonia!) there looms the possibility that other, more profound "special interests" are also hard at work here . . . attempting to eliminate not only public discussion of the subject, but the actual *verification* on Mars of what our eight years of research have so painstakingly revealed . . .

So, again, the question must be asked: why bother—if it's all "sheer nonsense?"

Is it possible that the *real* reason for the abrupt cancellation of NASA's *Hoagland's Mars* is grounded directly in our now actual *decoding* of the geometry we have discovered at Cydonia—

In our discovery, on Mars, of a literal "Message of Cydonia?"

*        *        *

For this is our penultimate triumph: after eight long, grueling years, and innumerable technical, financial and political obstacles, including the continuing, contradictory, downright bizarre behavior at this point of the one agency which should now be embracing what we've found—NASA—we can truthfully make the claim that we have finally "cracked the code" behind the *meaning* of the ruins at Cydonia.

A "meaning" that is literally stupendous—if we're right.

That Meaning seems to involve nothing less than "the fabric of Reality"—how Matter, Time, and Energy are woven into the tapestry of Everything . . . from stars . . . to planets . . . from atoms . . . to living systems . . . to Intelligence itself.

It is nothing less than the one appropriate Message that "someone" might just leave to "someone else" on the surface of a nearby planet . . . if they wanted to truly help . . . *without* interfering . . . without violating some actual Galactic "Prime Directive". . . .

**For—having now successfully decoded such a Message—what we *do* about it, as a society, . . . as an entire planet "in transition," is still strictly up to us. . . ."**

This extraordinary and eminently verifiable discovery—which we presented in detail (with appropriate caveats) in our *second* specifically-invited NASA-Lewis Briefing, in September, 1990 (which NASA *also*

*The Monuments of Mars*

videotaped)—involves nothing less, we now believe, than the fundamental *constructive energy* of the Universe . . . whose existence—and accessibility—was apparently *one* of the prime reasons for the very creation of the "ruins" at Cydonia:

**To demonstrate, through the reality of an actual, viable *community*—placed *deliberately* on the surface of a nearby "inhospitable" planet—how one can use this "energy and information" to transform Existence . . . if not literally to ensure it!**

The "Message" was left to us, we are coming to perceive, much as ancient Egyptian temples and other "sacred" structures here on Earth embodied a similar, "sacred," geometric "message"—as *living architectural examples* of their creator's understanding of a Universal Order.

But this "sacred architectural symbology"—initially, only a comparative example with the "familiar" here on Earth—through additional, stunning new discoveries by Myers, Munck, and others (as we stated in our Addendum to the Author's preface), now seems to be far more than merely a poetic *metaphor,* "borrowed" from this planet. . . .

"Cydonia" turns out to be: nothing less than an architectural affirmation of the fundamental *physics* of the Universe—the ultimate embodiment of a grand, "universal Architecture". . . at the most *archetypal* level—

For, in striking confirmation of the "terrestrial connection" that we initially proposed here in 1987 (Chapter XV), the Cydonia Investigation has now found multiple examples of The Message of Cydonia—

*Identically* "coded" elsewhere in the solar system . . . including, *here on Earth!*

*       *       *

The primary purpose—the real reason, we now believe—for the very existence of the complex and interlocking geometry we've deciphered at Cydonia:

**Nothing less than to provide whomever first deciphered it a deliberate communication of a whole new physical understanding of the Universe, if not the "ultimate" relationship of conscious beings—Intelligence—to all of it.**

The key to unlocking this multi-leveled interpretation (which we spent over six *years* trying to decode—before we *did* it!), now unmistakably centers on the geometric and geodetic properties of an *inscribed tetrahedron* (the primary geometric solid) surrounded by a sphere, communicated so redundantly through the interlocking geometry, angles, mathematical constants, and their associated trigonometric functions discovered so redundantly across Cydonia.

It all reduces to a simple tetrahedron—and what such a geometric figure *really* represents . . .

A "circumscribed tetrahedron" is a four-sided, four-cornered pyramid—the simplest regular "Platonic solid" that can exist—surrounded by a circumscribing sphere. If you place a tetrahedron in a *rotating* sphere, such that one vertex or "corner" sits on either the North or South polar axis of rotation, the other three "corners" of that tetrahedron will lie at either 19.5 degrees South or North latitude (relative to the equator), equally spaced—at 120-degree intervals—around the full 360-degree, 19.5-degree latitude-circle of that sphere (see Figure 29).

Some of the highly redundant geometric clues present at Cydonia now unmistakably point us toward a connection between this fundamental "inscribed tetrahedral geometry," and basic planetary geophysics; acting upon a suggestion made initially by David Myers, the distance—between one specific "benchmark" on the Face, the so-called "teardrop"—and another benchmark on the wedge-shaped front of the five-sided "D&M Pyramid" nearby, turns out, upon measurement, to be *exactly* 1/360th of the diameter of Mars! (see Figure 30)

Furthermore, it straddles (precisely bisects!) a *36-degree* angle within the structure of the "pyramid" (see Figure 31).

These (and several additional related clues) have now convinced us that whoever built Cydonia used our familiar 360-degree system of angular notation and, in fact, probably intended that we should learn of its existence (if we didn't already know and use it) from the specific layout and geometry at Cydonia itself (which, of course, ties in with our now almost unbelievable discovery of *specific* "Cydonia Geometry" distributed around the earth—as you will see below).

The basic "energy" discoveries alluded to above, deciphered through the "decoding" of these Cydonia geometries and mathematics, turn out to only truly work in our *360-degree notation system;* thus, like "kindergarten," it now seems apparent that, among other basic "lessons," the "builders of Cydonia" laid out the highly redundant Cydonia geometry *to lead us specifically toward this notation system.*

Further evidence of this comes from the fact that these same distances (the Face/Pyramid spacing, and an identical City/Face spacing) also measure *exactly* **19.5 arc minutes** on the Martian surface (an arc minute—in our 360-degree system—is 1/60th of one degree) (See Figure

32). This number, 19.5, which in this context can only be communicated if the 360-degree system is *specifically* intended—is, as we have demonstrated above, and elsewhere (8), *overwhelmingly* significant.

Being none other than *the precise latitude of three vertices of "an inscribed tetrahedron in a sphere."*

All this seems to be directing us—in the model that proposes that Cydonia is indeed "Galactic kindergarten"—to "place the inscribed tetrahedron in a *planetary sphere* such as Mars itself." We can do this so that one corner of the tetrahedron corresponds to the longitude and latitude of the largest geophysical feature (energy upwelling) on the planet—a vast shield volcano named Olympus Mons, located at 19.5 degrees North latitude on Mars. But when we place one vertex of the tetrahedron under Olympus Mons, the next vertex corresponds almost precisely to *the specific longitude of Cydonia itself*, one third of the way around the planet to the East (see Figure 33).

And of course, all of this is just "coincidence".

Following this "first lesson," we then noticed in rapid succession that ALL of the major "geophysical disturbances" across the solar system—from the latitude of the largest volcanic "upwelling" on Earth (the Hawaiian Shield Volcano), to the siting of the two major suspected active volcanic complexes on Venus, Alpha and Beta Regio, to the location of the Great Red Spot on Jupiter and its recently-discovered counterpart on Neptune[6]—the Great Dark Spot—ALL occur (within a degree or so) **At either 19.5 North or South—or both**!

Even the Sun seemed to obey this "circumscribed-tetrahedral" pattern; on the solar surface, the peak latitude of the 11-year sunspot cycle, and the peak latitude of solar temperature emission corresponding to that cyclic sunspot maximum,[25] remarkably, occurs at **~19.5 North and South**!

Could it be (we began to ask ourselves) that there is something *physically* significant in a rotating, *fluid* sphere, such as a spinning Jupiter-like planet; in the *liquid* core of a solid planet like Mars or Earth; or in a high-temperature, spinning gaseous body (an "idealized fluid"), like the Sun—regarding an "inscribed tetrahedral geometry" and the 19.5-degree latitude?" Could such a "new physics" be controlling both the energy flow deep inside such bodies . . . if not (somehow) the *generation* of the energy itself?

And could it be (we now realized enough to ask "the 64-dollar question") that Cydonia was *deliberately* built—

*To make us ask such questions!?*

We simply do not have room within this Update to describe adequately, or to substantiate *at all*, the detailed study and the calculations which have gone into attempting serious answers to these questions; that level of detail must await publication of *The Cydonia Papers,* and future issues of *Martian Horizons.* What follows, therefore, is at best merely a summary of extraordinary "work in progress.". . .

*       *       *

Through the independent work of Stan Tenen (*Mera Foundation*—see again the Author's Preface: Addendum), we have been introduced over the past few years to a parallel series of "redundant tetrahedral metaphors," expressed repeatedly, it turns out, in a wide-variety of ancient terrestrial texts. Tenen's twenty-one years of previously separate research[26] have now turned up *identical* expressions of "inscribed tetrahedral geometry" on Earth—in Ancient Egyptian hieroglyphic manuscripts, Ancient Hebrew scrolls, Ancient Greek texts, etc.—raising the obvious question: "what was so important to the ancients about "tetrahedra placed in circumscribing spheres?"

And how does this relate (if it relates at all) to the remarkable "tetrahedral" pattern we have been led to, through "decoding" the geometry of Cydonia—now demonstrably spread across the solar system in a series of major astrophysical phenomena?

From Tenen and his work on ancient texts, we were, in turn, led to a series of modern papers on the arcane subject of "mathematical topology"—which, it turns out, is probably now of *pivotal importance* in connecting these remarkable examples of astrophysics with the "Cydonia geometry"—as evidenced below.

*       *       *

For over a hundred years, a legion of *terrestrial* scholars (we must somehow get used to thinking in these expansive terms!) have created entire abstract universes out of numbers such as we have now found separately "coded" at Cydonia; the literature of this small and close-knit subculture of the terrestrial world—the "topological" world of pure mathematics—is *filled* with theoretical models, complex calculations and extended proofs of topologies and geometries which have no counterpart in our familiar three-dimensional existence. These are the mathematical modelings of "higher dimensionalities"—so-called "n-spaces" (where "n" is a number equaling a particular spatial dimension—3, 4, 5, etc.).

Just as a three-dimensional sphere, when projected back into a two-dimensional realm (as, for example, when its shadow is cast on a wall), can only be seen as a two-dimensional cross-sectional *circle or disc,* so "higher-order" geometric figures, when "rotated" or projected mathematically back from a "higher dimensionality" (4, 5, or 6 geometric dimensions, back into our familiar 3), can only produce three-dimensional geometric "shadows" of their true, higher-dimensional forms.

Following this logic, topologists have constructed elaborate three-dimensional geometric figures, actually "shadowgram representations," depicting abstract "n-space mathematical extensions" of familiar three-space objects—such as "hyperspheres," "hypercubes," and "hypertetrahedra," etc. One of the most classic "higher-dimensionality" geometric proofs is called "the problem of the 27 lines on the general cubic surface."[27]

**The solution to the mathematical puzzle posed by the "Message of Cydonia" is ultimately, it turns out, a deep connection to this previously *completely esoteric world of "pure mathematics!"***

According to these topological papers, and according to Tenen's actual *physical modelings* of the resulting figures—of the key geometric forms spelled out again and again, through the world's most illustrious, most ancient "sacred texts"—in a *rotating* physical reference frame (in this case, a planet), the previously abstract mathematical modeling of *higher-dimensional geometries* specifically predicts "vorticular rotation" of resulting inscribed "hypertetrahedral forms" **At 19.5 degrees North and/or South!**

**The precise latitude of *all* the major visible (or underlying) "vorticular forms" occurring in the solar system. It is these "internal vorticular forces" that in turn create the visible "spots" on Jupiter, on Neptune and on the Sun, and indirectly the massive volcanoes at the same latitudes on the solid "moons" and planets. . . .**

(Incidentally, one specific astronomical prediction of these "Cydonia Equations" pertains to the existence of a "Great Dark Spot" on Uranus—in the southern hemisphere, at 19.5 S. When *Voyager II* flew by in 1986, this hemisphere was deep in night; after 1991, Uranus' orbit of the Sun will move it into position for the *repaired* Hubble Space Telescope to see this *predicted* atmospheric feature, from Earth orbit. . . . So, stay tuned.)

These "higher-dimensional mathematical topologies" also predicted *another* specific set of atmospheric features, it turned out, around the *axis* of the rotating, planetary "hypersphere"—

**A precise, *hexagonal* pattern—produced by the 3-space projection, in a "working fluid," of two interlocking, rotating, 4-space "hyper-tetrahedra."**

Exactly like the baffling *hexagonal cloud* pattern the unmanned *Voyager* missions imaged in the 1980s around *Saturn's* north rotational pole[28]—the most "idealized," rotating fluid planet in the solar system! (See Figure 35.)

But there was one thing more . . .

ALL of these giant, rapidly rotating, fluid outer planets—Jupiter, Saturn, Uranus and Neptune—possess one other extremely significant attribute . . . now *totally* consistent with a "hypedimensional connection":

All radiate into space—inexplicably but significantly—*more energy than they receive from the Sun*; in the case of Neptune, almost *three times more!*[29]

Is this remarkable energy excess another "planetary signature" of a direct connection with a "higher" (more energy/information rich) spatial dimension?

*　　　　　*　　　　　*

Was it possible—we asked ourselves almost incredulously, when we realized that these planetary parameters might be actual *"hyperdimensional signatures"*—that the "Message of Cydonia" (and its newly-discovered, ancient terrestrial counterparts in the texts) could be intended to communicate—through *recording and preserving* for "whomever would come after"—the geometries of a demonstrable *physics* of such a "higher-dimensional connection" with another n-space?

And was that "higher n-space connection" then somehow *physically* responsible for the demonstrable upwelling of these "excess" energies from planetary (and possibly the solar) interiors, emerging specifically—as predicted by these topological models—as *large-scale vortices* **at 19.5 North and South**—or as "in-welling" forces from a higher n-space, as explicate *hexagonal features* around the rotating poles . . . on objects all across the solar system . . . and beyond?

The short answer seems to be a qualified "yes—with all its *stunning* implications!"

*　　　　　*　　　　　*

In addition to Tenen's astonishing discovery, and subsequent successful decoding of "tetrahedral geometry" amid a variety of ancient terrestrial sacred texts—whose very archaic letterforms (in which the texts were originally written) now seem geometrically to resolve to topological projections ("shadowgrams") of the higher mathematics of *a tetrahedron* we've discussed above. Other researchers have discovered *identical* geometry connected now with ancient "sacred" architecture and geodetic measurements . . . apparently *deliberately encoded* in a wide variety of ancient archaeological sites; these discoveries include work now decades old, made by completely independent workers, such as L. C. Stecchini and H. Harleston, Jr. (radically extended, recently, in new efforts carried out under the auspices of our research by *Mars Mission* associates David Myers, Carl Munck and Erol Torun.)

Stecchini, formerly at the University of Rome, and a well-known, world-class historian of science, became a specialist in ancient measurements and geodetic standards. In 1971, in an Appendix to Peter Tompkins' highly-successful compilation of material on the Great Pyramid of Gizeh, Egypt, Stecchini wrote:

> In order to describe [Africa] down to the equator, the [Ancient] Egyptians used a system of right triangles, in which one side was one of the three axes of Egypt and the other a perpendicular to it; the hypotenuse usually indicated the course of a segment of the east coast of Africa. The most important of these triangles was one obtained counting from Behdet *19 degrees 30 minutes [19.5] south along the central axis of Egypt and then 19 degrees 30 minutes [19.5] to the east* . . . [emphasis added].[30]

Harleston, a civil engineer-turned-archaeologist, working in Mexico beginning in the *1940s,* in 1974 gave a paper at the Forty-First International Congress of Americanists on a "sacred" archaeological complex, Teotihuacan—found almost a quarter of the way around the world from Ancient Egypt, northeast of Mexico City.

Harleston discovered, to his immense surprise, not only "e" and "pi" embedded in the fundamental architecture of this Ancient Mexican site (the two "tetrahedral" constants encoded so specifically and redundantly across Cydonia), he also discovered redundant examples of the specific circumscribed *tetrahedral angle*—19.5 degrees—indelibly imprinted in the *fourth-level-angle* of stones making up one of the two major pyramids within the Teotihuacan Complex: the famed "Pyramid of the Moon." This identical angle was then specifically *repeated* in several, carefully "pecked" geodetic markers found throughout the Complex.

And, if this were not astonishing enough, in the *other* stepped pyramid—the Pyramid of the Sun—Harleston found (in its precise *fourth-level* slope angle) 19.69 degrees—the *exact geodetic latitude* of Teotihuacan itself!

Harleston's own perspective on Teotihuacan's "self-embedded geometric 'messages'" are eerie echoes of our own, only recent, understandings of Cydonia . . .

> . . . the messages of Teotihuacan point to a new way of looking at time and space, and to *some new source of energy* from the cosmos, some new field fabric that our science has not yet isolated . . . The fundamental message conveyed by the Teotihuacanos is that *the physical universe is tetrahedral* from the microscopic level of the atom all the way up to the macroscopic level of the galaxies on a scale of vibrations in which man stands about the center . . . [emphasis added][31]

This, from a man exploring an ancient complex here on Earth, *decades* before *Viking* would leave for Mars.

Obviously, whoever laid out Teotihuacan not only knew and appreciated "inscribed tetrahedral geometry" *identical* to that discovered at Cydonia, they also knew (somehow!) and memorialized, *their specific location* on "a spherical Earth's" surface. How was that possible—in a "primitive, stone-age culture" (to whom archaeologists currently attribute the building of Teotihuacan, circa 1400 BC)—without (known) global surveying instruments or higher mathematics?

This very basic question must not only now be *re*-addressed by mainstream archaeologists, it must be *answered* . . . particularly when *identical* "geometric memorialization and mathematical constants" (if not the "Message of Cydonia") have now been discovered (and by completely *different* folks, who didn't even know Harleston existed, let alone the details of his work)—

In a Complex *on a completely separate planet!*

Which brings us to other, more recent, though equally astonishing archaeological examples . . .

Carl Munck (before he learned of, and then promptly joined, the *Mars Mission* Cydonia Investigation, in 1991), independently, over a decade, discovered literally dozens of additional examples of "Cydonia geometry" on Earth. Munck zeroed in on a repeating geodetic pattern, linking Teotihuacan to a wide variety of *other,* vastly different, archaeological sites and complexes, at diverse locations on the planet—stretching from England, to North America, to Meso-America and South America. He initially termed this work "Rediscovering the Global Grid. . . ."[32] (Recently, with the aid of David Myers, Munck's "grid" has been successfully extended to the Near East. And, through the work of another *Mars Mission* associate, Kent Watson, there are strong indications of a continuation in Japan. . . .)

*The Monuments of Mars*

Back in Britain, Munck determined that the *internal* layout of Stonehenge, England's premier "megalithic monument," was established according to precise, redundant, overwhelmingly "tetrahedral" geometric units (!); the very angle off True North of its famed northeast Avenue (as opposed to the current azimuth of the rising Solstice sun—see Chapter IV) is, astonishingly, *another* key "Cydonia angle" (see Figure 36)—

**49.6 degrees.**

**Identical not only to a key theoretical "tetrahedral" angular relationship [to within *0.2 arc seconds,* as measured by the Nineteenth-Century scientific giant, Sir Norman Lockyer[33]]— but also identical to another specific angle, expressed *twice* in the internal geometry of the D&M Pyramid itself!**

49.6 degrees is, of course, equivalent to e/pi *radians.* And "e/pi" is redundantly encoded across Cydonia innumerable times, in several ways—including, as we have reiterated above, in the very *siting latitude* of the D&M Pyramid upon the Martian surface. Its existence coded in perhaps the most famous ancient terrestrial monument of all (apart from the Great Pyramid), opens up profound questions as to "who," and "when". . . and "why."

And speaking of the Great Pyramid . . .

Myers himself has found "tetrahedral geometry" memorialized in the very "slope angle" of this remarkable structure: if you divide the long-estimated angle of the exterior (~51.9 degrees—see Chapter XV) by 60 degrees (an obvious "tetrahedral angle"), the result is again the "tetrahedral ratio": e/pi!

Now, remember that the *tangent* of this ratio—e/pi—equals 40.87 degrees—the D&M Pyramid's latitude *on Mars?;* OK, couple this to the fact that *another* of the "trig functions" associated with this ratio (e/pi = 0.865) equals the *sine* of slightly over 30 degrees, and you begin to see the reason for our developing excitement over Gizeh—

For, this second "e/pi latitude"—30 degrees—is none other than—

**The *latitude* of the Great Pyramid on Earth!**

But there is more.

Another key angle at Cydonia is the tilt of the entire Face off True North—22.5 degrees (a refinement now of our preliminary value, expressed in Chapter IV). Another critical internal angle—represented at least twice within the D&M, and several additional times within the larger Complex—is, of course, 19.5 degrees.

This "inscribed tetrahedral angle," 19.5 degrees, divided by the Face off-set angle, 22.5 degrees, is equal once again to 0.865, or e/pi! (See Figure 37.)

Myers then discovered that if you draw a "great circle" on the Earth, so that it passes through the Great Pyramid and intersects the Earth's equator at *19.5 degrees* East of Gizeh, it will be tilted, relative to the Equator, by precisely *60 degrees*—the most obvious "tetrahedral angle" of all! Then, another "great circle"—connecting Stonehenge and Gizeh—will intercept the Equator precisely at *22.5 degrees* East of Gizeh.

And 22.5, divided by 19.5, equals . . . e/pi.

Redundancy.

Munck had previously discovered totally different "interconnections" between the geodetic siting latitudes and longitudes of Stonehenge, the Great Pyramid, their internal measurements, etc., which made it clear to him (even before he discovered this research), that there *had to be* "some kind of ancient global, geodetic knowledge" encoded in this series of ancient terrestrial architectural constructions—knowledge strongly implying some kind of highly sophisticated, "unified, ancient global *culture.*"

When he and Myers began detailed geodetic collaborations in 1991, they discovered (to Carl Munck's profound surprise) not only that his previously separate "ancient global grid" is overwhelmingly "tetrahedral," but that it echoes *again and again* the specific angles and geometry we have now discovered on a completely *separate* planet . . . amid the "Monuments of Mars."

A couple of remarkable—and extremely recent—new pieces, in this rapidly developing *multi-planet* puzzle . . .

In late 1991, John Anthony West, a self-proclaimed "rogue Egyptologist" (Chapter XV), with the aid of colleagues in the geological community—particularly, Dr. Robert M. Schoch, a tenured professor in the College of Basic Studies at Boston University; and Dr. Thomas L. Dobecki, a seismologist with McBride-Ratcliff and Associates, a geophysical oil-drilling firm in Houston, Texas—presented a remarkable scientific paper to the annual Geological Society of America's 1991 convention, held this year in San Diego, CA. The subject of their paper:

**"Redating the Great Sphinx of Gizeh, Egypt."**

Traditionally, Egyptologists have long held that the Sphinx—this massive 240-foot long "half man/half lion," crouched a few hundred yards from the Great Pyramid itself—was sculpted out of the limestone of the Gizeh Plateau approximately 4500 years ago, during Ancient Egypt's so-called Fourth Dynasty of the Old Kingdom. It was created, so this scenario goes, in parallel with the construction of the *second* major pyramid on the Plateau, and like it, on the orders of the Old Kingdom Pharaoh, Chephren—in his likeness . . .

Critics of this mainstream Egyptological reconstruction of events on the Gizeh Plateau point to many inconsistencies about the story—starting with the obvious fact that the Sphinx does *not* resemble Chephren, and ending with fundamental questions of its true geological age.

Among the best-known of the critics of this "standard scenario," has been John West. On researching a book in the mid-seventiess about the philosophies and cosmologies of Ancient Egypt (34), West came upon a comment from another scholar, R.A. Schwaller de Lubicz: "the Great Sphinx shows evidence of *severe water erosion* . . ." To which West internally responded with the obvious question:

"In a *desert?!*"

West's curiosity, prompted by de Lubicz' casual reference to the "anomalous erosion" of the Sphinx, was based on the following hard facts:

To create "severe water erosion" on the Sphinx, the Sahara Desert must at one time NOT have been a desert. Readily available climatological data for Ancient Egypt make it abundantly clear that the Sahara has been in place from *seven to ten thousand years* . . . since the end of the last Ice Age. Meaning, that both the carving of the Sphinx *and* its "severe water erosion" had to have taken place sometime *before* . . . ! (See Figure 38.)

Such an age, if established, would, of course, automatically preclude an Egyptian Pharaoh by the name of Chephren from ordering the carving of the Sphinx only 4500 years ago—*if* the Sphinx's current state of weathering could a) be traced unambiguously, *geologically* to a period of massive *rainfall* on the Gizeh Plateau (as opposed to wind or sand erosion), and b) that period could be independently dated, by *geological* (as opposed to "Egyptological") techniques, to *before* Chephren's reign during Egypt's so-called "Old Kingdom."

In 1991, *both* of these results came together for West's privately-funded Sphinx Project Team—resulting in the Team's highly significant, radical conclusions:

Based on this chain of reasoning . . . we can estimate that the initial carving of the Great Sphinx (i.e. the carving of the main portion of the body and the front) may have been carried out ca. 7,000 to 5,000 B.C. This tentative estimate is probably a *minimum date;* given that weathering rates may proceed non-linearly . . . the possibility remains open that *the initial carving of the Great Sphinx may be even older than 9,000 years ago* . . . [emphasis added].[35]

Privately, West's geologists suspect an *even greater age* for this remarkable Egyptian effigy. . . .

This is required to produce the "advanced state of water weathering" they detect (in part, via seismological techniques)—not merely on the Sphinx—but in the walls, and under the "floor," of the carved "ditch" which separates the Sphinx from the limestone of the Gizeh Plateau proper. This pronounced state of deep erosion is also readily visible in the "Sphinx Temple"—the massive construction, sited a few hundred yards from the Sphinx, composed of 100-ton limestone blocks. It has long been presumed that these were excavated from the "ditch" at the time of the carving of the Sphinx itself.

ALL these features, according to Schoch and Dobecki, show evidence of such *severe* water weathering that eroded fissures *12 feet in depth* are visible inside the ditch; similar man-made excavations (to the ditch), in similarly hard limestone on other parts of the Plateau (for Old Kingdom tombs, dated by other methods to 5000 years), show literally *no erosion* . . .

The remarkable conclusion?

According to John West:

> If the Sphinx predates dynastic Egypt, this would have major archaeological implications. Quite simply, *we would have to rewrite the history of when advanced civilization began* . . . [emphasis added].[36]

Apropos this radical viewpoint is a comment by well-known mainstream historian, Will Durant, in his *Story of Civilization:*

> Immense volumes have been written to expound our knowledge, and conceal our ignorance, of primitive man . . . Primitive cultures were *not necessarily the ancestors of our own;* for all we know they may be the degenerate remnants of higher cultures that decayed when human leadership moved in the wake of the ice . . . [emphasis added].

Indeed.

And then . . .

As we were assembling the new figures for this Update, one of Carlotto's remarkable 3-D reconstructions was being routinely transferred from magnetic tape to a black and white transparency. It was while examining this image that we were able to *confirm* what—for over a year previous to this—has been only a developing suspicion. . . .

The "Face on Mars" is NOT merely the image of a "terrestrial hominid" ca. 500,000-300,000 BP, lying where it has no business being.

*One half* (the right—see Figure 39) is ALSO the perfect image of a cat; specifically, *a lion*—the "King of beasts". . . .

*The Monuments of Mars*

So—the "Martian Sphinx" (according to its now so apt, if not prophetic, designation by Vladimir Avinsky!—see Chapter XI)—is in truth also a *combination* of two "families". . . . a left (sunset) "hominid" half, and a right (sunrise) "feline" half—

**The *identical* symbolic "message" of its terrestrial counterpart . . . located at *another* equivalent "e/pi latitude"—on Earth . . . at Gizeh!**

The profound implications stemming from this newly-recognized symbology—the fusion of these two specific *terrestrial* forms . . . communicated so powerfully through *one* combinatorial artistic "monument" now appearing as two "Sphinxes," on *two* worlds—is truly beyond the scope of the space available here. But one thing must now be truly obvious: no more elegant expression of a *deep* "terrestrial connection" for this entire mystery could possibly be found . . .

The question is, of course:

What does it *mean!?*

A postscript to this amazing sequence of discoveries. . . .

Measuring a diagram of the layout of the Sphinx in relation to the major Pyramids at Gizeh (see Figure 40), Erol Torun discovered that the Great Sphinx itself *is also specifically sited*—north/south; east/west—*by none other than* . . . our familiar "Cydonia" ratio:

**e/pi!**

As alluded to above, these global "Cydonian" archaeological discoveries—*identical* geometric relationships, and now, complete duplicates of whole *symbolic images,* deliberately encoded in ancient ruins on *two* worlds—must elevate our entire previous discussion (Chapter XV) re "a possible terrestrial connection" for Cydonia, to a level of unprecedented implications . . .

Presentation of the stunning data and detailed discussion supporting these remarkable contentions *must,* however, await future issues of *Martian Horizons*, and the eventual publication (next year) of *The Cydonia Papers.*

*　　　　*　　　　*

But as stunning, and as overwhelmingly obvious, as this terrestrial archaeological "connection" to Cydonia has now become, the term does not adequately *begin* to describe what we would *next* discover. . . .

In the Summer of 1990, the now annual and literally worldwide mystery of the "crop circles" once again reappeared—as it has every Spring and Summer for fifteen years . . . since, in fact, *Viking* was launched to Mars.

This curious phenomenon—a series of baffling "swirled vortices" appearing suddenly, overnight, in wheat, rye, corn, and other cereal crops, most notably amid the gently rolling fields and ancient megalithic monuments of southern Britain—has for almost two decades mystified scientists, the media, and "ordinary" folks alike. Theories to explain it have ranged from the obvious (simple hoaxes) to the attempted scientific ("heretofore unobserved, swirling *electrified winds*") to the absurd ("hundreds of crazed hedgehogs, dancing in circles . . .").

Then, in 1990, the "circles"—as the phenomenon has become generally known—took on a radical new twist: not only did they rapidly expand *in a single summer* to over *a thousand* examples reported from all around the globe, they suddenly "evolved" internally, in England, into highly complex *geometric* forms—

Entire, highly intricate, "glyph-like" figures imprinted in the fields— not merely simple "swirls of flattened, horizontally-growing crop," but entire complex "symbols," made up of "circles and bars," "circles and rings," "circles with claws," and "circles with spurs"—stretched horizontally along a central, flattened axis . . . sometimes for up to *a quarter of a mile!*

Working from initial suspicions that these exotic figures *might*—somehow—be involved in our own mystery, occasioned by the appearance of the majority of these strange forms *in a huge triangle centered around Stonehenge*,[37] we began our own investigation of "crop circles" only in mid-1990; we reported our first successful analysis, of one of the most striking 1990 figures, in the first issue of *Martian Horizons*, that same year.[38] Again, for lack of space, the supporting documentation to this first inquiry cannot be included here, but the "bottom line" was simply this:

**At least one "crop circle"—unequivocally, if inexplicably— shared *an identical geometry* with the "Monuments of Mars"!**

Since then we've had an opportunity to subject a wider range of this remarkable phenomenon to this level of detailed, "Cydonia analysis"— including an overwhelmingly obvious example of *explicit* "Cydonia Geometry amid the crops," that dramatically appeared in 1991 (the exquisite, artistic figure represented on the back cover of this book and, in more detail, inside—see Figure 41).[39]

And the "message" is the same:

"Crop circles," somehow unquestionably "know," and are attempting to "communicate," *the same geometry* as that being communicated by the "Monuments of Mars." That much is simple now to demonstrate. The *implications* of this discovery are, however, anything but simple. . . .

The spectacular "glyph" which really "locked in" this interpretation for us (and, apparently, for a *lot* of other folks—see below) debuted July 17, 1991, in a field of ripening wheat, at a place in Central England termed Barbury Castle—after a nearby medieval ruin. According to accurately-surveyed maps (drawn by long-time British circle investigator, John Langrish, with additional on-site details subsequently furnished by European Director of *Mars Mission* Operations, David Percy), the figure measured over 300 feet to the outside of the three "satellite glyphs": two identical "rings," and an astonishing "stepped spiral," spaced at 120 degree intervals around the central, multiple-ringed structure. The inner converging "tetrahedral" lines, spaced also at 120 degree intervals around the center, measured a mean of 105 feet out to each of the three "satellites."

We've included here a detailed artistic sketch, with various aspects of the Barbury "glyph" appropriately marked, to facilitate our subsequent discussion (see Figure 42a and b).

A few comments.

First and foremost, as we have alluded to above, was this figure's blatantly obvious attempt—in two-dimensions—to represent none other than a *three-dimensional tetrahedron!*

This was so obvious to everyone, in fact, that after the Barbury Castle glyph's dramatic appearance, I received several calls from members of the press and other groups—previously highly skeptical that "circles" could *at all* be connected, *in any way,* with our previous discussions of Cydonia—exclaiming "Have you seen the new *tetrahedral* 'circle'!"

But, first impressions notwithstanding, when we elicited the appropriately surveyed maps, detailed *measurements* of this amazing glyph not only stunningly confirmed a redundant "tetrahedral message," they revealed mathematical and geometric subtleties—

Known at that time *only* to ourselves!

(And "no" we did NOT "hoax" the Barbury Castle "circle."

Not only didn't *we* do such a thing, neither did the two British "gentlemen" in the South London pub—who created quite a media stir in the late summer of 1991 when they claimed to have effectively hoaxed, over a fifteen year period, *most* of the world's "crop circles." I seriously doubt, after seeing BBC and "Good Morning America" interviews with both of them, on their methods and their motives—David Chorley and Douglas Bower, both self-proclaimed "landscape painters" in their sixties—that either could even spell "tetrahedron"—let alone create one *a football field* in length . . . or, *over a thousand others,* of similar mathematical subtlety and internal geometric complexity . . . literally, around *the world!*)

The most intriguing "tetrahedral" aspect of the Barbury Castle structure concerned both the "bull's-eye" of concentric rings in the center of the figure, as well as the two "bent lines" connecting two of the exterior "satellites." When David Myers and I put our heads together, during several long-distance telephone attempts to figure out this obviously "tetrahedral message," the conversation—which soon resulted in stunning affirmation of our own major "Cydonia discoveries"—went something like this:

"David," I prompted, "try a polar projection of *a planet*. See if the 'rings' and inner 'circle' on the top of the figure translate into some kind of key *latitudes* . . ." (See Figure 43.)

He agreed (he had the actual photos, "Federal Expressed" from England), and we hung up.

Minutes later the phone rang again.

"Listen to this," David said excitedly, "outside edge of the first (and largest) ring: Equator—equals 'zero.' Next, *inner* edge of that same ring: equals—

**"22.5 degrees—the *tilt angle* of the Face!"**

The room (if not my ear) literally rang with his triumphant exclamation.

"Next inner ring—outer edge: 45 degrees; inner edge—52 degrees.

"Outer edge of the inner circle itself: 69.4 degrees—the "back angle" of the D&M, and the central angle of "the City Square"—

"The whole damn thing is 'tetrahedral' and 'Cydonian'—through and through!"

David was correct. Not only was the "tilt angle" of the Face, 22.5 degrees, blatantly expressed by the "latitude projection" of the inner edge of Barbury Castle's first concentric "ring" (Figure 43b)—various *other,* both obvious and more subtle, mathematical combinations of "the Message" were reflected redundantly throughout this strikingly elegant geometry:

~52 degrees = the latitude (to two places) of the "tetrahedral circle"— exactly the same self-referential "message" as the D&M

45 degrees/52 degrees = 0.865 = **e/pi!**

60 degrees = the "tetrahedral" angles at each vertex of the "two-D tetrahedron"

~60 degrees/69.4 degrees = 0.865 = **e/pi!**

*The Monuments of Mars*

69.4 degrees = e/sq rt 5 radians, the most redundant mathematical relationship discovered at Cydonia—and the most "biologically" significant

Later, David Percy would measure *another* azimuth—connecting
the center of the "pinwheel satellite" (one of three, spaced equilaterally
around the glyph) with the linear "radius extension" of the "circle satellites." This line, it turned out, crossed the first "offset ring" at a latitude
equivalent to . . . 19.5 degrees (see figure). And 22.5, divided by 19.5,
equals, of course . . .

### ~e/pi!

And, when the angles of the "bent leg" of the "inner two-D tetrahedron" were measured by Percy, they turned out to be "aimed" directly at the centers of two of the three respective "satellite figures"—at
*another* highly specific "Cydonia" angle, the side angles of the D&M—
85.3 degrees . . .

Again, not bad . . . for a phenomenon skeptics emphatically insisted
*had* to be produced by an "electrified tornado," or "those two blokes
from a London pub . . . with a string and board, a funny cap, and a couple
of pints . . . under a full Moon."

But, David now reminded me, we weren't yet finished: the figure
wasn't "perfect." There was that very "bent," obviously deliberately distorted side of the "almost equilateral" inner triangle (see again, Figure 42).
With so much mathematical and geometric elegance represented elsewhere in the figure, and so redundantly as well, why this major "imper-
fection". . . ?

I looked once again at my own faxed copies of the photos. The
"bend" was necessary, I could plainly see, otherwise that side of the
"equilateral" triangle would have to pass directly *through* the second
ring of flattened crop; the bull's-eye ring" and circle seemed, in turn, to be
deliberately *offset* from the center of the figure . . . and *in the direction of
the bend.*

David," I asked slowly, "what latitude do you get if you straighten
that bent leg of the triangle, so that it goes *through* the inner ring?

I already strongly suspected what he'd find. There was a staticky
long-distance pause, while Myers drew the line—

"49.6 degrees," he finally, wonderingly replied.

"Bingo!" I cried, "Damn, these guys are elegant!"

Because . . . 49.6 degrees is not only the *specific* angle of the symmetrical *front* buttresses of the D&M (see above), it is also e/pi *radians*
. . . and . . . that critical "Avenue angle" leading out of *Stonehenge* . . . a few

miles distant from the Barbury Castle glyph itself.

Understand: the apparent success, if not the ease, of our "decoding" of these contemporary enigmas, is profoundly disconcerting—even to those of us who have lived and worked for many years with this potential: the discovery, at long last, of a bona-fide ET intelligence. If we truly have resolved the "crop circle mystery"—and discovered that, it is merely a "subset" of the larger Cydonia Puzzle—this is NOT a trivial new discovery; it comes with profound and *major* implications (see below).

For those not quite ready to confront—head on—these decidedly non-trivial implications, I would simply leave as a reminder to the reader (particularly, the intensely *skeptical* reader), that *none* of this aspect of our work is "necessary" to prove our central thesis: the unmistakable "intelligence" now apparently behind Cydonia itself.

Nor does it negate the *fact* that—NASA, the Congress, and (most important) the American *people* willing—in 1993, we all will KNOW what truly is lying on "the Sands of Mars."

Need I say more?

$$* \qquad * \qquad *$$

There is, in fact, a great deal more one could and should say—accepting the argument to this point—about the Barbury Castle figure (if not all the other "crop glyphs"): our decoding of the apparent "astrophysical," if not "biological" *meaning* behind these repeated, exquisite "Cydonian" geometric statements; but, as we have discussed this extensively before—both previously here (see above) and elsewhere[8]—I will simply reflect for a few moments on the larger *context* of these stunning new discoveries. . . .

$$* \qquad * \qquad *$$

"Someone," it should now be overwhelmingly apparent to any truly objective observer (as it has become to those of us working intimately with this material), is trying *very* hard to tell us something apparently *crucial* to our own past . . . if not to our future.

It is one thing, after eight long years, to "decode" successfully an "alien geometric message" on another planet, as we have apparently now done—at Cydonia, on Mars. It is quite another to have that *identical* "message" reappear—inexplicably and, now, repeatedly—in a bizarre fashion on *your own planet* . . .

First in your earliest archaeological remains of long-vanished "dead" terrestrial civilizations (including some, which until a few days ago, you never knew existed!—36) . . . and then, suddenly—contemporaneously—*in our very midst* . . . a message potentially involving the fundamental processes of life itself, appearing dramatically, across hundreds of square miles, amid your most productive fields and crops . . . !

Our discovery and verification now of the Cydonia Geometry "amid the crops" is affirming, then, of an extraordinary, very *current* Truth: "someone". . . who apparently knows the same geometry as those who, over *half a million years ago,* left their indelible geometric imprint on Cydonia . . . has *finally* come to Earth . . .

This can only mean one of *three* equally extraordinary things:

Either, whoever built Cydonia is *back* . . .

Or . . . *different* folks have now arrived, to remind us *once again* of our own heritage, which somehow encompasses not only *this* one planet, Earth—but also "Cydonia" and "Mars.". . .

Or—*everyone* "out there" knows what is depicted at Cydonia . . . and *someone* is attempting now to tell us *one more time,* through the "Message in the Crops". . . because only *we* . . . across those yawning half a million years . . . somehow . . . totally forgot.

And now . . . **it's Time that we Remember.**

*        *        *

Implicit in this "urgent" interpretation of the "Message" (particularly, its dramatic *reappearance* in the "crop glyphs"), we suddenly realized, might be the promise of a "Cydonia technology"—which, if feasible, would have a desperately-needed positive environmental impact. Could it be, we asked in growing wonder—as we gazed at what seemed more and more like highly intricate, *engineering* crop glyphs—that "someone," in our "hour of global environmental crisis," was trying very hard to "subtly instruct us" in how to build a set of *mechanical systems*—designed to harness the *benign* energies and forces implicit in the "ultimate" development of a "new physics!"

In other words:

> **If the Universe is actually operating according to this "hyperdimensional process"—and thus tapping a hitherto unknown, prodigious source of *non-polluting* energy, demonstrable in stars *and* planets—might it not just be possible to engineer a device (or series of devices) to achieve the same effect *on Earth?***

Shortly after realizing this astonishing central implication behind *all* the "geometric messages" we have now decoded—both on Mars and Earth—we methodically began uncovering an entire series of *precisely* such devices—and their inventors—which, for over a hundred years, have apparently been built and *operated* in laboratories all around the Earth . . . and resoundingly ignored!

These devices seem to function (as near as we can tell, from our *very* preliminary historical research and comparison calculations) remarkably close to our projections of the "technological possibilities" inherent in a new "Cydonia physics"—and with spectacular results.

The problem with such devices, of course, has been that they seem to violate the *current* laws of physics and so have been consistently dismissed—both by a narrowly-focused scientific community of physicists ("There can be NO SUCH THING as 'perpetual motion'!"), as well as a concerted group of "business interests" (and allied political representatives) with obviously very different priorities and concerns than the discovery and application of virtually unlimited new sources of non-polluting energy, which one day may be too *inexpensive* to even charge for. . . .

*       *       *

Anyone remember the sad saga of "Nicola Tesla"—the brilliant, "genius engineer" from Yugoslavia, who seventy years ago gave us the electrical world we now take so for granted (over Thomas Edison's vigorous objections)—the world of *alternating* current? Isn't the heart of Tesla's electrical system (which we still use in the United States in all commercial power applications) a current of precisely *60 hertz* (60 cycles per second)? And, in some electrical generator applications, isn't it produced by a "three-phase system . . . of *120 degrees?*" Against a backdrop of "tetrahedral hyperdimensional geometry". . . based on *60-degree* and *120-degree* angles . . . think about that. . . .

Tesla (who did far more than merely invent the quaint "Tesla Coil" we see shooting sparks in high school physics labs), not only singlehandedly produced much of what was to become the foundation of our current commercial electrical power infrastructure during the early Teens and Twenties of this Century, he also kept creating more and more arcane electrical devices whose immediate purposes were not so readily apparent . . . and mumbling something about "*unlimited* electrical energy"—until his financial benefactor, J. P. Morgan, got wind of Tesla's *real* objectives—

**To provide humanity ultimately with virtually unlimited energy "too inexpensive to charge for"—from the planet itself!**

It is common knowledge among historians of science and technology that when Morgan learned of Tesla's long-range intentions—and of his experimental "rigs" in Colorado which seemed to indicate that he was about to pull it off!—he immediately "pulled the plug" on all further financial support of Tesla's research.

It is our growing belief—backed by the "Cydonia Equations" and their now inescapable "hyperdimensional" implications—that Nicola Tesla was merely the *best-known* of an entire cadre of physicists, engineers and inventors who, over the past hundred or so years, have repeatedly stumbled on "something extraordinary," and built various devices which function according to its very *different* rules. . . .

Specifically: a phenomenon which *seems* to violate the current laws of physics, because it actually operates according to the currently unknown (and, with the exception of certain esoteric discussions among a handful of physicists—regarding a possible, theoretical "zero-point energy of the vacuum"—still totally unsuspected) laws of a radical new "hyperdimensional physics."

This premise is, of course, as eminently *testable* as it is seemingly incredible.

Which is exactly what we proposed, at NASA-Lewis—at our *second* specifically-invited Briefing.

At the official NASA AESP Conference we were invited by Dr. Lynn Bondurant to address on September 11, 1990 at NASA-Lewis, after we laid out our "cracking of the code"—the potential "hyper-dimensional physics" behind the "Cydonia Equations"—we featured one current technological device which we now believe *could* operate according to this physics:

The "N-machine."

This device is an offshoot of a 150-year-old mystery in physics initially discovered by Michael Faraday in England in 1831, that has remained unexplained for well over a century and a half. The mystery is this: that a *rotating magnetic field*, relative to *a conductor rotating at the same speed* (which means the field is *stationary* relative to the rotating conducting elements), can *also* create an electric current!

About twenty years ago, another physicist, Dr. Bruce DePalma, formerly of MIT, began a series of basic mechanical experiments with "rotating frames"—culminating in a device he termed the "N-Machine,"[40] which above a critical RPM (revolutions per minute), *appears to generate more electrical current out of a spinning "Faraday disc" than the input energy required to maintain its mechanical rotation!*

It is our growing belief that Dr. DePalma's "N-machine"—if it *is* successfully producing more energy than required for its input—*must* be operating according to the Cydonia "hyperdimensional physics" we have been so graciously "bequeathed"... by someone.

And we challenged NASA-Lewis, in our September 11, 1990 address—as NASA's official "energy and power research center"—to bring this device, and its inventor, into NASA-Lewis for a systematic, scientific, and very public test of this anomalous technology . . . and its stunning implications for current physics, if not for a radical new means to generate environmentally benign electricity for an entire Earth.

We have since learned that NASA-Lewis has, indeed, established a small office dedicated to "research into anomalous technologies," staffed currently by one physicist—Dr. Ira Myers. This came about, apparently, because again, we discussed our NASA-Lewis challenge on several subsequent media appearances, after our address to Lewis in the Fall of 1990. Some people who heard our discussion, and our description of DePalma's possible technology, immediately wrote their Senators and Congressman, asking them to look into the possibilities for this radical alternative to oil.

These Congressmen promptly replied to their constituents, and also fired off a list of questions to the Space Agency. These official Congressional queries then went, of course, directly to NASA Headquarters, which, in turn, passed them on to the new Director of the NASA-Lewis Research Center (remember, NASA's official energy and power research facility, etc.), who, in turn, passed them on to the appropriate division *within* the Lewis Lab itself. A special office was apparently soon set up to handle these (still growing!) numbers of official requests for information on 'free energy' technologies—so a credible answer eventually could be returned to the appropriate Congressional office.[41]

The "system," in fact, worked!

Stay tuned for further political, if not technological developments. . . .

Since our research into DePalma, we have discovered a host of *other* successful, completely independent inventors, engineers, and basic researchers—who have apparently also stumbled on devices and approaches which, we strongly now suspect, operate according to the same "hyperdimensional physics."

Among these are Troy Reed, inventor of the "Reed Magnetic Motor," which produces "anomalous torque" from a set of spinning magnets, which can then be harnessed to turn a conventional electrical generator; Dr. Paul LaViolette, originator of a new theoretical approach to quantum mechanics, termed "Subquantum Kinetics" (which is remarkably consistent with—but completely independent of—our own early astrophysical analysis of the "Cydonia Equations"); Dr. W. Lambertson, inventor of a solid-state, direct conversion process for producing electrical current from the "vacuum," termed the "WIN System"; Arthur Thiel, architect of another solid-state energy technology, based on a mathematically-wound wire coil—a "resonator"—"tuned" to 60-cycle alternating current; and many, many others . . . (see *The Cydonia Papers* and *Martian Horizons* for additional details).

Reed and Thiel (as opposed to LaViolette and others—who possess interesting *theoretical* models of possible "free energy" technologies), seem to have actual, working systems. Both, with no prior knowledge of our own Cydonia research, upon viewing the NASA video ("Hoagland's Mars"), *independently* proclaimed that their technology works—

**According to the same "Cydonia" geometry and mathematical constants!**

We are currently in the process of researching both these extraordinary claims.

And in a separate, striking, *political* development, this entire category of so-called free-energy devices (for so many years —since Tesla—literally "beyond the pale," in terms of mainstream engineering, if not fundamental physics) has now suddenly emerged to new "respectable" examination and discussion.

In the Fall of 1991 (August 4-9), the 26th Intersociety Energy Conversion Engineering Conference (IECEC '91) was held in Boston, Massachusetts. This is an annual event, sponsored by the half-dozen or so leading engineering societies in the United States with cooperating sponsorship from similar professional groups from as far away as Japan. The list of these sponsoring societies reads like a Who's Who of mainstream engineering: the American Nuclear Society; the Society of Automotive Engineers; the American Chemical Society; the American Institute for Aeronautics and Astronautics (AIAA); the American Society of Mechanical Engineers; the Institute of Electrical and Electronics Engineers (IEEE); the American Institute of Chemical Engineers (AlChE); and the Japan Society of Mechanical Engineers (JSME).

At their 1991 Boston meeting, however, in a special series of sessions titled "Innovative and Advanced Systems" (arranged for and chaired by Dr. Patrick Bailey, an MIT graduate nuclear physicist, with high DOD security clearances, and currently with the Power Systems, Space Systems Division of the Lockheed Missiles and Space Company, in Sunnyvale, California) for the first time the IECEC Conference *officially* invited papers from a series of "free-energy" researchers and inventors. . . including DePalma, Reed, Lambertson, and a score of other workers.[42]

Addressing a variety of theoretical "free energy" approaches, and presenting actual experimental engineering systems for "tapping the energy of hyperdimensional n-space" (as we now believe these systems function, based on our Cydonia research), the IECEC '91 Conference Proceedings, and the technical papers of individual researchers, will be made available to all members of *The Mars Mission* (because of the obvious "Cydonia connection") at significantly reduced rates. Before the Conference, Dr. Bailey himself called me—affirming that he had not only read *Monuments*, but that he had recently "seen and was impressed by the NASA-Lewis tape." It was in this conversation that he agreed to this cooperative arrangement on distribution of the Conference papers.

This, of course, is the stunning, major surprise of our successful "decoding" of the "Message of Cydonia":

That there is, indeed, an apparent *near-term engineering application* of the fundamental physics which awaits us . . . amid the "Monuments of Mars."

For it is now clear—even at this very preliminary stage in our investigation—that, if appropriately researched and then applied to many current global problems, the potential "radical technologies" that might be developed from the "Message of Cydonia" could significantly assist the world in a dramatic transition to a *real* "new world order". . . if not a literal New World. . . .

But, perhaps "some" would not be happy with that prospect . . .

*        *        *

Which prompts the crucial question:

Is the deep resistance we've been experiencing, to even the *possibility* of fundamental change (which would inevitably be spurred by verification of Cydonia's reality), what's *really* "driving" the seemingly contradictory events which still continue to swirl around the "off-again/on-again" NASA public responses to the "challenge of Cydonia?" Is this "internal bureaucratic indecision, in the face of significant opposition" behind the now sudden *rescheduling* (!) of NASA's increasingly controversial TV presentation: "Hoagland's Mars?"

For—

Dr. Robert Brown, the head of NASA's nationwide Educational Affairs Division (and Bondurant's boss, in Washington, you will remember), in the wake of the totally unexpected major public outcry over his last-minute censorship of the original Lewis "Cydonia program," abruptly (at least, publicly) reversed himself only a month later and began sending out *thousands* of identical copies of the following "official NASA response" to everyone—members of the public, and reporters—advising them that production of "Hoagland's Mars" would "now continue". . . .

One of these "Brown letters" was sent in response to an original inquiry[43] from Little Rock:

Dear Mr. & Mrs. Clinton:

Thank you for your recent inquiry on the status of the educational videotape being produced by NASA's Lewis Research center on the subject of possible life on Mars.

A NASA panel reviewed the preliminary version of the manuscript (sic) and video tape for the proposed presentation "Hoagland's Mars." Based on the panel's review, *NASA is continuing with the production of the video* that will include Mr. Hoagland and *his perspective,* as well as *other theoreticians* and scientists and their *opinions* about life on Mars [emphasis added].

To allow the producer time to fully develop the concepts and ideas about the complex and much-debated question of Martian life, the release data is presently targeted for the summer of 1991 . . .

Sincerely,
Dr. Robert Brown
Director, Educational Affairs Division

The Clintons, on March 3, 1991, sent back to NASA Headquarters their own[44] admirably direct response:

Dear Dr. Brown:

We have reviewed your form letter reply to our inquiry regarding the NASA "Hoagland's Mars" presentation, and have the following question and comments.

Why was it necessary for a NASA panel to make an unprecedented review of the manuscript and video tape of Mr. Hoagland's presentation to NASA scientists and educators?

Your description of the NASA version that it scheduled to be aired this summer has a fatal flaw. It labels as "opinion" the scientific findings of a dedicated team who has spent several years studying and measuring the NASA images of the Cydonia region.

As you should know, science was not founded upon "opinion," but upon testable or verifiable measurements and observations.

Mr. Hoagland and his team have said, "We have studied the NASA Cydonia images of the artifacts and believe those objects are: a) probably artificial in origin, and; b) have revealed not only a mathematical Rosetta stone, but a new physics based upon tetrahedral geometry as well."

All the measurements and claims made by Hoagland's team are testable, and the hypotheses presented are scientifically-based.

Not one NASA scientist [however] has bothered to refute or verify one element of these claims.

NASA has, instead, chosen to attack Mr. Hoagland and his team by producing a video presentation which from all appearances will be a one-sided attack arguing "opinions," not easily-verifiable facts . . .

If NASA's video presentation to be aired this summer is anything but a fair treatment and discussion of the facts presented by Mr. Hoagland and his team, we will actively campaign for *a Congressional investigation* [emphasis added].

While you and your colleagues may suppress the truth in this matter for a while, the facts will eventually be made public . . . what answer [to the charge of "cover-up" at that point] can NASA give that would be believable?

Dr. Brown, the ball is in your court.

> Sincerely,
> Lee F. Clinton
> Bonnie B. Clinton

> cc: Richard C. Hoagland
> Chairman George Brown

On October 1, 1991, Dr. Robert Brown abruptly resigned as Director of the Educational Affairs Division, NASA Headquarters.

To date, not only has he *not* responded to the above letter . . . his previous assurances that "the release date [for a reedited "Hoagland's Mars] is presently targeted for the summer of 1991" have not proved accurate; at this writing—late Fall, 1991—we are *still* waiting to see what NASA plans as its "official and revised" PBS release: *Hoagland's Mars.*

*      *      *

But, is this continuing back-and-forth delay, on a mere *television program,* based on merely "bureaucratic indecision" when confronted with "a highly controversial subject"?

Or—

Is the *real* reason why "someone" doesn't want the American people to see "Cydonia Unedited" on NASA television—

Because It, if not its "physics/energy implications,"could *literally*—Change The World?

*      *      *

Consider this . . .

When NASA Headquarters abruptly terminated Bondurant and NASA-Lewis' original plan to broadcast *Hoagland's Mars,* in addition to the literally thousands of calls and letters received by NASA in those first few days—which apparently soon forced Brown to *publicly* reverse course (at least for appearances of fairness)—Chairman Roe's *House Committee* also got its share.[45] All demanded that "Roe and the Committee step in and investigate this highly unusual and suspicious last-minute NASA action!"

Then . . . *a few days later* (as we noted above), on January 3, Robert Roe, suddenly, and with no prior official announcement, stepped down as Chairman of the House Committee on Science, Space and Technology.

Roe gave up "space" (after *publicly acknowledging there might well be actual alien artifacts on Mars* (!) over which, as head of the key Congressional Committee with NASA oversight, he would have dominant control in terms of how they were explored) to take up the Chairmanship of the Commitee on Public Works and Transportation, a committee which has federal oversight of bridge construction, dams, highways, and federal garbage landfills.

Was Robert Roe's previously vocal and highly visible support for "seriously looking at Cydonia," as head of the critical congressional Committee also suddenly being asked by *thousands* of angry American citizens to look into NASA's actions over "Hoagland's Mars," suddenly a political "embarrassment" to someone . . . ?

How was Roe convinced to suddenly vacate the most pivotal committee in the Congress if, in Roe's own words, **"it looked like they [the 'face' and 'pyramids'] had to be fashioned by some intelligent beings . . ."**[1] a mere two years before *Mars Observer* finally tells us (and Bob Roe!) *if he is right!?*

And if NASA will "take steps" to insure that even *discussion* of this subject "is not legitimized" in 1991 through *television,* what assurances do we now have—with Roe suddenly absent from this key Congressional Committee—that NASA will still live up to its commitment—to him and to the American people—that, in Roe's own words, **"all information [in 1993] about these unique features will be provided to the public?"**

Again, the question looms: is there a more broadly-based political effort here—far outside of merely "NASA"—to suppress all real investigation of an extraordinary scientific possibility . . . one with, among other major "upsets and implications," a foreseeable, imminent, major technological and economic impact on an already (to some) too-rapidly changing world order . . . ?

Remarkably, on January 17, Roe sent a letter[46] to a constituent who was inquiring regarding NASA's handling of *Hoagland's Mars,* in which—

fourteen days *after* he gave up the Chairmanship of the House Science Committee—he still expressed the following:

> Dear Mr. Reinauer:
>
> Thank you for your recent communication concerning the film presentation of "Hoagland's Mars." As you may know, I have strongly supported further exploration of Mars and I believe we should undertake whatever efforts are practicable to understand the origin of landforms *such as the Cydonia region.* For this reason, *I have asked NASA to provide assurances* that the *Mars Observer* mission *will include* this as one of its imaging objectives. . . .
>
> Although I am no longer the Chairman of the Committee on Science, Space and Technology, having assumed the chairmanship of the Committee on Public works and Transportation, *be assured that I maintain my strongest interest regarding this matter* . . . [emphasis added].
>
> Sincerely,<br>Robert Roe<br>Member of Congress

But if his professed "continuing interest" was *that* strong, if he truly feels that Cydonia is truly *that* important—*why* did Robert Roe leave the one committee in the Congress which would assure him pivotal control of "the greatest scientific discovery of modern times"—*public confirmation,* via *Mars Observer,* that the Human Race is NOT Alone!?

What were the *real* reasons Robert Roe suddenly "switched" Congressional committees, and successfully removed himself (and at the *height* of public controversy over the NASA cancellation of its own Cydonia program, *Hoagland's Mars*) from "the line of fire," from the "hot seat of public and political responsibility," from the *Chairmanship* of this key Congressional Science Committee, as we come up on the pivotal date of *Mars Observer's* 1992 return to Mars . . . ?

One grave possibility, indicated by the sudden timing of his hastily-arranged departure, and then his continuing, publicly-expressed, seemingly contradictory opinions regarding the "importance of Cydonia," even after he has left:

That Roe's "decision" was not completely voluntary.

That, as a well-connected Washington politician, Roe was gently informed during the "flap" surrounding *Hoagland's Mars* (perhaps, through his membership on the House Intelligence Committee) that "there will be no 'public pictures' of Cydonia in 1993"—or *anytime* soon after. . . .

*        *        *

*The Monuments of Mars*

In August, 1991, our Director of Operations, David Myers, specifically wrote to David Evans, NASA-JPL Manager of the *Mars Observer* Project. On behalf of all our *Mars Mission* members, Myers requested clarification of NASA's specific intentions toward Cydonia—in the wake of Chairman Roe's abrupt departure, and the conflicting claims [see Vera Hirshberg's statements, above] re NASA's mission objectives with regard to Cydonia, and the continuing ambiguity surrounding its commitment to Roe and the full House Committee on behalf of new Cydonia photography.

On September 4, Evans wrote a three-page "technical response" to Myers' inquiry,[47] which stated in part:

> Dear Mr. Meyers (sic):
>
> ... as Project Manager, it is my responsibility to insure that the objectives of the [*Mars Observer*] mission are met. To this end, the objectives for the mission have been used to develop a set of functional requirements. Both the objectives and functional requirements were established *early during the mission's evolution in CY [calendar year] 1985.* These functional requirements have never included the systematic imaging of any region of Mars or *the targeting of specific features* [emphasis added].
>
> ... this requirement was not a part of the [original 1985] mission ... for a variety of reasons.
>
> (1) Mars Observer was (and is) to be developed using existing Earth orbiting spacecraft designs ...
>
> (2) The gravity field and atmospheric density of Mars are non-uniform, leading to both a down-track (largely latitude) predictive uncertainty, and a cross-track (largely longitude) predictive uncertainty ...
>
> (3) The orbit of the spacecraft in mapping will be *non-repeating* in order to map the entire planet [in low-resolution mode] and *provide uniform sampling in longitude* ... [this] implies that for most places on Mars, except at very high latitude, there will be at most *one or two chances to photograph a given target* [emphasis added]. ...
>
> MOC [*Mars Observer* Camera] is actually two cameras, both are of the line-scan type (not framing) and both are rigidly mounted on the spacecraft and point directly downward. The wide-angle camera samples approximately 300 m per pixel [meters per picture element] and provides limb-to-limb coverage with lower resolution toward the limb. The narrow-angle camera samples approximately 1.5 m per pixel at nadir [directly below the spacecraft], and with 2048 pixels has a maximum field-of-view of about 3 km in the cross-track [at right angles to spacecraft orbital motion] direction.
>
> ... During the mission we will image a specific point repeatedly at lower resolution and we will probably image it once at 300 m per pixel.

However, we cannot be certain of imaging any specific point, such as the *Viking* Landers [or, the "face on Mars"], at the 1.5 m per pixel resolution. . . .

I hope this explanation is useful to you and it may provide at least a partial answer to your question 'What is going on here? Pardon us, if we cannot figure it out.' It is not possible [given the above technical constraints of *Mars Observer*] to guarantee high resolution images of any target, including the *Viking* Landers, but attempts will be made including targets in Cydonia. I believe any apparent uncertainty you observe has stemmed from our inability to *assure* we can obtain the images you desire. . . .

> Best Regards,
> David D. Evans, Manager
> *Mars Observer* Project

Setting aside, for the moment, the *technical merits* of Mr. Evans' "official NASA response" to our request for clarification—

*On the same day*—September 4, 1991—*another* "official" letter was also being drafted on the same subject[48]—to Mars Mission member, Mr. John R. Howell, of West Des Moines, Iowa—

This, *from none other than Roe's successor*—the *new* Chairman of the House Committee on Science, Space and Technology, Mr. George E. Brown. Said Chairman Brown, in part:

Dear Mr. Howell:

As the new Chairman of the Committee on Science, Space and Technology, I appreciate the opportunity to respond to your suggestion that NASA should investigate the "Monuments of Mars". . . .

NASA plans to launch in 1992 the *Mars Observer* to follow up on the discoveries of *Mariner 9, Viking I* and *Viking II*. The *Mars Observer* is equipped with both a wide-angle camera and a narrow field of view camera. . . . It is my understanding that NASA *does* intend to try to capture, with the narrow angle camera, the Cydonia region *including the unique features* you have referred to as the "Monuments of Mars" (emphasis added) . . . ⁻

Once again, thank you for your interest in the space program.

> Sincerely,
> George E. Brown, Jr.
> Chairman

Not only was this letter almost exactly, *word for word,* what Chairman Roe had been telling media inquiries[1] on this subject for over two years (and writing to *constituents*), it affirms Brown's obvious assumption that NASA will be securing new images of "the unique features" in

Cydonia in 1993; the idea of NASA "attempting" anything—and *failing* (if it's serious in its attempt)—is, after thirty years of seeing the Space Agency routinely accomplish the "impossible," almost patently unthinkable. Or, so any average American (or member of Congress) could be forgiven for believing—the former, *especially* after being reassured of this fact by the latter, and none other than the *head* of a major science committee in the U.S. Congress!

So—which is it?:

*Can we or can't we* assure the American people that NASA will indeed acquire these vital new close-ups of Cydonia? Is Brown's "official" letter, in truth, merely a cruel hoax being perpetuated on the American people by himself, former Chairman Roe, and other members of this House committee—who know "behind the scenes" that NASA *can't deliver* what the House Committee's promising?

Or, is Chairman Brown (and Chairman Roe before him) as "misled" on this issue as Evans implies that we are—in expecting to see a NASA "can-do" effort at complying with both the wishes of the Congress, *and* the American people, on this vital issue . . . ?

And, is this official NASA response to us, re the subject of "new pictures of Cydonia," merely NASA's carefully-worded, carefully-planned "official out"—when 1993-1994 finally comes around?—

And the space agency suddenly announces: "Regrettably, we just couldn't effectively *target* the 'City,' the 'Face,' and the other 'unique objects' in the Cydonia region, for *technical* reasons beyond our control . . . as we attempted to inform you two years ago. . . ."

With this "excuse" as an effective "cover," will NASA, in fact, go on *in secret* to secure these vital new Cydonia images . . . and, on instructions from others in this government—

*Never let us know!!*

And, do I sound just a touch paranoid re this entire subject . . . ?

Or, did I just *imagine* that NASA Headquarters recently stepped in and *censored* a television show *its own people* had produced, which attempted *merely to describe* some of our research . . . not *physically go back to Mars* . . . and secure effective photographic *proof* of the hypothesis!

Again, in a society now inured to tales of Presidential misdeeds during "Watergate," "Iran-Contra" and the like—issues which *pale* before the political and social implications (see above) of confirmation of a genuine set of ET *ruins* right next door!—I submit that nothing, and no "scenario," can be considered a priori "too paranoid"—in view of what has *already happened vis-à-vis* "Cydonia," and NASA's fifteen years of less-than-forthcoming behavior on this subject—

**And the fact that "someone"—** *demonstrably not from Earth—* **is now attempting to drive home the "Message of Cydonia," as a "message in the crops," before our very eyes right here on Earth!**

Against that backdrop, to assume that everything will proceed "as planned" with regard to vital new images of "the Monuments of Mars," would, in our opinion, not merely be "naive". . . it would be *irresponsible.*

*        *        *

Which brings us to my own response[49] to Mr. Evans' September 4th "official NASA comments," *vis-à-vis Mars Observer* NOT being able "to *assure new images of the Monuments of Mars."*

Dear Mr. Evans:

. . . As someone with a bit of experience in space mission planning, I fully appreciate the technical constraints under which *Mars Observer* will be forced to operate, in securing new science and new imaging of Mars during the 1993-1995 (one full Martian year) 'planned nominal mission.' But, in your detailed response to Mr. Myers' queries, I am now a bit confused, in particular, by some of your assumptions—beginning with the scale of the specific targets we would like to see reimaged in Cydonia.

You referred several times in your letter to Mars Observer's inability to image "any specific point, such as the Viking Landers, at the 1.5 m per pixel resolution . . ."

With all due respect, Mr. Evans, this is a technical "straw man;" the Landers are relatively minuscule objects on the Martian surface—merely a few meters—and are sitting "somewhere" inside "landing CEP ellipses" which measured over 50 km in down-range errors, and perhaps 30 km across.

Even allowing for substantial downward revision of these 1976 uncertainties, the major task of finding—and then targeting—the "microscopic" Viking Landers for new Mars Observer's images is totally irrelevant to our request: simply a sincere effort to reimage massive structures, such as "the Face"—and collections of structures, such as "the City"—measuring several kilometers in length and width!

By your own reckoning, certainly these latter objects fall well within your own error uncertainties for new *Mars Observer* narrow-angle images; thus, I cannot help but infer that, not to give the American people (or the Congress) any assurances regarding these specific objects we've identified (for which, incidentally, we possess extremely precise geodetic coordinates—perhaps better than any other features currently known on Mars!) is much more determined by the politics of NASA, than by the engineering parameters of the *Mars Observer* mission you describe.

Which brings me to your second, curious "assumption."

You state "both the objectives and functional requirements [of Mars Observer] were established early during the mission's evolution in CY 1985." I find this a remarkable statement—for it means, in essence, that none of the solid, diversified (and now published) scientific research into the "ET hypothesis," re the "anomalous objects" at Cydonia— which has taken place outside NASA since this CY 1985 mission plan was initially crafted—has apparently made a "lick of difference" in modifying—even slightly—the science objectives of a publicly-funded mission—Mars Observer—

Which is the *only* means the American people will possess to test the "intelligence hypothesis" *on Mars* . . . for the foreseeable future.

This, despite repeated public statements now by *two Chairmen* of the pivotal House Committee on Science, Space and Technology, assuring Americans and the media that "NASA *does* intend to try to capture, with the narrow-angle camera, the 'unique formations' in the Cydonia region . . . referred to as 'the Monuments of Mars.'"

I guess it all comes down to your definition of that curious word (for NASA): "try."

I believe that the Committee Chairmen are assuming on this issue what most Americans would rightly assume at this point, having seen *repeatedly* what NASA can accomplish when NASA "truly" tries: three men sent audaciously "where no one had gone before"—*to the Moon,* in an epic Christmas journey over twenty years ago—sent there on an essentially untested, 6-million-pound *bomb*—the Saturn 5—and then returned safely "to the Earth"; three *other* men brought safely back to Earth from almost certain death, during the harrowing experience of Apollo 13; a 12-year unmanned voyage to the outer solar system, with a spacecraft designed for only a fraction of that time, and all depending on *one* faulty back-up radio receiver . . . or, an unprecedented *three astronauts in one EVA,* standing on a "Rube Goldberg" trusswork above the open payload bay of the Space Shuttle, reaching up with *only their gloved hands,* in an unrehearsed attempt to *manually* grapple in a 100 million dollar satellite, seen *live* around the world. . . .

That was "trying," Mr. Evans—part of the proud tradition that NASA has itself carefully inspired through the decades.

Frankly, telling the American people you cannot, for "technical reasons," "assure" new *Mars Observer* images—of potentially artificial structures *the size of the island of Manhattan!*—does NOT exactly inspire confidence in NASA's *political* forthrightness on this highly sensitive issue. I, for one, would have much higher confidence in your "intended efforts"—if you had taken some of your lengthy explanation and detailed your *specific* technical plans "to try to capture" images of the objects in question in Cydonia—in simple compliance with the stated wishes now of both the American people, and their duly-elected representatives in Congress.

That's how the system is *supposed* to work, Mr. Evans.

My erstwhile friend, Carl Sagan, has often said "Extraordinary claims demand extraordinary evidence"—to which I can merely add the obvious corollary: "Extraordinary *implications* demand equally extraordinary *efforts* . . ."

Mr. Evans, there is nothing "extraordinary"—by your own description—surrounding *any* of NASA's published efforts to "secure one image in a full Martian year" of potential ET *ruins* in Cydonia; no discernible "extra effort" to *fairly* test potentially one of the most important scientific problems of our time, one with literally "off-scale" cultural, philosophical, if not *geopolitical* implications . . .

For, we are *not* discussing here merely "another type of Martian crater," or "a profile of an ancient Martian river valley"; we are discussing nothing less than a possible *extraterrestrial civilization,* which *eight years* of independent, published (and, incidentally, thoroughly discussed and video taped, in several *NASA* seminars) scientific analyses—from comparative fractal scaling to geometric modeling—now strongly indicates *just might well exist!*

Not to make some kind of "good faith effort"—*not* to "give up" one or two of the previously established (in 1985) *low-priority Mars Observer* scientific objectives—to fairly test this non-trivial, *extraordinary* possibility, not only seems to us at this point a breech of simple scientific (if not institutional) common sense—it's beginning to seem more and more like "someone" in NASA is not quite leveling with the rest of us out here . . .

Or, Mr. Evans, is the problem simply much more "human": a classic one of "Not Invented Here"—that this work, and this hypothesis have, from the beginning, originated *outside* the "planetary science community" of NASA . . . a hypothesis, therefore, that that community is simply not about to *test* (at least, in public . . .)?

The American people are waiting for an answer . . .

*      *      *

Since the above correspondence was written, several critical new developments—both on the political and research fronts—have come to pass. We would be derelict if we did not report some of these developments and their potentially crucial impact on the increasingly imminent opportunity in 1993-94—via Mars Observer—for definitive verification of the Intelligence Hypothesis.

On February 27, 1992 we accepted an earlier invitation by two United Nations staff members, Susan Karaban and Mohammad Ramadan, and addressed a special meeting of delegates, staff and invited special guests at the Dag Hammarskjold Auditorium at UN headquarters. During our two standing-room-only, back-to-back three-hour Cydonia presentations, we discussed for the first time before the international community several of the latest research developments regarding the ongoing Cydonia Investigation, as now reported in this volume. (These are also reported on in *Martian Horizons,* Vol. I, No. 3.) A 90-minute video, *The U.N. Briefing* (a follow-on to *The NASA-Cydonia Briefing*) was also recorded at the meeting, and is available through *The Mars Mission.*

Subsequent to these two major presentations, several ambassadors from member states in the United Nations, as well as official news agencies headquartered at the U.N. (including TASS), requested private, one-on-one continuing discussions regarding the ongoing Cydonia Investigation and its global implications.

The purpose of the U.N. Briefing and discussions: to place the problem of Cydonia verification squarely in the lap of the United Nations through, among other means, belated implementation of "General Assembly Decision 33/426" (enacted, but never enforced, in 1978, regarding "extraterrestrial life . . . and the scientific evidence thereof")—as another means (in addition to demanding action by the U.S. Congress) of politically pressuring the U.S. Administration to "take and immediately publish new images of Cydonia from NASA's up-coming *Mars Observer* Mission."

The individual meetings with U.N. delegations, regarding "Decision 33/426" and its relevance to the *Mars Observer* Mission, currently continue.

\*            \*            \*

Newly published official documents based on internal *Mars Observer* Project memoranda have now come to our attention, relevant to *technically* answering the many questions which have arisen concerning *Mars Observer's* specific capabilities for reimaging the "Monuments of Mars." These documents now present *The Mars Mission* with *specific engineering evidence,* from JPL itself, concerning the exact nature of *Mars Observer's* orbital capabilities. This will allow a new level of informed outside analysis of the impact of these capabilities on the continuing controversy over reimaging specific targets in Cydonia: e.g., the "City," the "Face," the "D&M," etc., during the nominal 2-year duration of the *Mars Observer* Mission.

As suspected, based on our own earlier discussion (see response to David Evans, *Mars Observer* Project Manager at JPL, above), these documents, authored by JPL and GE (the prime contractor for the spacecraft) management and engineering personnel, reaffirm our fundamental questions regarding David Evans' and Dr. Michael Malin's (the *Mars Observer* Camera Principal Investigator's ) continued negative assertions as to the capability of the *Mars Observer* Project to "guarantee" re-imaging of Cydonia by *Mars Observer*. With specific reference to Dr. Malin's latest comments:[50]

> . . . we will have, at best, *one* or *maybe two* opportunities to photograph any 3 km square piece of Mars [such as Cydonia—during the entire Mission]. In addition, given the day to day uncertainty of the orbit (resulting from atmospheric drag because the spacecraft is so low), *it is unclear whether we will be able to predict when Mars Observer will be likely to fly over a specific location on Mars* [such as Cydonia]. Finally, even if we are able to predict when appropriate images should be taken, Mars may not cooperate—clouds, fogs, hazes, and dust storms frequently obscure the surface. So, I hope you understand that I cannot *guarantee* the images of interest [of Cydonia] will be taken . . . (emphasis added).
>
> Sincerely,<br>Michael Malin, Mars Observer<br>Camera Principal Investigator

As previously stated, the technical basis for these consistently discouraging comments re new, high-resolution imaging by *Mars Observer* of the "Monuments of Mars," primarily concern the "bolted-on" aspects of the *Mars Observer* Camera (MOC) and its *inability to swivel:* i.e., to be deliberately aimed toward objects of specific scientific interest as *Mars Observer* orbits overhead. Because the camera is rigidly constrained to point *straight down,* the only time an image can be taken is when the spacecraft is passing *directly overhead.* According to Evans and Malin, the exact times when these events will happen at Cydonia (in fact, if they'll occur *at all!),* due to "large uncertainties in the predicted orbit of the *Mars Observer* spacecraft," are currently totally unknown, making it impossible to "guarantee the images of interest will be taken."

The JPL documents we now have in our possession cast significant doubt on these continued disingenuous representations of *Mars Observer's* capabilities as "limited" *vis-à-vis* targeted photography of specific objects on the Martian surface, like the "Face." They reveal, instead, a highly sophisticated, highly flexible spacecraft design and mission *technically* capable of acquiring precisely the confirmation of Cydonia we

*The Monuments of Mars*

need . . . provided 1) NASA *intends* seriously to seek such confirmation, and 2) the Agency intends to tell us if they get it!

The new engineering evidence is part of an extended *Mars Observer Mission* Overview initially published in the September-October 1991 issue of the *Journal of Spacecraft and Rockets,* a publication of the American Institute for Aeronautics and Astronautics (AIAA).[51] Several papers, written specifically by *Mars Observer* engineers, now reveal *substantial reductions* (by factors up to 26!) in the "estimated uncertainties of *Mars Observer's* final mapping orbit." These "orbit uncertainties," we have been informed repeatedly by Evans and by Malin, are "the main constraints in planning (or 'guaranteeing') any targeted high-resolution photography of 'the D&M,' 'the City,' or 'the Face.'"

The newly-published, *official* JPL computer estimates of these "uncertainties," direct from the navigation engineers assigned to carry out the *Mars Observer Mission,* translate directly into equally dramatic reductions in the *photographic* uncertainties involved in taking new images of the anomalous objects in Cydonia. How, then, do these actual calculations correlate with the much *larger* "orbit errors and photographic uncertainties" consistently depicted in Evans' and Malin's repeated statements to the press and public on this controversial subject?

Are we, in fact, being *deliberately and systematically misled* on "technicalities" regarding this entire, vital issue . . ?

All the engineering information one would require in order to come to an informed conclusion on these matters is now available, to anyone, in any library. All one has to do is read the relevant JPL planning and engineering documents, in this highly informative AIAA 1991 "*Mars Observer* Special Issue." These documents reveal the extremely narrow financial and project-management environment in which the Evans/Malin estimates of the "impossibilities" of reimaging Cydonia have heretofore been made. The "uncertainties" so often quoted, it turns out, seem to be based as much on a worst-case interpreatation of NASA's "pre-Cydonia" *Mars Observer Mission* management Policy, as on any realistic engineering specifications. . . .

**Which—to keep the costs of Mission Operations down—was originally designed to discourage constant, real-time correction of the accumulating errors in the spacecraft position in Mars orbit [through constant analysis of spacecraft orbital position, and real-time uplinking of commands for on-board thruster firings]. In the current "low cost" Mission Plan, orbital errors will deliberately be allowed to slowly accumulate, until regularly-scheduled orbital corrections can be carried out—no oftener than once every two weeks! However, there is nothing in the space-**

**craft design which technically prohibits much more timely orbital corrections—with consequent dramatic reductions in "imaging uncertainties" for any given target.**

Under existing Mission Rules, in 14 days, the spacecraft positional uncertainties can, in certain worst-case scenarios, accumulate to almost 30 km—regarding the uncertainty of where *Mars Observer* actually will be, somewhere along its orbit!

No wonder that Malin is insisting that he "cannot *guarantee* new images" of the "Monuments of Mars" IF he also insists on sticking to this existing, highly rigid, "low cost" Mission Plan for tracking and correcting *Mars Observer's* orbit—both of which directly affect pre-knowledge of the *exact* location of all images.

To illustrate the dramatically negative effect this *single* operational constraint (if rigidly adhered to) will have on knowing and correcting the orbital position of the *Mars Observer* spacecraft (in relation to Cydonia), let us refer to two computer simulations on this subject (orbital uncertainties), that were carried out at JPL in 1989, described in some detail in the AIAA Special Issue.

These studies refer to a "December, 1993 analysis period" (just after the spacecraft has been placed in its intended "mapping orbit," but *before* critical information on Mars' gravity field has been derived); and several months later, an epoch "in April, 1994"—*after* this crucial gravity field information has been secured (from long-term tracking of the *Mars Observer* spacecraft.) The latter case involves convolving this critical "Mars gravity information" into all future computer models of "the forces affecting *Mars Observer's* reconstructed and predicted orbits."

The dramatically differing results of just these two comparative analyses, are quite enlightening:

> The Orbit Determination results for the two cases, due to analysis of Doppler under the above conditions, are given . . . For the December 6, 1993 epoch [just after the spacecraft is placed in the Mapping Orbit], the largest position errors for reconstruction are in the crosstrack and downtrack directions [1.34 km, and 1.30 km, respectively] . . . Prediction errors are given for 7, 14, and 21 days past the analysis epoch. As shown, the errors in the downtrack [along the orbit] component of position rise dramatically [ 7.3 km, in 7 days; 24.8 km, in 14 days; 53.6 km, in 21 days] and are due almost exclusively to atmospheric drag . . .
>
> Position errors are also given for a second case identified by the analysis epoch, April 4, 1994. The reconstructed errors [in instantaneous position of the spacecraft—0.05 km] are *much smaller* than those of the previous case [1.3 km] because we have used smaller gravity field

errors. This is due to the Gravity Calibration results [obtained in the first 7 days of tracking in the low-orbit Mapping Phase of the Mission] . . . Downtrack errors show a similar growth pattern [7.1 km, in 7 days; 29.4 km, in 14 days; 66.8 km, in 21 days] when compared with the December 6, 1993 case. This is because a priori [atmospheric] density errors are larger for April 4, 1994 [because Mars is then closer to the Sun, and the atmospheric density, and thus atmospheric drag, at *Mars Observer's* orbital altitude, increases with increased solar heating] . . . [emphasis added].[52]

If Malin and the *Mars Observer* Navigation Team are allowed to uplink new thruster-firing commands to *Mars Observer* based on *immediately-processed* tracking information, then (according to JPL's own internal *Mars Observer* tracking and navigation documents) the entire problem of re-imaging Cydonia rapidly becomes quite academic; the "predicted instantaneous uncertainties in reconstructing the position of the spacecraft" will be less than 0.25 km in "crosstrack error" (at right angles to the orbital path), and less than 0.05 km—*50 meters*—in "downtrack error" (along orbit).

This latter error is, in fact, *less than the size of one pixel* on the original *Viking* Cydonia frames!

The amount of fuel required for a change in spacecraft velocity in orbit depends on the magnitude of the orbital correction needed. Note that the exitsing spacecraft orbit would already place *Mars Observer* relatively near the "Face," the "D&M," etc. (ten to twenty km miss-distance is envisioned). For small orbit corrections such as we are hypothesizing here, the amount of velocity correction needed to alter such a close pass to one which would take the spacecraft directly over the "Cydonia Complex" in a *predictable* fashion, could be (according to the data in the *Mars Observer* documents) as little as "a couple of hundred *millimeters per second!*"[53]

Since the spacecraft carries fuel equivalent to about *45 meters/sec* (total on-board velocity correction capability,[53] it is easy to see that to adjust the *Mars Observer* orbit—to trim the initial "close pass" so that the spacecraft goes into a repeating, 88 revs/every 7 Martian days orbit,[54] *right over Cydonia;* then, carry out subsequent small thruster corrections in order to passover Cydonia in *multiples* of that repeating 7 "sols" (for complete imaging, at several different geometries and lighting); then, to rephase the orbit (by more thruster firings), so as to resume the nominal mapping mission schedule—all this would require, *at most,* perhaps *one meter per second* of total velocity correction capability (if that!) out of the available 45 meters/sec on-board capability.

Now—whether NASA and the *Mars Observer* Project will *permit* this "massive intrusion" on the carefully planned, existing mapping mission schedule (because of overtime costs; availability of key mission personnel; tracking antenna use by other missions; computer time availability, etc.), is an entirely different matter, one with major *political* implications for the Agency, if not for Malin.

But, the engineering *capability* to do this (and thereby *guarantee* new Cydonia high-resolution images from *Mars Observer)* is unquestionably there . . . right in the JPL's own, now-published Mission documents!

But this is not, by any means, the total *"Mars Observer-Cydonia"* picture.

If we are more realistic, if we don't expect or demand a *major* revision in the existing *Mars Observer* schedule (designed for "synoptic mapping, at low-resolution, of all of Mars,") what then? Can *Mars Observer still* secure new high-resolution images of the "Monuments of Mars?"

You be the judge:

Take the "reconstructed errors" from the first (worst-case) "pre-Martian Gravity Map" simulation:[52] instantaneous "sideways" (crosstrack) errors and downtrack (along the orbit path) errors come out about the same (1.3 km—see again, above). But even *these* major orbital positioning errors for the spacecraft are *smaller than the crosstrack and downtrack dimensions of the "City" at Cydonia,* which measures approximately 7 x 15 km!

This means that, with *no new information* concerning the Martian gravity field (beyond *Viking's* current information), *if* Dr. Malin received this caliber of tracking data even *a few days* before an actual predicted *Mars Observer* crossing of the "City"—*without changing the orbit of the spacecraft one iota*—he could *still* effectively time his Cydonia high-resolution images to occur as *Mars Observer* flew directly overhead. Even *with* these errors he would likely get extraordinary close-ups of the "City"— an *extended* target underneath the spacecraft, literally bigger than the island of Manhattan!

But *can* the good Doctor rapidly re-program his *Mars Observer* Camera—even if he *has* access to this up-to-date- tracking information—to take advantage of this kind of opportunity? And, critically important, can he do this without negatively affecting not only the overall health of the rest of the *Mars Observer* spacecraft, but *all the other* scientific instruments on-board as well?

In the "good old days" of *Viking,* the answer would have been a resounding "no." Science acquisition during the *Viking* Orbiter and Lander missions was distinctly a team effort: most observations by one instru-

*The Monuments of Mars*

ment had the potential to affect all the others, as well as the overall condition of the spacecraft. This was why *Viking* was essentially a "billion dollar mission:" it took literally *hundreds* of crewmen to constantly and safely fly it in Mars orbit by remote command from Earth.

By contrast, the core of the *Mars Observer* concept—the key (we have repeatedly been told) to radically reducing the cost of this and later unmanned planetary missions—lies in the "radical decentralization of technology," and the "new philosophy" of Mission management this new technology currently allows. To quote from another Mission document:

> The fixed mounting of the payloads on the nadir-pointing platform [of the *Mars Observer* spacecraft] provides nearly *independent control* for the science instruments. There is no articulating platform shared by instruments that would *limit observation opportunities.* Instrument commanding in general *does not require inter-instrument coordination,* and the spacecraft's command system is invisible to commands addressed to the instruments.
>
> Consequently, the instruments can operate autonomously and can sequence themselves with *minimum information from the spacecraft bus.* The spacecraft bus function becomes decoupled from payload operation, and *the scientists can interact directly (via data terminals in their home institutions) with the command uplink process.* Science instruments will be controlled primarily by such noninteractive *real-time commands* . . . the largest portion of which [will be] utilized *for the control of the Mars Observer Camera* . . . [emphasis added].[54]

In other words, according to these *official JPL Mars Observer Mission* documents, Dr. Malin, without consulting either the spacecraft managers at JPL or any of the other science experimenters (who will be simultaneously sending autonomous real-time commands to *their* own science instruments), *will* be able to uplink directly from his own computer terminals, either at his home or office, a series of new imaging instructions *direct* to the *Mars Observer* Camera. Via these commands, during the course of the nominal mapping mission, he may *unilaterally* target objects *anywhere on Mars,* including (when the orbital opportunity arises) in Cydonia itself. The latter, of course, as soon as ongoing orbital analyses at JPL provide a warning of an upcoming Cydonia pass.

And, *even if the first such opportunity occurs very early in the mapping mission* (before the new "gravity map" is effectively secured, analyzed and implemented), the odds are that the MOC (operating within the orbital uncertainties already analyzed in detail by *Mars Observer* navigation specialists) could (according to their published data) *still* secure "a frame or two" of something as extended as the "City". . . if Malin *truly* wants to do it.

But that is only the beginning—again, according to JPL's *own Mission documents.*

The previous description of a hypothetical *Mars Observer* Camera Cydonia imaging opportunity has been based solely on the assumption that any images acquired over Cydonia *would have to be recorded prior to being transmitted to Earth;* however, this may not always be the case: The MOC has a solid-state buffer capacity of about 12 megabytes,[55] and periodically dumps this stored imaging data (a few images worth of digital information) to one of the several on-board spacecraft digital tape recorders for later playback to Earth. As images are acquired and shunted over to the tape recorders, the buffer is cleared and a new image may be stored. Thus, the rapidity of image-taking is determined by the rate (in bits per second—bps) at which the camera buffer may be read out to an on-board spacecraft tape recorder. Under these conditions the total number of images acquired in any one 2-hour orbit of the planet is thus ultimately determined by *the amount of data space* allocated on the spacecraft tape recorders, and the *rate at which this imaging data can be recorded,* when compared to that required by other spacecraft engineering and science experiments at the time.

In this scenario, the likelihood of catching a high-resolution image of an important feature in Cydonia depends critically on the *rate of camera shuttering,* the *rate of the readout of the camera buffer* to the spacecraft tape recorders and the *rate of re-shuttering the next image.* If the readout rate is too slow, images may be taken too soon (before the spacecraft passes over an important target) or too late (after it has passed "the Fort," "the Face," etc.). The rate of imaging individual frames is thus quite critical to actually capturing a specific target, particularly when the orbital uncertainties are factored in.

But, present in the new JPL documents is a little-known additional feature of the MOC and *Mars Observer:* its ability to send *a continuous series of "live" images—directly to Earth!*

> The project has included a real-time data rate of 40 ksps (34.9 kbps) to permit the return of some high-bandwidth data *that would otherwise be constrained to the lower record rates.* The project policy is that *an additional tracking pass will be scheduled approximately every three days* over the mapping phase to return data at the real-time rate. This additional real-time data . . . augments the recorded data returned *every day* at the available playback rate. The 40-ksps rate can be returned at the 34-m (HEF) antennas over most of the mapping phases, and only a

*The Monuments of Mars*

brief period of 70-m tracking support is required in January and February 1994, when Mars is farthest from Earth.

A strategy for collecting the real-time data adds some complexity to the mission design. The recorded data modes provide complete coverage over each orbit and around the planet on each mapping cycle. However, the real-time data can only be collected when the Earth is in view, this is, *primarily on the dayside of the planet, and when an additional tracking pass is scheduled.* Only data from TES [Thermal Emission Spectrometer] *and MOC [Mars Observer Camera],* along with spacecraft engineering data, will be returned during real-time coverage . . . [emphasis added].[56]

Given the above, there is now *no known technical reason* (Evans and Malin's anticipated protests notwithstanding) why Dr. Malin cannot adequately anticipate *any and all* imminent *Mars Observer* crossings of Cydonia, and (via the now-established, *real-time* capability of the Mars Observer Camera) record, on Earth, a 3-km wide, almost 9600-km long "facsimile image" of *an entire hemisphere of Mars* . . . including, obviously (if the geometry on any individual orbit permits), a real-time, north/south image strip *across the entire "Cydonia Complex" itself!*

Two hours later, *another* real-time swath, 3 km wide, and 9600 km long, could again be relayed from the MOC and recorded here on Earth. Two hours later another strip, and so on . . . for as long as Mars remains above the horizon at the tracking stations spaced at roughly 120-degree intervals around the Earth. Since from these three stations Mars will be in view *continuously,* only Mission costs (and Mission personnel fatigue!) can ultimately limit what can be relayed to Earth in this real-time mode when the spacecraft is in view—which, in even the most limiting geometry, will still be approximately *one hour* out of each 2-hour Martian orbit!

The fact that the Mission *already* is planning for such real-time activity *every third day* means that there are some "high-priority surface features" that Malin is already anticipating and obviously doesn't want to miss through "limited spacecraft recording capability." One can only wonder what they are . . .

And if all else fails, if Malin *truly* intends, as he insists, "to try and capture images across Cydonia," there is, of course, the Ultimate Last Resort:

*The entire spacecraft orbit can be deliberately modified, repeatedly,* with Cydonia *specifically* the target, as we described before.

Again, the new JPL documents provide a treasure-trove of insight into the exact on-board *Mars Observer* capabilities for exercising this last option. But at this point "Cydonia orbit-changing" remains only a feasible *theoretical* consideration—at total odds with Malin's (and Evans') firm insistence that they are sticking with the "nominal global mapping mission plan" (which will afford, according to Malin, at best, *one or maybe two* opportunities [in the nominal 687-day mission] to photograph any 3 km square piece of Mars . . .)

*   *   *

The third recent development directly addresses this last outstanding question: the possibility of a *deliberate orbit change* for securing high-resolution images of the "Monuments of Mars."

A few days ago a new document from an authoritative source outside of JPL also came to our attention. This document directly implies (by its strong, positive language *vis-à-vis* acquiring new imagery of Cydonia, and specifically imagery apparently to be acquired at "a variety of geometries and sun angles") that this most controversial aspect of the entire Cydonia issue—the *deliberate alteration* of the *Mars Observer* orbit—is now, in fact, being quietly . . . perhaps even secretly . . . *officially* considered, as the one *certain* means of securing definitive new information on Cydonia.

This confirmation is contained in a recent letter from Congressman Paul Kanjorski,[57] representing the 11th District in upper, central Pennsylvania. Kanjorski is a member of several committees in the Congress related to banking, finance and urban affairs, and is Chairman of the Subcommittee on Human Resources.

On May 8, 1992, Congressman Kanjorski wrote to Ian Richardson, a constituent in Shavertown, PA, the following remarkable new information on the *Mars Observer Mission* . . .


May 8, 1992

Dear Ian:
    Thank you for contacting me regarding the *Mars Observer* program. I appreciate knowing of your concerns on this matter and, like you, *am impressed by the picture you enclosed.*
    As a result of your letter I spoke with Congressman George Brown, the Chairman of the House Committee on Science and Technology about the *Mars Observer Mission* during the floor consideration of the legislation authorizing the National Aeronautics and Space Administration (NASA). Chairman Brown assured me that the *Mars Observer Mission* was still on schedule and that the Cydonia region of Mars *was* included in the photography schedule. Apparently the previous missions to Mars (including the one which produced the picture you for-

warded to me) have only looked at the Cydonia region in limited detail. *The Mars Observer Mission currently plans to photograph the region extensively, from numerous angles and at different times to allow for different angles of the sun* [emphasis added].

Thank you again for your letter; I hope that my conversation with Chairman Brown has resulted in satisfactory assurances. If there is anything more I can do for you on this or any other matter, I hope that you will not hesitate to contact me.

Sincerely,<br>
Paul E. Kanjorski<br>
Member of Congress

Which is VERY interesting in light of Malin's continued insistence on "at best, *one* or *maybe two* opportunities [during the whole Mission] to photograph any 3 km square piece of Mars . . ."

Just *what* is going on? Has Chairman Brown in fact seen the *real* (secret?) *Mars Observer Mission* Plan—which he's quoting from to a fellow Member of the House?

And how can you have "extensive" reimaging of Cydonia, "from *numerous* angles and at *different* times"—given the rigid requirements of the currently planned "global mapping grid"[54]—Without *deliberately changing the orbit of the entire spacecraft?*

So, just what's going on here?

Needless to say, this recent communication, from *another* Member of the U.S. House of Representatives re the subject of reimaging Cydonia, raises some profound new questions *vis-à-vis* NASA's (if not Malin's and Evans') forthrightness all along on this entire issue.

For, *Mars Observer* will carry into Mars orbit approximately *twice the amount of on-board hydrazine fuel* estimated by JPL's own engineers to be required for the nominal Mars mapping mission. This excess fuel could 1) extend the global Mars mapping mission beyond the nominal 687 days (in which case, Cydonia will automatically pass once again beneath the spacecraft orbit sometime in this "extended mission"), or, 2) be used to alter *Mars Observer's* orbit, in order to *deliberately* fly across Cydonia early in the initial 687-day mission—a scenario just about as we described above and just as Congressman Kanjorski's independent information now *strongly* reinforces!

Either way, the crucial orbit-changing capability—according to JPL's own planning documents[53]—already exists, with a *2-to-1 fuel reserve* now planned for flight aboard the *Mars Observer* spacecraft. The critical question still outstanding then becomes: *will* this "excess" fuel be used *specifically* to target the anomalous features in Cydonia . . . and will Evans,

Malin, JPL, or anyone at NASA even tell us (and the Congress!) if they try . . . ?

Based on the above, we are forced to draw only one rational conclusion:

These published capabilities of the *Mars Observer* spacecraft (fewer orbital uncertainties, independent command of picture-taking, live transmission capability, almost twice the amount of on-board fuel required . . .) now make it *virtually certain* that *Mars Observer can,* one way or another, secure high-resolution imaging of the "Monuments of Mars" . . . *despite* what we've consistently been told by high officials on the Project, and at NASA.

Whether we will ever *know* if it succeeds, seems now to depend on NASA's (and Dr. Malin's ) "good intentions." On this latter point, in the recent letter from which I previously quoted,[50] Dr. Malin specifically remarks:

> . . . No one at NASA has ever attempted to dissuade me from acquiring images in the Cydonia region. No one at NASA has encouraged me to take such images, either, *but this is because the choice of areas to photograph has been mine from the start* . . . [emphasis added].

And I was under the impression that, in this Republic, that kind of power, on an issue of *this* magnitude, carried out with *publicly appropriated funds,* ultimately rests with the Congress of the United States of America, acting on behalf of the People—and of their sovereign Right to Know . . .

I guess I was mistaken.

*       *       *

Again, the inconsistencies—between the public proclamations and the private, contradictory actions and events—are *almost* definitive enough to reach a firm conclusion . . . given what the evidence now shows is waiting at Cydonia (if not right here on Earth!) . . . but not quite.

One thing however is now extremely clear:

Without *overwhelming and sustained public interest and political support*—between now and 1993-1994, when *Mars Observer* is finally in position to rephotograph Cydonia . . .

We will simply never know.

*       *       *

# Notes

1. D. P. Myers, "Face-Saving Mission to Mars," *Washington Post,* July 13 (1991); B. Cox, "Mysteries of Mars," USA Today Sept (1990); *ibid., Florida Today,* July 15 (1990).

2. E. Torun, personal communication, July (1988).

3. E. Torun, "Preliminary Investigation of the Geometry of the D&M Pyramid," unpublished paper (1988).

4. *Op. cit.*

5. E. Torun, personal communication, September (1988).

6. C. Sagan, *Cosmos,* New York: Random House, (1980).

7. M. J. Crowe, *The Extraterrestrial Life Debate 1750-1900: The idea of a Plurality of Worlds from Kant to Lowell,* Cambridge: Cambridge University Press, (1988).

8. R. C. Hoagland and E. O. Torun, "The 'Message of Cydonia': First Communication from an Extraterrestrial Civilization?," reprints available through *The Mars Mission* (P.O. Box 123, Danville, CA 94526-0123; and electronically through CompuServe, "Issues Forum," Section 10, August (1989).

9. J. Zeleznik, personal communication, March (1989).

10. T. M. Stat, S. M. Thesis, "A Numerical Method for Shape from Shading from a Single Image" (Dept. of Electrical Engineering and Computer Science, MIT, Cambridge, MA, January, 1979); D. Terzopoulis, "Computing Visible-Surface Representations," Memo AIM-800 (MIT Artificial Intelligence Laboratory, March, 1985).

11. M. J. Carlotto, "Digital Imagery Analysis of Unusual Martian Surface Features," *Applied Optics,* Vol. 27, (1988); *The Martian Enigmas: A Closer Look,* M. J. Carlotto, North Atlantic Books (1991).

12. B. Cox, "Mysteries of Mars," *Florida Today,* July 15 (1990).

13. M. J. Carlotto and M. C. Stein, "A Method for Searching for Artificial Objects on Planetary Surfaces," *Journal of The British Interplanetary Society,* Vol. 43 (1990); M. C. Stein, "Fractal image models and object detection," *Society of Photo-optical Instrumentation Engineers,* Vol. 845 (1987); B. B. Mandelbrot, *The Fractal Geometry of Nature,* W. H. Freeman, New York (1983).

14. B. Cox, on-the-record interview with Robert Roe, Chairman, House Committee on Science, Space and Technology, January (1990); B. Cox, "Mysteries of Mars," *USA TODAY,* Sept (1990); *ibid., Florida Today,* July 15 (1990); personal communication from Cox to Hoagland re substance of Roe interview, July (1990).

15. Communication from James G. Osborn, Science Editor, *The World*

*and I—Washington Times,* to Walter Gelles, free-lance journalist September (1990).

16. Reply from Walter Gelles, to James G. Osborn, *Washington Times,* September (1990).

17. Communication from Walter Gelles, to Robert Roe, Chairman, House Committee on Science, Space and Technology, September (1990).

18. Reply from Robert Roe, to Walter Gelles, October (1990).

19. Audio recording in NASA-Lewis Auditorium of John Klineberg's unedited Hoagland introduction; Klineberg's on-the-record confirmation of his NASA-Lewis introductory remarks, with "explanation": B. Cox, "Mysteries of Mars," *USA TODAY*, Sept (1990); *ibid., Florida Today,* July 15 (1990).

20. *Op. cit.,* B. Cox, "Mysteries of Mars," *USA Today,* Sept (1990); *ibid., Florida Today,* July 15 (1990).

21. V. DiPietro, G. Molenaar and J. Brandenburg, *Unusual Martian Surface Features,* Fourth Edition, Mars Research, PO Box 284 Glen Dale, MD (Copyright 1982, 1988).

22. E. Torun, "The Cydonia Complex: A Geological Perspective," *Martian Horizons,* Quarterly Journal of *The Mars Mission,* Vol. 1, No. 1, Summer (1991).

23. B. Cox, private communication, January (1991).

24. Off-the-record NASA Headquarters source, January (1991).

25. E. N. Parker, "The Sun," *Scientific American Special Issue,* "The Solar System," September (1975); J. R. Kuhn, et al "The surface temperature of the sun and changes in the solar constant," *Science,* 242, 908 (1988).

26. S. Tenen, personal communication (1990); "Geometric Metaphors of Life," 108 minute videotape (VHS or Beta) with supporting materials, *The MERU Foundation,* P. O. Box 1738, San Anselmo, CA 94960.

27. H.S.M. Coxeter, "Polytope 2[21], Whose Twenty-seven Vertices Correspond to the Lines on the General Cubic Surface," pp. 457-479, Vol. 62, *The American Journal of Mathematics* (1940); H.S.M. Coxeter, *Regular Polytopes,* Dover, (1963); D. Hilbert and F. Cohn-Vossen, *Geometry and the Imagination,* Chelsea Press (1952).

28. D. A. Godfrey, "A Hexagonal Feature around Saturn's North Pole," *Icarus,* 76 (1988).

29. B. Conrath, et al. "Infrared Obvservations of the Neptunian System," *Science,* 246, December 15 (1989).

30. L. C. Stecchini, "Notes on The Relation of Ancient Measures to The Great Pyramid," Appendix to P. Tompkins,' *Secrets of the Great Pyramid,* New York: Harper and Row, (1971).

31. H. Harleston, Jr., "A Mathematical Analysis of Teotihuacan," Mexi-

co City: XLI Internationalist Congress of Americanists (October 3, 1974); quoted in P. Tompkins,' *Mysteries of the Mexican Pyramids,* New York: Harper & Row (1976).

32. C. P. Munck, *Reestablishing the Grid* (aka, *The Pyramid Matrix: A Program from The Past—for The Future*), (Copyright 1991) [P.O. Box 147, Greenfield Center, NY 12833; also available through *The Mars Mission*].

33. Lockyer, Sir Norman, *Stonehenge and other British Monuments Astronomically Considered,* 2nd Edn., London: Macmillan (1909).

34. J. A. West, *The Serpent in The Sky: The High Wisdom of Ancient Egypt,* New York: Harper & Row (1979).

35. R. M. Schoch, "Redating the Great Sphinx of Giza, Egypt," paper presented at the Geological Society of America's Annual Meeting, October 23 (1991).

36. J. A. West, quoted by T. Friend, "The Latest Mystery of the Sphinx: His Age," *USA TODAY,* October 10 (1991).

37. P. Delgado and C. Andrews, *Circular Evidence: A Detailed Investigation of the Flattened Swirled Crops Phenomenon,* London: Bloomsbury Publishing Ltd. (1989); *The Cereologist: The Journal of the Crop Circle Studies,* Editor, John Michell, Vol. 1, No. 1, Summer (1990).

38. R. C. Hoagland, "The 'Crop Glyphs' And The 'Message of Cydonia,'" *Martian Horizons,* Quarterly Journal of *The Mars Mission,* Vol. 1, No. 1, Summer (1991).

39. R. C. Hoagland, "Tetrahedral Crop Glyph Supports 'Message of Cydonia,'" *Martian Horizons,* Quarterly Journal of *The Mars Mission,* Vol. 1, No. 2, Fall (1991).

40. *The DePalma Research Papers,* R. C. Hoagland, Ed., (1990), published by "For the People," 3 River Street, White Springs, FL, 32096. Also available through *The Mars Mission.*

41. Dr. I. Myers, personal communication, January (1991).

42. "IECEC '91: Preliminary Program" (26th Intersociety Energy Conversion Engineering Conference), August 4-9, 1991, Boston, MA. Patrick G. Bailey, Program Chairman. Reprints of this and the final Conference Program, and presented papers available from American Nuclear Society, 55 North Kensington Avenue, La Grange Park, Il 60525.

43. Dr. R. Brown, NASA form letter "reversal of cancellation" to the Clintons, of NASA-Lewis TV production, "Hoagland's Mars," February (1991).

44. Mr. and Mrs. L. F. Clinton, personal communication and enclosures from Dr. Brown, March (1991).

45. B. Cox, personal communication, January (1991).

46. B. Franklin Reinhauer II, personal communication and enclosures from Roe, January (1991).

47. Communication from David D. Evans, Manager, *Mars Observer* Project, NASA-JPL, to *The Mars Mission*, September 4 (1991).

48. Communication from George E. Brown, current Chairman, House Committee on Science, Space and Technology, to *Mars Mission* member, John R. Howell, September 4 (1991).

49. Excerpt from *Mars Mission* Founder and President, Richard C. Hoagland's reply to David D. Evans, Manager, *Mars Observer* Project, NASA-JPL, November 7, 1991.

50. Excerpt from *Mars Observer* Camera (MOC) Principal Investigator, Dr. Michael Malin's letter to *Mars Mission* Member, Mark P. Archambault, January 2, (1992).

51. E. L. McKinley, "Introduction to Special Section: The *Mars Observer Mission*," *Journal of Spacecraft and Rockets,* Volume 28, Number 5 (1991).

52. P. Esposito, D. Roth and S. Demcak, "*Mars Observer* Orbit Determination Analysis," *Journal of Spacecraft and Rockets,* Volume 28, Number 5 (1991).

53. C. A. Halsell, and W. E. Bollman, "*Mars Observer* Trajectory and Orbit Control," *Journal of Spacecraft and Rockets,* Volume 28, Number 5 (1991).

54. S. Palocz, "*Mars Observer Mission* and Systems Overview," *Journal of Spacecraft and Rockets,* Volume 28, Number 5 (1991).

55. F. O. Komro, and F. N. Hujber, "*Mars Observer* Instrument Complement," *Journal of Spacecraft and Rockets,* Volume 28, Number 5 (1991).

56. W. H. Blume, S. R. Dodd and C. W. Whetsel, "*Mars Observer Mission* Plan," *Journal of Spacecraft and Rockets,* Volume 28, Number 5 (1991).

57. Communication from Congressman Paul Kanjorski to a constituent, Ian Richardson, re Kanjorski's discussion with Chairman George Brown (House Committee on Science, Space and Technology) on the subject of "*Mars Observer* plans for 'extensive' reimaging of Cydonia," May 8, (1992).

# INDEX

ABC News, 330, 338–340, 344–347
*see also:* Morgan, Keith
absolute chronology 271, 272,
celestially anchored, 271
*see also:* dating, archaeology
Abu Simbel, 50, 270
Abydenus, 262
*see also:* Berossus
"account number," 340
*see also:* NASA-Lewis Cydonia Briefing; Klineberg, Dr. John
additional "unusual Martian surface features,"
225, 317–322
Administrator, NASA
*see also:* Goldin, Daniel S.; National Space
Council; Truly, Richard H.; Quayle, Vice
President Dan
Agrest, M. M., 262, 294
Air Force Cartographic and Mapping Service, 324
*see also:* Defense Mapping Agency
Akkadians, 263, 272, 273
Alexander the Great, 262
Alexandria, Library of, 17, 18
algae, blue-green, 118, 123, 170, 173, 175
*see also:* life, origin of
alignments, astronomical:
of terrestrial temples and monuments, 51, 56,
242, 268, 280–282, 306
of City and Face, 51, 57, 58, 62, 63–65, 124,
125, 220, 241–244, 268, 359–361
*see also:* Cydonia, Geometric Relationship
Model for; Geometric Relationship Model,
of Cydonia
alignments, geometrical, 88, 151, 152, 156,
194–196, 218–220, 226–230, 237, 319,
351–355, 359
*see also:* Cydonia, Geometric Relationship
Model for; Geometric Relationship Model,
of Cydonia
Alpha Centauri, 250, 255, 257
Amateur Space Telescope, 189
*see also:* Independent Space Research Group
[ISRG]
American Institute for Aeronautics and Astronautics (AIAA), 373, 387–388
*see also:* Evans, David; *Journal of Spacecraft
and Rockets;* Malin, Michael; *Mars Observer*
amino acids, 118, 173, 306
*see also:* life, origin of
Anasazi, 51, 66–69, 74
*see also:* Chaco Canyon
ancient astronauts, 211–213, 266
*see also:* Von Daniken, Erich

ancient cratered terrains, 112–117, 120, 121, 177
ancient global grid, 360
*see also:* ancient terrestrial monuments; inscribed tetrahedral geometry; Munck, Carl;
terrestrial connection, to Cydonia
ancient oceans, on Mars, 78, 175–179
*see also:* water, on Mars
ancient terrestrial monuments, 351, 357–365
*see also:* ancient global grid; Great Pyramid;
Lockyer, Sir Norman; mounds; Munck,
Carl; Stonehenge; terrestrial connection,
to Cydonia
ancient texts, 354–356
*see also:* inscribed tetrahedral geometry; mathematical topology; Tenen, Stan
anomalous objects, 87, 88, 140, 146, 158, 186,
192, 225, 317–322
anthropologist, 135, 164, 169, 232, 264
*see also:* anthropological analysis
anthropological analysis, 135, 156, 158, 159,
264–266,
Anunnaki, 294
*see also:* Sumer; myth, as history
Apkallu, 262, 263
*see also:* Berossus
Apollodorus, 262, 298
*see also:* Berossus
*Applied Optics, Journal of,* 333–334
archaeological sites, connection to Mars, 272–
273, 274–277, 281–283, 357–363, 368
*see also:* Munck, Carl; Myers, David
archaeology:
aerial analysis techniques, as applied to 35A72,
16, 17, 22–28, 66, 67, 69–72, 74–78, 135
as applied to analysis of human artifacts, 19,
20, 21, 65, 66, 164, 168, 263, 268–284,
294
as applied to Martian "ruins" by future missions,
309, 310
Archbishop Ussher, 107
*see also:* geology
architect, 27, 88, 147, 159, 237, 269, 275, 283,
312, 314
*see also:* architecture
architecture, 83, 89, 156, 160, 195, 218, 237,
242, 244, 245, 247, 262, 275, 277–279, 282,
283, 306,
arcologies, 147, 237, 238, 243, 244
as applied to Martian "pyramids," 147, 148,
238, 239, 242, 243, 246, 247, 259, 312, 313,
319, 322
as applied to "the Fort," 244, 246, 305
*see also:* Soleri, Paolo
Ares, 297
*see also:* "grove of Ares"
Argo, 296, 297
*see also:* ark; ārq; myth, as history
Argonauts, 296, 297
*see also:* Anunnaki; myth, as history

ark, 296
  *see also:* ārq; myth, as history
ārq, 296
  *see also:* hieroglyphs; ark
ārq ur, 298
  *see also:* hieroglyphs; Sphinx
artifacts, on Mars, 194, 195, 197, 199, 202, 251, 257, 299, 303, 328
Artificial Intelligence Laboratory, 333
  *see also:* MIT
"artificial mountains," pyramids as, 49, 270, 312
  *see also:* "lofty dwelling places"
Asimov, Isaac, 92, 248, 252, 260, 285, 302
astronomy, 30, 41, 48, 51, 60, 124, 271, 277, 280, 286, 301, 306
  description of Earth's orbit, seasons, and effect upon sun's motion, 52–55
astrophysical phenomenon, 354–356, 368
  *see also:* evolution, planetary; hyperdimensional physics; inscribed tetrahedral geometry; solar system
Astronomer Royal, of Scotland, 279
  *see also:* Smyth, Piazzi
Ascending Passage, 280, 281, 284
  *see also:* Great Pyramid
Assyrians, 273, 291
Athene, 297
  *see also:* Argo; Cadmus
atmosphere, of Mars, 14, 22, 29, 32, 33, 241, 253
  composition of, 149, 170–174, 178, 180, 307, 308
  greenhouse properties of, 80, 129, 177, 307, 308
Atum-Ra, 287
averaging techniques, as applied to computer imaging enhancements, 153, 154
  *see also:* Independent Mars Investigation
*Aviation Week [ Space Technology,* 200, 304, 315
Avinsky, Vladimir, 195–199, 201–204, 221, 268, 363
  background on, 211, 212, 214,
  *see also:* Hoagland, Richard C.; "half man/half lion"; hominid; Sphinx; Martian Sphinx; *Soviet Life* Magazine, "Pyramids on Mars?;" terrestrial connection, to Cydonia; West, John A.

Babylon, 262, 298
Babylonians, 263, 273, 275, 291, 295
Badewy, Alexander, 275
  *see also:* architect
Barbury Castle, 365–368
  *see also:* "Crop Circles"; inscribed tetrahedal geometry; Cydonia; terrestrial connection to Cydonia
Bateson, Gregory, 232–234, 244, 246, 260
  *see also:* anthropological analysis
Beatty, Bill, 169, 175, 197, 317–320
  *see also: Independent Mars Investigation*

Becquerel, Henri, 109, 110
  *see also:* geology
Bel-Marduk, 262
  *see also:* Berossus
Bell Helicopter, 158
  *see also:* Young, Arthur
Bell Laboratories, 164
Bergstrom, Joyce, 340–345
Berossus:
  background of, 262
  "contact myth" related by, 262–264, 298, 299
  mentioned, 271
  *see also:* Oannes, legend of
Bethe, Hans, 110
  *see also:* sun
*Bible,* 262, 279, 281, 292, 294
  *see also:* myth, as history
Bicentennial, as year of Viking Mission, 4
biological contamination, prevention of on Mars, 306, 308–310
"black-footed people," 293
  *see also:* Egypt, ancient
"black-headed people," 293, 294
  *see also:* Sumer; myth, as history
Bohemian Club, 331
  *see also:* Cydonia, censorship of
Bondurant, Dr. Lynn, 341–342, 346, 349, 371, 374–376
  *see also: Hoagland's Mars;* NASA-Lewis Research Center
"Boulder Conference," 181–187, 197
  *see also: Independent Mars Investigation*
"Boulder Paper," 206, 211, 212, 244
  *see also: Independent Mars Investigation*
Bower, Douglas, 365
  *see also:* "Crop Circles"
Bracewell Probe, 102, 187
  *see also:* SETI
Bracewell, Ronald, 102, 103
  *see also:* SETI
Bradbury, Ray, 3, 36, 47, 65, 96, 142, 301
  *see also:* "Martian Chronicles"
brain case, volume of hominid, 19, 28
  *see also:* human evolution
Brandenburg, John, 148, 158, 169, 170, 173, 175, 177, 178, 182, 186, 207, 224, 244, 319
  *see also: Independent Mars Investigation*
Breck, Ren, 137, 139, 152
  *see also:* InfoMedia, Inc.
Briareus, 296
  *see also:* Hercules; myth as history
Brin, David, 252, 260
  *see also:* interstellar flight
British Interplanetary Society, 98, 187, 289, 335
  *see also: Applied Optics, Journal of;* Carlotto, Dr. Mark
Brown, Rep. George E., 304, 376, 380–381, 394–396

see also: House Committee on Science, Space and Technology; *Mars Mission,* congressional lobbying efforts; *Mars Observer;* Space Cooperation Initiative

Brown, Dr. Robert, 346–347, 374–377
  see also: *Hoagland's Mars;* Bondurant, Dr. Lynn; NASA Headquarters

Budge, Wallis, 275, 286–288, 296, 302
  see also: Egypt; Sumer

Burgess, Eric, 45, 46, 98, 289,
  see also: Pioneer 10 Plaque

Burroughs, Edgar Rice, 36, 96, 179
  see also: "dead sea bottoms of Barsoom . . ."

Bush, President George, 346
  see also: Cydonia Briefing, NASA-Lewis; Hoagland, Richard C.; Klineberg, Dr. John; NASA-Lewis Research Center

Cadmus, 297
  see also: "spring of Ares"

Cairo, 269, 277, 279, 288
  derivation from "Mars," 289, 298
  see also: Egypt; Mars

"Cal" International Program in Applied Systems Design, 160, 163
  management of "Mars Investigation Group (MIG) by, 184, 186, 203, 204, 206
  "Cal Proposal" for further Mars research by, 204, 206, 207, 210
  joint Soviet participation in MIG, 194, 202–205, 210, 211
  fate of, 213, 214

camera, Malin, 331–332, 336
  see also: Malin, Michael; *Mars Observer*

cameras, Viking, 16, 17, 61, 151, 186, 245, 246

canals, on Mars, 30, 31–33, 45, 46, 48,
  see also: Lowell, Percival

carbon-14 dating, 19

carbon dioxide, 34, 37, 40–44, 46, 80, 122, 172, 172, 307, 308
  see also: atmosphere, of Mars

Carlotto, Dr. Mark, 305, 327, 332–337, 349, 362
  see also: Face on Mars, three-dimensional modeling of

Carr, Michael, 73, 74, 126, 127, 134, 138, 140, 141, 188
  see also: Viking Orbiters

Carter, Howard, 311
  see also: Tutankhamen

cartographic data, for Cydonia, 141
  see also: Cydonia, martian coordinates for

cartouche, 269

"Case for Mars" Conferences, 13, 145, 181–183, 185–187

casing stones, 276, 277, 280
  see also: Great Pyramid

CBS News, 185, 250

celestial equator, 53
  see also: astronomy

celestial sphere, 53
  see also: astronomy

Center for Research in Management, 213, 214
  see also: "Cal" International Program in Applied Systems Design

Chaco Canyon, 66–69
  see also: satellite-assisted archaelogy; Anasazi

Chaldean, 298
  see also: Apollodorus

Challenger, 304
  see also: space shuttle

Chamberlin, T. C., 109
  see also: geology

chance, 88, 220, 230, 244
  see also: statistical analysis, of Cydonia

channels, Martian, 38, 114–117, 121, 185
  see also: water, on Mars

Channon, Jim, 166–169
  see also: "work of art"

*Chariots of the Gods?,* 211, 264
  see also: Von Daniken, Erich

Cheops, 269
  see also: Khufu; Greeks

Chephren, 270, 286
  see also: Sphinx

*Childhood's End,* 301
  see also: Clarke, Arthur C.

chlorine, 179
  see also: ancient oceans, on Mars

Chorley, David, 365
  see also: "Crop Circles"

Christo, 23
  see also: "work of art"

Church, 91, 279

Churchman, C. West, 160, 161, 210, 214, 224, 234, 333
  see also: University of California (Berkeley)

CIA, 160
  see also: Rautenberg, Tom

Circê, 297
  see also: "falcon of Horus"

City, mentioned, 3, 4, 48
  author's discovery of, 25–27
  scale of, 25, 26, 71, 72
  description of, 26, 27, 70, 88
  relationship to Face, 26–28, 70, 88, 101, 229–231, 237
  initial analysis of, 47, 49, 70–72
  comparison with terrestrial "ceremonial centers," 70
  habitation hypothesis, 101, 127, 147, 148, 160, 182, 242, 247, 254, 259, 301, 305, 312, 313, 314, 319, 322
  architectural analysis of, including 3-D reconstructions, 134, 138, 147
  unnatural destruction of, 148, 305
  as discussed during the Independent Mars Investigation, 153–155, 157–159
  possible underground explosion of, 159, 243

*The Monuments of Mars*

possible connection with City, 26, 27, 48–52, 56–59, 62–64, 71, 75–79, 101, 229–231, 237
detailed measurements of, 88, 134
possible connection with D [ M, 151, 152, 218–220
"simian" aspects of, 165, 169
epistemological problems presented by, 217–220, 224, 225, 236, 237, 261, 273
new 3-D image-processing of, 305
explored by future Martian expeditions, 312–315
as "heru-sa-agga" . . . *hero* of the hill, 313
tilt of Face off North of 22.5 degrees, 359, 366
new 3-D modeling of, 334–336
image processing of, 333–334
*see also:* M. Carlotto; R. Jastrow
"falcon of Horus," 287, 297
*see also: heru*
Faraday, Michael, 371
*see also:* Faraday Disc; hyperdimensional technology; DePalma, Dr. Bruce
Faraday Disc, 371
*see also:* Faraday, Michael; hyperdimensional technology; DePalma, Dr. Bruce
"feline half," of Face on Mars, 363
*see also:* Cydonia Investigation; "half man/half lion"; Hoagland, Richard C.; "Martian Sphinx"; Sphinx, at Gizeh; terrestrial connection to Cydonia
"First Martian Expedition," 214, 311
*see also:* Von Braun, Wernher
Fort:
discovery of and initial impressions, 14, 15
as part of City, 26, 71, 72, 88
detailed analysis of, 81, 83, 157, 224, 244, 246, 305
evidence for massive destruction of, 148, 305
geometric integration into Cydonia Complex, 219, 228
new image-processing of, 305
"For the People" radio show, 347, 349
*see also:* Harder, Chuck; *Hoagland's Mars*
fossils, 18, 19, 28, 105–107, 110
*see also:* evolution
Forward, Robert, 251, 260
*see also:* Interstellar flight
four dimensions, of space,
*see also:* hyperdimensional physics; inscribed tetrahedral geometry; mathematical topology; "Message of Cydonia"; n-spaces
Frankfort, Henri, 275, 290, 302
*see also:* Sumer
"free energy," 372–374
*see also:* Baily, Dr. Patrick; hyperdimensional physics; hyperdimensional technology; IECEC Conference; "Message of Cydonia"; Myers, Dr. Ira, office of, at NASA-Lewis; Tesla, Nicola

Friedman, Louis, 205, 206
*see also:* Planetary Society
Freud, Sigmund, 145
Fuller, R. Buckminster, 237
*see also:* contained environments, on Mars

galaxy, 97–99, 125, 128, 182, 222–224
colonization of, 252, 256
"Earth-like" planets in, 254–256
"Mars-type" planets in, 256, 257
alive with "a million alien cultures," 314
Galileo mission, 190
*see also:* Jupiter
Gauss, 328
*see also:* Cydonia, Geometric Relationship Model for
Gelles, Walter, 338–339
*see also:* Hoagland, Richard C.; Roe, Chairman Robert; *Washington Times, The*
General Assembly Decision 33/426, 385
*see also: Hoagland's Mars: The U.N. Briefing;* Karaban, Susan; Ramadan, Mohammad; United Nations
genetics:
as possible key to interstellar flight, 251, 252, 300
involved in human evolution, 265, 300
in determining possibilities for "relatedness" of *Homo sapiens* to "the Martians," 306, 307
geology:
Martian, 6, 8, 10, 11, 13, 24, 27, 33–46, 64, 78–80, 82, 83, 87, 89, 90, 101, 111, 157, 170, 171, 318
Martian, pre-space, 30–33, 38, 39, 48
terrestrial, 105–111, 171
general dating methods of, 105–113
Geometric Relationship Model, of Cydonia, 147, 224–233, 227, 325, 341, 352, 364–367
*see also:* Cydonia, Geometric Relationship Model for; Hoagland, Richard C.; inscribed tetrahedral geometry; Torun, Erol
geometry:
as exemplified in layout of City, 26, 27, 70, 88, 89, 192, 218–220, 226–233, 237, 301
as exemplified in purported Martian "canal network," 31, 32
as Sagan criteria for detection of planetary intelligence, 70
in Avinsky's independent analysis of Cydonia, 195, 196, 199
from Torun geometric analysis of D&M, 326–327
and Hoagland re-analysis of Cydonia Complex, 331, 352, 364–367
geomorphology, 89, 164, 225–226, 305, 325–326
*see also:* epistemology, of Mars investigations
Gilgamesh, 296
*see also:* Sumer; myth, as history

Gizeh, 269, 278, 280, 282, 357–363
   *see also:* Egypt; Great Pyramid; Martian Sphinx; "Message of Cydonia," mathematical connection to; Sphinx, Gizeh; Sphinx, revised age of; terrestrial connection, to Cydonia; West, John A.
global grid, 358, 360
   *see:* ancient global grid
"God of War," 148
   *see also:* nuclear holocaust, possible evidence for on Mars
Goddard Colloquium Committee, 330
   *see also:* Goddard Spaceflight Center; Cydonia Investigation, official NASA invitations to
Goddard Spaceflight Center, 5, 16, 148, 150, 158, 191, 192, 330
   *see also:* Goddard Colloquium Committee; Cydonia Investigation, official NASA invitations to
Godfrey, D. A.
   *see also:* hexagonal cloud pattern; hyperdimensional physics; inscribed tetrahedral geometry; pole, of rotation; Saturn
gods, 50, 165, 207, 261, 264, 267, 269, 281, 286–288, 290, 291, 296, 302, 315
"God's quarantine regulations," 250
   *see also:* interstellar distances
Golden Fleece, 296, 297, 299
   *see also: heru*
Golden Section, 151, 156, 283
   *see also:* da Vinci; Dipietro and Molenaar Pyramid [D & M]
Gould, Stephen J., 217, 261
government censorship, of Cydonia, 331, 345–349
   *see also:* Bohemian Club; Brown, Dr. Robert; Cydonia Investigation; NASA Headquarters; *Hoagland's Mars;* National Press Club
"great circle," 360
   *see also:* ancient global grid; Gizeh; inscribed tetrahedral angle; Munck, Carl; Myers, David; Stonehenge; terrestrial connection, to Cydonia
gravity:
   Martian, 25, 94, 95, 174, 238
   of the sun, 97
   as ejection technique from the solar system, 97
   biological effects of, 217, 258, 260, 300
   used in archaeology, 284
   *see also:* extended lifespan
Great Dune Sea, on Mars, 177, 178
   *see also:* environment, Martian
"great ending . . . ," 298
   *see also:* Sphinx; ārq ur
Great Pyramid, 137, 268–270, 298, 301, 302
   scale, compared to Martian analogs, 72
   construction of, 269, 276, 278, 281, 306
   statistics of, 269, 276, 280, 282
   precision measurements of, 276–278, 280–283, 306

"Pi controversy," 278, 280, 281, 283
   claims of arcane "knowledge" in, 278, 279, 281, 282
   builders of, 279, 281
   age of, 281, 306
   possible new chambers of, 284
   e/pi relationship to, 359–361
   *see also:* ancient global grid; great circle; Munck, Carl; terrestrial connection, to Cydonia
Great Dark Spot, on Neptune, 353
   *see also:* energy upwelling; evolution, planetary; hyperdimensional physics; inscribed tetrahedral geometry; solar system; Uranus; *Voyager,* spacecraft
Great Dark Spot, on Uranus, prediction of, 355
   *see also:* "energy upwelling"; Hubble Space Telescope; inscribed tetrahedral geometry; solar system; *Voyager,* spacecraft
Great Red Spot, on Jupiter, 353
   *see also:* energy upwelling; evolution, planetary; hyperdimensional physics; inscribed tetrahedral geometry; solar system
Greeks, 207, 269, 281, 287, 288, 291, 293, 296–299
   *see also:* myth, as history
Greenbank, 221
   *see also:* SETI
greenhouse effect, 177
   in Martian "arcologies," 241
   *see also:* atmosphere, of Mars
Grossinger, Richard, 264, 265, 301
   *see also:* anthropological analysis
"grove of Ares," 297
   *see also:* "spring of Ares"

"half man/half lion," 360
   *see also:* Hoagland, Richard C.; hominid; Martian Sphinx; Sphinx, Gizeh; "terrestrial connection" to Cydonia; West, John A.
Hall, Asaph, discoverer of Martian moons, 30
Hapgood, Charles, 20, 21, 28
   *see also:* "high-tech" civilizations, possible previous terrestrial
Harder, Chuck, 347, 349
   *see also:* "For the People," radio network; *Hoagland's Mars*
Harleston, Jr., Hugh, 355–358
   *see also:* ancient global grid; Teotihuacan; inscribed tetrahedral geometry; hyperdimensional physics; Munck, Carl; terrestrial connection, to Cydonia
Hartmann, William, 30
   *see also:* Mariner, unmanned interplanetary spacecraft series
Harvard University, 164, 169, 174
Hatshepsut, 272
   *see also:* Egypt
Hawaiian Shield Volcano, 353

# ORDER FORM

| QUANTITY | TITLE AND PRICE | TOTAL |
|---|---|---|
| | The Monuments of Mars: A City on the Edge of Forever (2nd edition) @ $6.95 | |
| | The Monuments of Mars: A City on the Edge of Forever (hardcover) @ $25.00 | |
| | The Monuments of Mars: Evidence of a Lost City on Mars? (audiotape) @ $8.00 | |
| | Hoagland's Mars: The NASA-Cydonia Briefings (VHS video) @ $23.95 | |
| | Hoagland's Mars: The U.N. Briefings (VHS video) @ $23.95 | |
| | The Martian Enigmas: A Closer Look @ $12.95 | |
| | The Martian Enigmas: A Closer Look (hardcover) @ $24.95 | |
| | The Face on Mars: Evidence for a Lost Civilization? @ $11.50 | |
| | Folio of 12 research-quality 8" by 10" glossy prints @ $120.00 | |
| | Planetary Mysteries @ $9.00 | |
| | The Night Sky: The Science and Anthropology of the Stars and Planets @ $8.95 | |
| | Waiting for the Martian Express @ $3.95 | |
| | Sky Signs: Aratus's Phaenomena @ $2.95 | |
| | Forbidden Science (hardcover) @ $22.95 | |

Postage ($2.50 for first book, $1.00 for each additional one) _____

(California residents please add 8.25% sales tax) _____

Total _____

**Please send card with payment to:**
**North Atlantic Books, 2800 Woolsey Street, Berkeley, California 94705**

NAME _____

ADDRESS _____

CITY _____ STATE _____ ZIP _____

# Special Discount Card (tearing from book will not affect other pages)

☐ **The Monuments of Mars: A City on the Edge of Forever** by Richard C. Hoagland (2nd edition). This version contains some 50 pages of writing and the original photograph section not in the current edition. $14.95 in paper, with this card: $6.95. Ten copies for $50.00 in hardcover.

☐ **The Monuments of Mars? by Richard C. Hoagland and Daniel Draisin.** Audiotape docudrama, includes chart of nine photos. Voices of many scientists discussing Cydonia, ending with dramatic simulation of U.S./Soviet manned landing on Mars. 60 minutes. $10.95; with this card: $8.00.

☐ **Hoagland's Mars: The NASA-Cydonia Briefings.** Full-color VHS format video. Be in the audience at NASA's Lewis headquarters as Richard Hoagland presents his stunning research on Mars to astonished engineers and scientists. Includes never-before-released, custom-relief fly-over of Cydonia, official NASA footage of the "face" and "pyramid." 2-color foldout reference map and speculative decoding of some aspects of the "monuments." 83 minutes. $24.95; with this card: $23.95.

☐ **Hoagland's Mars: The U.N. Briefings—The Terrestrial Connection.** Full-color VHS format video. Be in the audience at the United Nations in February, 1992, as Richard Hoagland describes startling relationships between Cydo-nia and sites on earth, compares the Martian and terrestrial sphinxes, and proposes a whole new four-space energy source suggested by the rebus at Cydonia. 93 minutes. $29.95; with this card $28.95.

☐ **The Martian Enigmas: The Face, Pyramids and Other Unusual Objects on Mars — A Closer Look** by Mark J. Carlotto. This large-format photo album summarizes the results from the few technical papers that have been written about the Face and other enigmatic objects on Mars and presents the best-quality images, enhancements, and quantitative analyses of these, controversial Viking photos, specifically illustrating the processes used to digitally restore and clarify the images — also striking three-dimensional renditions. Glossy-paper oblong, 130 pages. $29.95 in hardcover, with this card: $24.95. $16.95 in paperback; with this card: $12.95.

☐ **The Face on Mars: Evidence for a Lost Civilization?** by Randolfo Rafael Pozos. The original account of the Mars Investigation. 160 pages oversize, highly illustrated. $12.95, with this card: $11.50.

☐ **A folio of 12 research-quality 8" by 10" glossy prints** of the Martian enigmas, reconstructed and clarified by Dr. Mark Carlotto, made directly from original first-generation negatives. $120.00.

☐ **Planetary Mysteries**, edited by Richard Grossinger. This anthology includes a 45-page interview with Richard C. Hoagland. 272 pages. $12.95 in paper; with this card: $9.00.

☐ **The Night Sky: The Science and Anthropology of the Stars and Planets** by Richard Grossinger (revised 1988 edition). Includes a history of Western astronomy from the Greeks through quasars and black holes. Sections on ancient and non-Western astronomies, science fiction, UFOs and extraterrestrial life, the planets and moons of the Solar System, the artifacts at Cydonia, plus the complete text of the talk "Giving Them a Name" (re: aliens and UFOs). 512 pages, $12.95 in paper; with this card: $8.95.

☐ **Waiting for the Martian Express** by Richard Grossinger. Further discussions of UFOs and extraterrestrials plus "A Critical Look at the New Age." 196 pages, $9.95 in paper; with this card: $3.95.

☐ **Sky Signs: Aratus's Phaenomena** translated by Stanley Lombardo. A 270 B.C. astronomy book with illustrations. 82 pages. $7.95 in paper; with this card: $2.95.

☐ **Forbidden Science** by Jacques Vallee. The autobiography of the famous UFO scientist portrayed in the movie *Close Encounters of the Third Kind*. 500 pages. $24.95 in hardcover; with this card: $22.95.